ELEMENTS
OF QUANTUM
CHEMISTRY

ELEMENTS OF QUANTUM CHEMISTRY

by
RUDOLF ZAHRADNÍK
J. Heyrovský Institute of Physical Chemistry and Electrochemistry,
Czechoslovak Academy of Sciences, Prague
RUDOLF POLÁK
J. Heyrovský Institute of Physical Chemistry and Electrochemistry,
Czechoslovak Academy of Sciences, Prague

PLENUM PRESS • NEW YORK AND LONDON

SNTL • PUBLISHERS OF TECHNICAL LITERATURE, PRAGUE

Distributed throughout the world with the exception of the Socialist countries by

Plenum Press,
a Division of Plenum Publishing Corporation,
227 West 17th Street, New York 10011
ISBN 0-306-31093-7
Library of Congress Catalog Card Number 77-85610

© 1980 Rudolf Zahradník, Rudolf Polák
Translated by Jiří Horký
English edition first published in 1980 simultaneously by Plenum Publishing Corporation and
SNTL – Publishers of Technical Literature, Prague

Printed in Czechoslovakia

CONTENTS

6

8

1. INTRODUCTION

The post-war generation of chemists learned to handle a blow pipe at the university as thoroughly as modern chemistry students learn to write computer programmes. Even after World War II the rule of three was considered to be sufficient mathematical knowledge for chemists and the short course of "higher mathematics" at technical universities was the test most feared by chemistry students. However, even then some envisaged the theoretical derivation of information on the properties of molecules from knowledge of the bonding of the component atoms.

During the last quarter of this century, amazing changes have occurred in chemistry, some of them almost incredible. Dirac's famous clairvoyant statement* has been partially realized. Incorporation of quantum mechanics into chemistry encountered numerous difficulties. After all, the reserve of experimental chemists is not surprising. For decades the hydrogen and helium atoms and the hydrogen molecule belonged among the systems most frequently investigated by theoreticians. Later these systems were supplemented by ethylene and benzene. The authors of this book can therefore recall with understanding the words of the late Professor Lukeš: "Well, when they succeed in computing a molecule of some alkaloid by those methods of yours...". Unfortunately, the calculations on calycanin were not completed before his death.

Now there is no need to convince even the members of the older generation of the usefulness of quantum chemistry for chemists. Even the most conservative were convinced after the introduction of the Woodward-Hoffmann rules.

---

* "The underlying physical laws necessary for the mathematical theory of a large part of physics and the whole of chemistry are thus completely known, and the difficulty is only that the exact application of these laws leads to equations much too complicated to be soluble. It therefore becomes desirable that approximate practical methods of applying quantum mechanics should be developed, which can lead to an explanation of the main features of complex atomic systems without too much computation". [P. A. M. Dirac: *Proc. Roy. Soc.* (London) **123**, 714 (1929).]

This book is concerned, on the one hand, with an introduction to the theory of the chemical bond to a degree necessary for active understanding of quantum chemical semi-empirical methods (Chapter 10, which completes the methodical part) and, on the other hand, with the study of the relationships between the structures of molecules and their properties. Among these properties, both the static characteristics (thermochemical, electric, magnetic, optical) and the dynamic characteristics, chemical reactivity characterized by the equilibrium and velocity constants, will be discussed. It is necessary to define the meaning of the term "structure" more precisely. In a narrow sense structure means the way in which atoms are bonded in molecules or the arrangement of molecules in a crystal lattice. In recent years, structure in this sense has often been determined directly using X-ray analysis. Here, as a rule, for the probable structure of a compound the theoretical characteristics will be determined by computation and afterwards will be compared with the experimental results.

An attempt will be made to acquaint the reader with these comparisons in such a way as to enable him not only to perform similar comparisons himself but also to open new possibilities. In comparing theoretical and experimental quantities, both a more profound qualitative explanation of the studied properties and processes and quantitative interpretation of experimental data will be necessary. This approach will help in generalizing the knowledge obtained and in condensing large groups of experimental data into empirical formulae, in which, of course, quantities appear resulting from quantum-chemical calculations. These relationships will be used as interpolation formulae and will permit estimation of the values of experimental characteristics in substances not yet prepared, whose properties are of interest. Moreover, there is also the very attractive possibility of using the quantum theory of the chemical bond not only for the interpretation, but also for the prediction of properties.

2. A BRIEF COMMENT ON THE DEVELOPMENT OF THE THEORY OF THE CHEMICAL BOND

It is admirable that, as early as in the nineteenth century, chemists succeeded in defining concepts of the structure of substances that are in remarkable agreement with modern knowledge of the quantum theory of the chemical bond and with direct structural determinations using electron or neutron diffraction and X-ray analysis. Only in the theory published in 1916 by Kossel and Lewis did electrons assume a decisive role in concepts of the origin of the chemical bond. (The electron was discovered by Thomson only 19 years earlier, and 5 years earlier Rutherford proposed the planetary model of the atom.) The basic concepts of this very successful and innovative theory are based on the ideas of electrovalency and covalency, which are still accepted at the present time. This theory of the chemical bond forms a basis for the theory of mesomeric and inductive effects which contributed considerably to the rationalization of organic and inorganic chemistry (Robinson, Ingold, Arndt, Eistert). The work carried out by their predecessors (Kekulé, Cooper, Butlerov, Werner, and in spatial structure Le Bel and van't Hoff) is of essential importance.

The difficulties encountered in classical mechanics will be mentioned in another context. Here, however, it should be noted that classical Newtonian mechanics is useful for the description and prediction of phenomena in the middle and macro cosmos. The growing need to describe the motion of particles forming molecules and atoms led to the establishment of a new mechanics, quantum mechanics, in the twenties of this century.

The fundamental equation of this new mechanics, the Schrödinger equation, can be obtained in two ways. The method given by Schrödinger is apparently less complicated, proceeding from the concept that electron motion can be described in terms used for the description of wave motion, leading to the term "wave mechanics".

Quite independently, the same result was achieved by Heisenberg, who made use of matrix properties. Although the two approaches are formally quite different, the results have been shown by Born and Jordan to be equivalent.

Later, Dirac and von Neumann formulated quantum mechanics more generally and showed that Schrödinger's and Heisenberg's approaches are special cases of a more general theory.

3. THE TIME-INDEPENDENT SCHRÖDINGER EQUATION

3.1 Introduction of the equation

It is important to remember that the Schrödinger equation, similar to the principal thermodynamic laws, cannot be derived from the general principles of physics. It is true that we can proceed from the classical law of conservation of energy and, through a number of modifications (some of them inconceivable from the point of view of classical mechanics), arrive at the Schrödinger equation ("derive it"). However, this procedure does not possess the character of derivation by deduction that is considered normal in classical physics. The only method of determining whether the equation obtained has physical significance, i.e. whether it gives a true picture of the real behaviour of particles, will lie in comparison of values for quantities calculated using this equation with experimentally obtained values.

In classical physics two fundamental objects are investigated, namely the particle and the wave. A particle can be localized in space and time and characterized by dynamic characteristics such as its linear momentum p and energy E. A wave originates in connection with a disturbance in a continuous medium and can be assigned kinematic characteristics, such as wavelength λ and frequency v. Although a wave can assume certain dynamic characteristics reminiscent of particle properties (i.e. momentum density, energy density), it is apparently an object quite different from a particle.

For the sake of simplicity we shall, first of all, discuss a point particle of mass m, moving in a constant (i.e. time-independent) external field along the x-axis. The principle of the conservation of energy holds for such a system and it is therefore denoted as a conservative system. The principle of the conservation of energy can be expressed by the equation

$$E = T + V, \tag{3-1}$$

where E is the total energy of the given point particle, T is its kinetic energy and V is its potential energy. The relationship

$$T = \frac{1}{2} m\dot{x}^2 \tag{3-2}$$

is also valid, where $\dot{x} = \mathrm{d}x/\mathrm{d}t$ (x is the trajectory, t is the time).

Since the potential energy of a point particle in an external field is a function of its coordinates, it then follows that

$$E = \frac{1}{2} m\dot{x}^2 + V(x) \tag{3-3}$$

If in equation (3-3) the expression for the linear momentum p is introduced in the form

$$p = m\dot{x}, \tag{3-4}$$

it then follows that

$$E = \frac{p^2}{2m} + V(x) \tag{3-5}$$

As is familiar from classical mechanics, for a conservative system the total energy can be identified with the corresponding Hamiltonian function H, and thus

$$H(p, x) = E \tag{3-6}$$

Equation (3-5) is a first-order differential equation; considering the initial conditions it is then possible, by integration, to derive the equation for the trajectory of the particle, $x = x(t)$. It thus follows that the solution of the equation of motion in classical mechanics (represented here by the principle of the conservation of energy) provides fully defined functions describing the dependence of the dynamic quantities on time, thus permitting calculation of values of these dynamic quantities at each instant.

The discussion of the behaviour and properties of "classical" particles can be extended to microparticles, using the electron as a representative example. It has been experimentally demonstrated that the electron behaves as a particle: its charge has a discrete value (cf. Millikan's experiment) and it can be localized (a track in the Wilson chamber). However, if an attempt were made to localize the electron by giving its position in a given instant of time (with an arbitrary precision), it would be found that this cannot be achieved. Experiments have even been carried out in which electrons behave as waves. In interference or diffraction phenomena (the experiments of C. J. Davisson, L. H. Germer, and E. Rupp), electrons must be treated as waves with a wavelength given by the de Broglie relation

$$\lambda = \frac{h}{p}, \tag{3-7}$$

where p is the momentum of the electron and h is Planck's constant.

The electron, similar to other microparticles, is therefore an object which in the classical sense resembles neither a particle nor a wave. This complication always appears in the character of laws describing the behaviour of microparticles.

In his five papers published in the first half of 1926 in the journal *Annalen der Physik*, E. Schrödinger proposed a new system of dynamics for the description of microparticles, in which the *wave function*, Ψ, assumes a leading role.

The Schrödinger wave function is a quantity that characterizes the state of the particle in a particular way. By solving the wave equation, a function is obtained giving the dependence of this quantity on the spatial coordinates of the particle (and possibly also on time). The position of an electron is given by the probability function, which is a function of the coordinates, usually written ϱ (x, y, z), and is denoted as the probability density. Its value increases with increasing probability of electron occurrence in a given area in space. It appears that this probability density can be expressed by the wave function, Ψ. The physical significance of the wave function, if real, is such that its square (Ψ^2) gives the probability distribution function for the particular coordinate system and permits calculation of the physical quantities of the given particle. It is necessary to add, however, in the general case, the wave function can be complex, so that instead of the square of the function the product of Ψ and its complex conjugate, $\Psi^*\Psi$, is employed. It is preferable to choose the multiplication constant for the wave function so that the equality $\varrho(x, y, z) = \Psi^*(x, y, z)\,\psi(x, y, z)$ holds. The probability of finding a particle in a volume element $d\tau$ $(d\tau = dx\,dy\,dz)$ whose centre has coordinates x, y, z is given by the expression $\Psi^*\Psi\,d\tau$. By summing all possible contributions of this type throughout the entire space, that is by integration, unity results, i.e. the particle must be located somewhere in the given space. If this condition is fulfilled, then function Ψ is said to be normalized. The physical meaning of the wave function is ensured when Ψ is continuous, single-valued and finite.

A detailed explanation of the fundamental postulates of quantum mechanics will be resumed later. Here, only the quantum-mechanical formulation for a moving point particle will be described; the classical formulation was discussed in the introductory part of this section.

In this way, the relationship between the two types of mechanics will become apparent.

Conversion of the classical formulation of the problem [expressed by equations (3-5) and (3-6)] to the quantum formulation can be effected in two steps:

a) Linear momentum p in the classical formulae will be replaced by the operation "differentiation with respect to a trajectory" multiplied by the constant $h/2\pi i$ (where h is Planck's constant and $i = \sqrt{(-1)}$; the symbol $\hbar$ is sometimes used in place of the expression $h/2\pi$). This process (representing a first step that, from the point of view of classical mechanics, is rather unexpected) can be symbolically described as follows:

$$p \rightarrow \frac{h}{2\pi i} \frac{d}{dx}$$

Similarly, for p^2,

$$p^2 \rightarrow \frac{h}{2\pi i} \frac{d}{dx} \left(\frac{h}{2\pi i} \frac{d}{dx} \right) = - \frac{h^2}{4\pi^2} \frac{d^2}{dx^2}$$

In the conversion from classical to quantum mechanics the x-coordinate remains unchanged.

As in classical mechanics any physical quantity of a system can be expressed in terms of its coordinates and its momentum — for instance the Hamiltonian function in equations (3-5) and (3-6) — it is possible, on the basis of such a "surprising" procedure, to form a corresponding expression for each quantity which will be called the operator of the given quantity. The operator of a physical quantity will later be denoted by the symbol of this quantity printed in school script. Consequently,

$$\mathscr{p}_x = \frac{h}{2\pi i} \frac{d}{dx}$$

$$\mathscr{V}(x) = V(x)$$

The operator of the Hamiltonian function, called the *Hamiltonian operator* (or simply the *Hamiltonian*), for a particle moving along a straight line in the direction of the x-axis can then be expressed by the relationship

$$\mathscr{H} = - \frac{h^2}{8\pi^2 m} \frac{d^2}{dx^2} + \mathscr{V}(x) \tag{3-8}$$

Simultaneously, the Hamiltonian can be considered to consist of two partial operators, i.e. the kinetic energy operator,

$$- \frac{h^2}{2\pi^2 m} \frac{d^2}{dx^2},$$

and the potential energy operator,

$$\mathscr{V}(x)$$

b) The wave function Ψ will be sought as a solution of the Schrödinger wave equation in the form

$$\mathscr{H}\Psi = E\Psi \tag{3-9}$$

This equation can be obtained formally from classical equation (3-6), according to paragraph a), by replacing the Hamiltonian function H with the Hamiltonian operator (conversion "to operator form") and both sides will then be multiplied by function $\Psi(x)$ (this multiplier is always written on the right-hand side). Thus, for this particle it follows that

$$\left[-\frac{h^2}{8\pi^2 m}\frac{\mathrm{d}^2}{\mathrm{d}x^2} + \mathscr{V}(x) \right]\Psi(x) = E\Psi(x) \tag{3-10}$$

This equation is often written in the form

$$\frac{\mathrm{d}^2\Psi}{\mathrm{d}x^2} + \frac{8\pi^2 m}{h^2}(E - \mathscr{V})\,\Psi = 0, \tag{3-11}$$

where, for the sake of simplicity, instead of $\mathscr{V}(x)$ and $\Psi(x)$, only $\mathscr{V}$ and Ψ are written.

Extension to a three-dimensional system encounters no difficulties. For the kinetic energy it then holds that

$$T = \frac{1}{2}m(\dot{x}^2 + \dot{y}^2 + \dot{z}^2) = \frac{1}{2m}(p_x^2 + p_y^2 + p_z^2),$$

where p_x, p_y and p_z are the components of the momentum which will be replaced by the corresponding operators:

$$\not{p}_x \rightarrow \frac{h}{2\pi i}\frac{\partial}{\partial x}; \quad \not{p}_y \rightarrow \frac{h}{2\pi i}\frac{\partial}{\partial y}; \quad \not{p}_z \rightarrow \frac{h}{2\pi i}\frac{\partial}{\partial z}$$

The Schrödinger equation now assumes the form

$$\frac{\partial^2\Psi}{\partial x^2} + \frac{\partial^2\Psi}{\partial y^2} + \frac{\partial^2\Psi}{\partial z^2} + \frac{8\pi^2 m}{h^2}(E - \mathscr{V})\,\Psi = 0 \tag{3-12a}$$

This type of equation can further be simplified by introducing the Laplace operator (symbol Δ)

$$\frac{\partial^2}{\partial x^2} + \frac{\partial^2}{\partial y^2} + \frac{\partial^2}{\partial z^2} \equiv \Delta$$

18

$\Bigg[$ in place of symbol Δ, ∇^2 can be written, where ∇ is the nabla vector operator $\left(\dfrac{\partial}{\partial x}, \dfrac{\partial}{\partial y}, \dfrac{\partial}{\partial z} \right) \Bigg]$. Thus the Schrödinger equation can be rewritten as

$$\Delta \Psi + \frac{8\pi^2 m}{h^2} (E - \mathscr{V}) \Psi = 0 \qquad (3\text{-}12\text{b})$$

3.2 Formulation of the Schrödinger equation for simple systems

3.2.1 A particle in a one-dimensional potential box

Let us consider a particle moving in the direction of the x-axis inside a so-called one-dimensional potential box. It is assumed that the particle has the same potential energy at any place in the box; this energy can conveniently be set equal to zero. It is further assumed that the energy of the particle everywhere outside the box is infinitely high (Fig. 3-1).

The Schrödinger equation for a particle in a box assumes the form

$$\frac{d^2 \Psi}{dx^2} + \frac{8\pi^2 m}{h^2} E \Psi = 0 \qquad (3\text{-}13)$$

(as $\mathscr{V} = 0$).

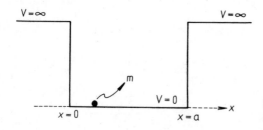

Fig. 3-1. A particle of mass m in a one-dimensional potential box of length a.

3.2.2 The harmonic oscillator

A particle of mass m is moving along the x-axis alternately in the positive and the negative direction and its equilibrium position is $x = 0$ (Fig. 3-2). The force F acting on the particle is directed against the displacement and is proportional to the magnitude of the displacement, x. Thus it follows that

$$F = -kx,$$

where k is a proportionality constant, called the force constant. For the potential energy it then follows that

$$V = -\int_0^x (-kx)\,dx = \frac{1}{2}kx^2$$

By substituting this expression for $\mathscr{V}$ into Eq. (3-11) the required Schrödinger equation is obtained:

$$\frac{d^2\Psi}{dx^2} + \frac{8\pi^2 m}{h^2}\left(E - \frac{1}{2}kx^2\right)\Psi = 0 \qquad (3\text{-}14)$$

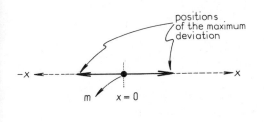
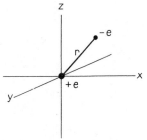

Fig. 3-2. A harmonic oscillator of mass m.

Fig. 3-3. Model of the hydrogen atom: nucleus $(+e)$ and electron $(-e)$.

3.2.3 The hydrogen atom

Even for the hydrogen atom (which is a proton-electron system, Fig. 3-3), the formulation of the Schrödinger equation poses no difficulties. The electron is considered to move in a three-dimensional space about the nucleus which is at rest. The potential energy of this system can be expressed as

$$V = \frac{e(-e)}{4\pi\varepsilon_0 r} = \frac{-e^2}{4\pi\varepsilon_0 r},$$

where e is the elementary charge, r is the distance between the electron and the nucleus and ε_0 is the permittivity of a vacuum. By substituting this expression into Eq. (3-12b), the Schrödinger equation is obtained in the form

$$\Delta\Psi + \frac{8\pi^2 m}{h^2}\left(E + \frac{e^2}{4\pi\varepsilon_0 r}\right)\Psi = 0 \qquad (3\text{-}15)$$

3.2.4 The hydrogen molecular ion, H_2^+

Now a system of two protons and one electron (Fig. 3-4), which is an ionization product of the simplest molecule, will be considered.

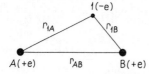

Fig. 3-4. The hydrogen molecular ion: nuclei (*A*, *B*), electron (1). Distances are denoted by *r*.

For the potential energy of this system it holds that

$$V = \frac{1}{4\pi\varepsilon_0}\left(-\frac{e^2}{r_{1A}} - \frac{e^2}{r_{1B}} + \frac{e^2}{r_{AB}} \right),$$

so that the Schrödinger equation [obtained by substituting for $\mathscr{V}$ in Eq. (3.12b)] has the form

$$\Delta\Psi + \frac{8\pi^2 m}{h^2}\left(E + \frac{e^2}{4\pi\varepsilon_0 r_{1A}} + \frac{e^2}{4\pi\varepsilon_0 r_{1B}} - \frac{e^2}{4\pi\varepsilon_0 r_{AB}} \right)\Psi = 0 \qquad (3\text{-}16)$$

3.3 Examples of the solution of the Schrödinger equation

3.3.1 The free particle

A free particle is defined as a particle that is moving in a constant potential field (i.e., it has the same potential energy everywhere). Then, without losing generality, it is possible to set $V = 0$ and the Schrödinger equation for a particle moving along a straight line then assumes the form

$$\frac{d^2\Psi}{dx^2} = -\frac{8\pi^2 mE}{h^2}\Psi \qquad (3\text{-}17)$$

throughout the whole *x*-coordinate region, in contrast to the limited validity of Eq. (3-13).

This is a second-order differential equation with constant coefficients, whose solution consists of two functions, as can be verified by substitution:

$$\Psi_1 = N_1 \exp\left[\frac{2\pi i}{h}\sqrt{(2mE)}\,x \right] \qquad (3\text{-}18)$$

$$\Psi_2 = N_2 \exp\left[-\frac{2\pi i}{h}\sqrt{(2mE)}\,x \right] \qquad (3\text{-}19)$$

(N_1 and N_2 are constants).

In addition, both these functions satisfy the relationships

$$\hat{p}\Psi_1 = \sqrt{(2mE)}\,\Psi_1 \qquad (3\text{-}20)$$

$$\not{p}\Psi_2 = -\sqrt{(2mE)}\,\Psi_2, \qquad (3\text{-}21)$$

agreeing with Eq. (3-9) in one significant property: applying the operator to a function yields the same function multiplied by a constant. An equation possessing this property is called a characteristic equation; function Ψ, which satisfies such an equation, is then a characteristic function (or eigenfunction), and the corresponding constant is denoted as the characteristic value (or eigenvalue). As will be shown later, characteristic values in equations of this type are measurable values of physical quantities which are represented by the corresponding operator. More specifically, from Eq. (3-20) it follows that, if a particle is in state Ψ_1 (with energy E), it is moving in the positive direction of the x-axis with the linear momentum $\sqrt{(2mE)}$. A particle in state Ψ_2 and with the same energy moves with an equally large momentum but in the opposite direction. The expression for the magnitude of the momentum will be readily understood by considering that the classical value for the total energy is given by the expression

$$E = \frac{1}{2}mv^2,$$

and thus

$$\sqrt{(2mE)} = m\,|\,v\,|$$

It is quite sufficient here to confine the discussion to a particle moving in the positive direction of the x-axis. Some interesting physical consequences result from the form of wave function (3-18): First, energy E cannot assume negative values, as for $E < V$ the exponential factor would become a real number and function Ψ for $x \to \infty$ would become infinite, thus losing physical meaning.

Wave function (3-18) can be employed for calculation of the probability density of the particle:

$$\Psi_1^*(x)\,\Psi_1(x) = N_1 \exp\left[\frac{2\pi i}{h}\sqrt{(2mE)}\,x\right] N_1^* \exp\left[-\frac{2\pi i}{h}\sqrt{(2mE)}\,x\right]$$

$$= N_1^* N_1 \qquad (3\text{-}22)$$

Hence it follows that the probability density of the particle is independent of the x-coordinate, so that the particle can be found with equal probability at any point in the one-dimensional space within which it moves. Thus, the uncertainty in the position of the particle is infinitely large, in agreement with the Heisenberg uncertainty principle, according to which the more accurate the determined value for the particle coordinates, the less accurate is the determined value of its momentum and vice versa.

[In the given case the particle has a quite definite exact momentum value, $\sqrt{(2mE)}$, so that the uncertainty in its position is infinitely large.]

Wave function (3-18) is sometimes written in the form

$$\Psi_1 = N_1 \exp(ikx), \tag{3-23}$$

where the expression $\dfrac{2\pi}{h}\sqrt{(2mE)}$ is replaced by a new quantity k, termed the wave vector (in a multidimensional case it would actually be a vector). This quantity is related to the energy by the expression

$$E = \frac{h^2}{8\pi^2 m} k^2 \tag{3-24}$$

The meaning of the wave vector will follow from comparison of relation (3-24) with the classical expression for the energy

$$E = \frac{1}{2} mv^2 = \frac{1}{2m} p^2, \tag{3-25}$$

into which the de Broglie relation (3-7) can be substituted:

$$E = \frac{1}{2m} \left(\frac{h}{\lambda} \right)^2 \tag{3-26}$$

Comparing this relation with Eq. (3-24), it follows that

$$k^2 = \frac{8\pi^2 m}{2m\lambda^2} = \frac{4\pi^2}{\lambda^2}$$

or

$$|k| = \frac{2\pi}{\lambda} \tag{3-27}$$

3.3.2 A particle in a potential box; the solution and its consequences

Let us return to the study of the behaviour of a particle in a potential box (Fig. 3-1). The same differential equation as for a free particle must be solved, except that wave function Ψ must satisfy the boundary conditions describing the fact that the particle cannot be present in some regions (i.e. outside the box) as infinitely high energy would be needed to transfer it to these regions. For these regions the x-coordinate has the values

$$x \geqq a$$

and

$$x \leqq 0$$

Thus, the probability of finding the particle anywhere outside the box equals zero, so that the wave function whose square is proportional to this probability must also have zero value. For regions outside the box it therefore holds that

$$\Psi = 0$$

It will be seen that, because of this condition, the particle can no longer assume any energy value in the interval $\langle 0, \infty \rangle$, but can have only certain allowed energy values; that is, the particle energy is quantized.

Here again, it is necessary to find a function $\Psi(x)$ that, when differentiated twice, yields the same function multiplied by a constant. From the theory of linear second-order differential equations it follows that the general solution of Eq. (3-13) can be found in the form

$$\Psi(x) = A \sin (\alpha x) + B \cos (\alpha x) \qquad (3\text{-}28)$$

It can easily be shown that, for function Ψ in this form,

$$\frac{d^2 \Psi}{dx^2} = -\alpha^2 \Psi \qquad (3\text{-}29)$$

This equation is identical to Eq. (3-13), which is to be solved, provided that

$$\alpha^2 = \frac{8\pi^2 mE}{h^2} \qquad (3\text{-}30)$$

So far, nothing has been involved in the solution that would limit the value of E.

Certain boundary conditions must be introduced: $\Psi(x)$ must equal zero at the edge of the box; hence

a) $$\Psi(0) = 0,$$

b) $$\Psi(a) = 0.$$

Condition a) is satisfied by expression (3-28) only if

$$B = 0$$

Moreover, condition b) requires that

$$\alpha a = n\pi,$$

where n is an integer called *the quantum number*. It then follows that

$$\alpha = \frac{n\pi}{a} \qquad (3\text{-}31)$$

By comparing equations (3-30) and (3-31) it is found that

$$\frac{8\pi^2 mE}{h^2} = \frac{n^2\pi^2}{a^2},$$

so that, for the allowed energy values

$$E = n^2 \left(\frac{h^2}{8ma^2} \right), \tag{3-32}$$

where $n = 1, 2, 3, \ldots$.

Now the value of constant A [Eq. (3-28)] must be found; function Ψ must be normalized, i.e.

$$\int_{-\infty}^{+\infty} \Psi^2 \, d\tau = 1 \tag{3-33}$$

Here this condition has the form

$$\int_0^a \left(A \sin \frac{n\pi x}{a} \right)^2 dx = 1 \tag{3-34}$$

Calculation of the given integral results in the condition

$$A = \sqrt{\frac{2}{a}}$$

The solution is therefore obtained in the form

$$\Psi_n = \left(\sqrt{\frac{2}{a}} \right) \sin \frac{n\pi x}{a} \tag{3-35}$$

$$E_n = \frac{n^2 h^2}{8ma^2} \tag{3-36}$$

In Fig. 3-5 the result of the calculation (for $n = 1, 2, 3$) is represented graphically. Wave functions Ψ_n are given at levels corresponding to the respective values of E_n. In addition to Ψ_n, Ψ_n^2 is also plotted as a function of x.

For a given quantum number n, the energy is inversely proportional to the mass of the particle and to the length of the box. The heavier the particle and the longer the box, the closer together the values of E_n lie. For example, with $m \approx 1$ g and $a \approx 1$ cm, the levels are so close together that they appear as a continuum. Therefore quantization occurs only when $ma^2 \approx h^2$ (where $h = 6.625 \cdot 10^{-34}$ J s). On the other hand, if

$$ma^2 \gg h^2$$

the quantum mechanical treatment leads to the classical result, i.e. to energies that are not quantized. Furthermore, from Fig. 3-5 it is apparent that Ψ_n changes sign at every nodal point (a point where $\Psi = 0$); the number of nodal points equals $(n - 1)$. In general, the larger the number of nodal points (or nodal planes) under otherwise constant conditions, the higher is the energy of the corresponding state.

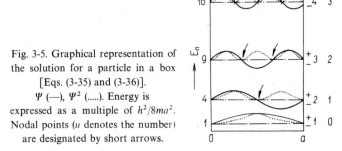

Fig. 3-5. Graphical representation of the solution for a particle in a box [Eqs. (3-35) and (3-36)]. Ψ (—), Ψ^2 (.....). Energy is expressed as a multiple of $h^2/8ma^2$. Nodal points (u denotes the number) are designated by short arrows.

A particle in a three-dimensional box provides a very instructive illustration. In this connection, a certain technique for the solution of the Schrödinger equation, which will be used later, will be introduced.

The Schrödinger equation (3-12b) for a particle in a three-dimensional box assumes the form [cf. Eq. (3-13)]

$$-\frac{h^2}{8\pi^2 m}\Delta\Psi = E\Psi \tag{3-37}$$

This equation can be rewritten to give

$$\frac{\partial^2\Psi}{\partial x^2} + \frac{\partial^2\Psi}{\partial y^2} + \frac{\partial^2\Psi}{\partial z^2} = -\frac{8\pi^2 mE}{h^2}\,\Psi \tag{3-38}$$

To solve this equation an attempt can be made to separate variables x, y and z, i.e. the solution must be found in the form

$$\Psi = X(x)\,Y(y)\,Z(z), \tag{3-39}$$

where each of the functions X, Y and Z depends exclusively on a single variable.

If the expression for function Ψ from the last equation is introduced into Eq. (3-38), then, after partial differentiation, the expression

$$YZ\frac{\partial^2 X}{\partial x^2} + XZ\frac{\partial^2 Y}{\partial y^2} + XY\frac{\partial^2 Z}{\partial z^2} = -\frac{8\pi^2 mE}{h^2}\,XYZ \tag{3-40}$$

is obtained. Dividing this equation by the product XYZ and modifying it leads to the relationship

$$\frac{1}{X}\frac{\partial^2 X}{\partial x^2} + \frac{1}{Y}\frac{\partial^2 Y}{\partial y^2} + \frac{8\pi^2 mE}{h^2} = -\frac{1}{Z}\frac{\partial^2 Z}{\partial z^2} \qquad (3\text{-}41)$$

In order that this equation hold for any set of values x, y, z, both sides must equal a constant; this constant can be expressed in the form $8\pi^2 mE_z/h^2$, where the value of constant E_z is still undetermined. Therefore,

a)

$$-\frac{1}{Z}\frac{\partial^2 Z}{\partial z^2} = \frac{8\pi^2 mE_z}{h^2} \qquad (3\text{-}42)$$

b)

$$\frac{1}{X}\frac{\partial^2 X}{\partial x^2} + \frac{1}{Y}\frac{\partial^2 Y}{\partial y^2} + \frac{8\pi^2 mE}{h^2} = \frac{8\pi^2 mE_z}{h^2} \qquad (3\text{-}43)$$

The same procedure as used in Eq. (3-41) can be applied to Eq. (3-43), which can be rearranged to give

$$\frac{1}{X}\frac{\partial^2 X}{\partial x^2} + \frac{8\pi^2 m}{h^2}(E - E_z) = -\frac{1}{Y}\frac{\partial^2 Y}{\partial y^2}, \qquad (3\text{-}44)$$

where both sides can be set equal to a constant; this constant can then be expressed as $8\pi^2 mE_y/h^2$:

a)

$$-\frac{1}{Y}\frac{\partial^2 X}{\partial y^2} = \frac{8\pi^2 mE_y}{h^2} \qquad (3\text{-}45)$$

b)

$$\frac{1}{X}\frac{\partial^2 X}{\partial x^2} = -\frac{8\pi^2 m}{h^2}(E - E_z - E_y) = -\frac{8\pi^2 mE_x}{h^2} \qquad (3\text{-}46)$$

Equations (3-42), (3-45) and (3-46) have the same form as the equation for a particle in a one-dimensional box [cf. Eq. (3-13)], the solution of which is already known. If the three-dimensional box has dimensions a, b, c, we can write

$$X = \left(\sqrt{\frac{2}{a}}\right)\sin\frac{n_x \pi x}{a}; \qquad E_x = \frac{n_x^2 h^2}{8ma^2} \qquad (3\text{-}47)$$

$$Y = \left(\sqrt{\frac{2}{b}}\right)\sin\frac{n_y \pi y}{b}; \qquad E_y = \frac{n_y^2 h^2}{8mb^2} \qquad (3\text{-}48)$$

$$Z = \left(\sqrt{\frac{2}{c}}\right)\sin\frac{n_z \pi z}{c}; \qquad E_z = \frac{n_z^2 h^2}{8mc^2} \qquad (3\text{-}49)$$

Then, for the total wave function Ψ and the total energy E, it holds that

$$\Psi = XYZ = \left(\sqrt{\frac{8}{abc}}\right)\sin\frac{n_x \pi x}{a}\sin\frac{n_y \pi y}{b}\sin\frac{n_z \pi z}{c} \qquad (3\text{-}50)$$

$$E = E_x + E_y + E_z = \frac{h^2}{8m}\left[\frac{n_x^2}{a^2} + \frac{n_y^2}{b^2} + \frac{n_z^2}{c^2}\right] \qquad (3\text{-}51)$$

One aspect of the solution, which can be illustrated for a particle in a three-dimensional box where $a = b = c$, is worth mentioning. Then, for the total energy

$$E = \frac{h^2}{8ma^2}(n_x^2 + n_y^2 + n_z^2) \qquad (3\text{-}52)$$

Hence, the lowest energy level ($n_x = n_y = n_z = 1$) is given by

$$E(1, 1, 1) = \frac{3h^2}{8ma^2}$$

The next energy-richer state can be described by three combinations of quantum numbers, where two quantum numbers are set equal to 1 and one is equal to 2:

$$E(2, 1, 1) = E(1, 2, 1) = E(1, 1, 2) = \frac{3h^2}{4ma^2}$$

A level with the same energy but characterized by different combinations of quantum numbers and consequently by different wave functions is designated as a degenerate level. The number of states with the same energy is given by the order of the degeneracy. The second energy level (i.e. the first excited level) of a particle in a cubic box is, therefore, threefold degenerate.

The described procedure is the basis of the free electron method (briefly denoted as the FEMO method – free electron molecular orbital method) which was proven useful in the study of simpler conjugated compounds. This method is by far not as important as the LCAO-MO method (see Chapter 10); nevertheless, it deserves attention and not only from a pedagogical point of view.

3.3.3 The harmonic oscillator

A particle executing simple harmonic motion, called a harmonic oscillator, is another simple system that is of interest because of its important role as a model in molecular spectroscopy.

In the Schrödinger equation of the harmonic oscillator [Eq. (3-14)] new symbols are introduced for expressions that contain only constants:

$$a = \frac{8\pi^2 mE}{h^2}; \qquad b = \frac{2\pi\sqrt{(mk)}}{h}$$

Thus the equation assumes the form

$$\frac{d^2 \Psi}{dx^2} + (a - b^2 x^2)\, \Psi = 0 \qquad (3\text{-}53)$$

Next, a new dimensionless variable $\xi = (\sqrt{b})\, x$ is introduced; therefore $d^2/dx^2 = b\, d^2/d\xi^2$. Equation (3-53) can then be written in the form

$$\frac{d^2 \Psi}{d\xi^2} + \left(\frac{a}{b} - \xi^2\right) \Psi = 0 \qquad (3\text{-}54)$$

Wave function Ψ must necessarily fulfil the following conditions: it must be continuous, single-valued and finite. For the sake of simplicity, Eq. (3-54) will first be solved for $|\xi| \gg \sqrt{(a/b)}$ (the asymptotic solution). The equation

$$\frac{d^2 \Psi}{d\xi^2} - \xi^2 \Psi = 0 \qquad (3\text{-}55)$$

is then obtained, which is satisfied by the solution

$$\Psi = e^{\pm \xi^2/2} \qquad (3\text{-}56a)$$

The positive sign in the exponent has, of course, no physical meaning, since for $\xi \to \infty$ function Ψ tends to infinity; therefore, only the solution

$$\Psi = e^{-\xi^2/2} \qquad (3\text{-}56b)$$

is valid.

Let us now return to the original differential equation, (3-54). From mathematical experience it follows that the expression, $e^{-\xi^2/2}$, will act as a factor in the solution:

$$\Psi = f(\xi)\, e^{-\xi^2/2} \qquad (3\text{-}57)$$

It would then remain to establish the form of function $f(\xi)$. By substituting for Ψ and $d^2\Psi/d\xi^2$ into Eq. (3-54) the equation

$$f'' - 2\xi f' + \left(\frac{a}{b} - 1\right) f = 0 \qquad (3\text{-}58)$$

is obtained, where $f'' = d^2 f/d\xi^2$ and $f' = df/d\xi$.

Equation (3-58) can be solved in the form of an infinite power series:

$$f = \sum_{k=0}^{\infty} a_k \xi^k = a_0 + a_1 \xi^1 + a_2 \xi^2 + \dots \qquad (3\text{-}59)$$

The first and second derivatives can then be written as

$$f' = \sum_{k=1}^{\infty} k a_k \xi^{k-1} \qquad (3\text{-}60)$$

$$f'' = \sum_{k=2}^{\infty} k(k-1) a_k \zeta^{k-2} \tag{3-61}$$

Substituting these power series into differential equation (3-58) gives

$$\sum_{k=2}^{\infty} k(k-1) a_k \zeta^{k-2} - 2 \sum_{k=1}^{\infty} k a_k \zeta^k + \left(\frac{a}{b} - 1\right) \sum_{k=0}^{\infty} a_k \zeta^k = 0 \tag{3-62}$$

The term $l = k - 2$ can be substituted into the first summation and the original symbols can be retained, so that k appears instead of l. The second summation can be augmented by the term corresponding to $k = 0$, as this term vanishes anyway. Then the equation

$$\sum_{k=0}^{\infty} (k+2)(k+1) a_{k+2} \zeta^k = \sum_{k=0}^{\infty} a_k \zeta^k \left(2k + 1 - \frac{a}{b}\right) \tag{3-63}$$

is obtained. This equation must be valid for any value of ξ, which is, of course, possible only when the coefficients of identical powers of ξ are equal on both sides of the equation. By comparing the expressions for the coefficient for the same power of ξ, the recursion formula

$$a_{k+2} = a_k \frac{2k + 1 - \dfrac{a}{b}}{(k+2)(k+1)} \tag{3-64}$$

is obtained, permitting formation of a set of even coefficients a_k from the initial value a_0 and a similar set of odd coefficients from the initial value a_1. No conditions are imposed on the initial values of a_0 and a_1, so that they may be chosen arbitrarily.

Thus, the solution of differential equation (3-58) has been found in the form

$$\Psi(\xi) = f(\xi) e^{-\xi^2/2}, \tag{3-65}$$

but it has not yet been determined whether Ψ fulfils the requirements imposed on the wave function, in particular, whether it is finite for all values of ξ. To this end, the behaviour of $\Psi(\xi)$ will be compared with the behaviour of the exponential function e^{ξ^2}, which evidently diverges for large values of ξ. This comparative function can be expressed in the form of an infinite series:

$$e^{\xi^2} = 1 + \frac{\xi^2}{1!} + \frac{\xi^4}{2!} + \dots + \frac{\xi^v}{\left(\dfrac{v}{2}\right)!} + \frac{\xi^{v+2}}{\left(\dfrac{v}{2} + 1\right)!} + \dots$$

$$= b_0 + b_2 \xi^2 + b_4 \xi^4 + \dots + b_v \xi^v + b_{v+2} \xi^{v+2} + \dots \tag{3-66}$$

For the quotient of two successive coefficients the relationship

$$\frac{b_{v+2}}{b_v} = \frac{1}{\dfrac{v}{2} + 1} \tag{3-67}$$

is obtained, which, for large values of v, can be replaced by the expression $2/v$. An analogous coefficient ratio in infinite series (3-59) is given for large values of k — as follows from Eq. (3-64) — by the expression:

$$\frac{a_{k+2}}{a_k} \approx \frac{2}{k} \tag{3-68}$$

The Schrödinger Equation and its Solution for the Rigid Rotator, Harmonic Oscillator and

Problem	Schrödinger equation (in Cartesian coordinates)	Schrödinger equation (in polar coordinates)
rigid rotator	$\Delta\Psi + \dfrac{8\pi^2\mu E}{h^2}\,\Psi = 0$	$\dfrac{1}{\sin^2\Theta}\dfrac{\partial^2\Psi}{\partial\Phi^2} +$ $+ \dfrac{1}{\sin\Theta}\dfrac{\partial}{\partial\Theta}\left(\sin\Theta\dfrac{\partial\Psi}{\partial\Theta}\right) +$ $+ \dfrac{2IE\Psi}{h^2} = 0$
harmonic oscillator	$\dfrac{d^2\Psi}{dx^2} + \dfrac{8\pi^2\mu}{h^2}\left(E - \dfrac{kx^2}{2}\right)\Psi = 0$	
hydrogen atom	$\Delta\Psi + \dfrac{8\pi^2 m}{h^2}\left(E + \dfrac{e^2}{4\pi\varepsilon_0 r}\right)\Psi = 0$	$\dfrac{\partial}{\partial r}\left(r^2\dfrac{\partial\Psi}{\partial r}\right) + \dfrac{1}{\sin^2\Theta}\dfrac{\partial^2\Psi}{\partial\Phi^2} +$ $+ \dfrac{1}{\sin\Theta}\dfrac{\partial}{\partial\Theta}\left(\sin\Theta\dfrac{\partial\Psi}{\partial\Theta}\right) +$ $+ \dfrac{2mr^2}{h^2}\left(E + \dfrac{e^2}{4\pi\varepsilon_0 r}\right)\Psi = 0$

It follows that both the series have the same asymptotic character. To examine the asymptotic behaviour of function Ψ, function $f(\xi)$ in Eq. (3-65) will be replaced by comparative function e^{ξ^2}:

$$\Psi \approx e^{\xi^2} e^{-\xi^2/2} = e^{\xi^2/2} \tag{3-69}$$

Hence it follows that function Ψ diverges for large values of ξ and consequently does not have the properties required of the wave function. It is evident that no solution containing $f(\xi)$ in the form of an infinite series is suitable. Thus another similar solution must be found that, in contrast to the former, does not diverge. To obtain a non-divergent finite

Hydrogen Atom		Table 3-1

E	Ψ	
$E_l = \left[\dfrac{l(l+1)\hbar^2}{2I} \right]$ $(l = 0, 1, 2, \ldots)$	$\Psi \equiv Y_{l,m}(\Theta, \Phi)$ $Y_{l,m}(\Theta, \Phi) = \underbrace{\left[\dfrac{2l+1}{2} \dfrac{(l-\vert m\vert)!}{(l+\vert m\vert)!} \right]^{1/2} P_l^{\vert m\vert}(\cos\Theta)}_{\underbrace{\qquad\qquad\qquad\qquad}_{\mathrm{Th}_{l,m}(\Theta)}} \underbrace{\dfrac{1}{\sqrt{(2\pi)}} e^{im\Phi}}_{F_m(\Phi)}$ $\underbrace{\hphantom{xxxxxxxxxxxxxxx}}_{\text{normalization factor}}$	associated Legendre polynomial (spherical harmonics) of degree l, of order m
$E_n = \left(n + \dfrac{1}{2}\right)h\nu =$ $= \left(n + \dfrac{1}{2}\right)h\omega$ $(n = 0, 1, 2, \ldots)$ $(\omega = 2\pi\nu)$	$\Psi_n(\xi) = \underbrace{\left[\dfrac{\sqrt{(b/\pi)}}{2^n (n!)} \right]^{1/2}}_{\substack{\text{normalization} \\ \text{factor}}} H_n(\xi)\, e^{-\xi^2/2}$ Hermite polynomial	
$E_n = -\dfrac{e^2}{8\pi\varepsilon_0 a_0 n^2}$ $(n = 1, 2, 3, \ldots)$	radial function angular function $\Psi = R_{n,l}(r)\, Y_{l,m}(\Theta, \Phi);$ spherical harmonics $R_{n,l}(r) = - \underbrace{\left[\dfrac{4(n-l-1)!}{n^4[(n+l)!]^3} \right]^{1/2} \left(\dfrac{Z}{a_0} \right)^{3/2}}_{\text{normalization factor}} \left(\dfrac{2\varrho}{n} \right)^l e^{-\varrho/n} L_{n+l}^{2l+1}\left(\dfrac{2\varrho}{n} \right)$	

solution, the infinite series can be terminated, that is, replaced by a polynomial. Then a_k will assume a value of zero after a certain term, i.e.

$$a_0 \neq 0, \; a_1 \neq 0, ..., a_n \neq 0, \tag{3-70}$$

and

$$a_{n+1} = a_{n+2} = a_{n+3} = ... = 0 \tag{3-71}$$

Considering Eq. (3-64), it must then hold that

$$a_{n+2} = a_n \frac{2n + 1 - \dfrac{a}{b}}{(n + 2)(n + 1)} = 0, \tag{3-72}$$

which is possible only when

$$2n + 1 - \frac{a}{b} = 0 \tag{3-73}$$

Substituting for quantities a and b and after appropriate rearrangement, the condition assumes the form

$$E_n = \frac{h}{2\pi}\left(\sqrt{\frac{k}{m}}\right)\left(n + \frac{1}{2}\right) = \left(n + \frac{1}{2}\right)h\nu, \tag{3-74}$$

where n is zero or a positive integer and the expression $(1/2\pi)\sqrt{(k/m)}$ has the meaning of frequency ν.

It then follows that the requirement that wave function Ψ be finite, which was realized by terminating the infinite series, leads to quantization

Some Special Functions Employed in Quantum Chemistry[a]   Table 3-2

Function	Symbol	Differential equation	Occurrence
harmonic	e^{imx}	$(d^2f/dx^2) + m^2f = 0$	translation motion
Legendre	$P_l(x)$	$(1 - x^2)(d^2f/dx^2) - 2x(df/dx) + l(l + 1)f = 0$	angular motion
associated Legendre	$P_l^m(x)$	$(1 - x^2)(d^2f/dx^2) - 2x(df/dx) + [l(l + 1) - m^2/(1 - x^2)]f = 0$	angular motion (hydrogen atom)
Laguerre[b]	$L_n(x)$	$x(d^2f/dx^2) + (df/dx) - (1/2 + x/4 + n)f = 0$	radial motion
Hermite	$H_n(x)$	$(d^2f/dx^2) - 2x(df/dx) + 2nf = 0$	harmonic oscillator

[a] Reproduced from J. M. Anderson: *Mathematics in Quantum Chemistry*, Benjamin, New York, 1966.

[b] Polynomial of the $L_k^s(x)$ type appearing in the solution of the wave function for the hydrogen atom is related to the Laguerre polynomial $L_k(x)$ by the relationship $L_k^s(x) = \dfrac{d^s}{dx^s} L_k(x)$.

of the energy values. For the sake of completeness, it should be added that differential equation (3-58) is called the *Hermite equation* and the corresponding solutions are denoted Hermite polynomials (cf. Tables 3-1 and 3-2).

The result of the quantum-mechanical treatment of the harmonic oscillator as represented by Eq. (3-74) is interesting in that it demonstrates the inadmissibility of zero energies (for $n = 0$, $E_0 = h\nu/2$, Fig. 3-6). This is related to the uncertainty principle; if the oscillating particle possesses zero energy, it would have zero momentum and would be located exactly in the equilibrium position characterized by the potential energy minimum.

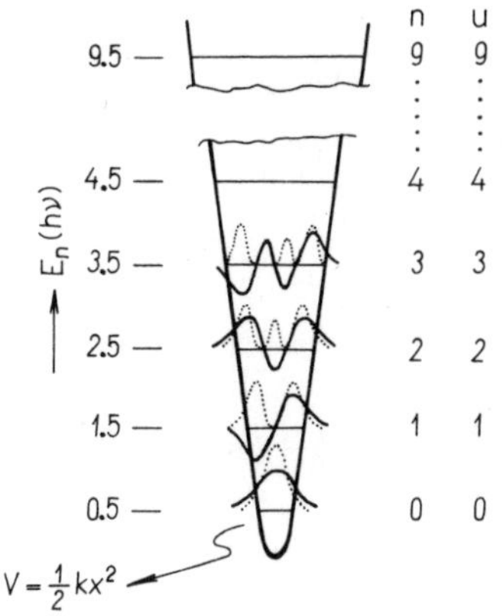

Fig. 3-6. Graphical representation of the solution for the harmonic oscillator [Eqs. (3-65) and (3-74)]. Ψ (———), Ψ^2 (.....). Energy is expressed in multiples of $h\nu$; u denotes the number of nodal points.

The particle in a box, for which the series of allowed quantum numbers begins with unity and not with zero (cf. p. 24), is analogous. The rotator (see the next section) is, however, different, as it can assume an infinite number of equilibrium positions in the plane and consequently its ground state can possess zero energy.

3.3.4 The rigid rotator

In the previous instances the complete solution of the problem was given. For the rigid rotator and the hydrogen atom the solution will be outlined to an extent sufficient for further discussion.

The theory of the rigid rotator is important in the analysis of the rotational spectra of diatomic molecules. The rigid rotator (Fig. 3-7) is

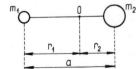

Fig. 3-7. The rigid rotator.

a system consisting of two point particles of mass m_1 and m_2, which are held a constant distance apart by a massless bond. This system rotates around the (O) axis passing through the centre of mass of the system and lying perpendicular to the projection plane. Here, translational motion of the rotator is not considered and therefore the centre of mass of the rotator is considered to be at rest, fixed in the origin of the coordinate system.

In classical mechanics the following relation [cf. Eq. (3-2)] holds for the kinetic energy of a two-particle system:

$$T = \frac{m_1 \dot{r}_1^2}{2} + \frac{m_2 \dot{r}_2^2}{2}, \qquad (3\text{-}75)$$

where r_1 and r_2 are radius vectors giving the positions of the two particles. The distance between the particles is given by vector r, for which it holds that

$$r = r_1 - r_2 \qquad (3\text{-}76)$$

If vector r is introduced into the relation expressing the assumed location of the centre of mass in the origin of the coordinate system (i.e. into the relation, $m_1 r_1 + m_2 r_2 = 0$), then

$$r_1 = \frac{m_2}{m_1 + m_2} r ; \qquad r_2 = -\frac{m_1}{m_1 + m_2} r \qquad (3\text{-}77)$$

Substitution of these equations into the expression for the kinetic energy (3-75) leads to the relationship

$$T = \frac{\mu \dot{r}^2}{2}, \qquad (3\text{-}78)$$

where

$$\mu = \frac{m_1 m_2}{m_1 + m_2} \qquad (3\text{-}79)$$

denotes the reduced mass of the system.

Since for a rigid rotator $|r| = a$ (a is a constant), it follows from Eq. (3-78) that the system is mathematically equivalent to a system in which a particle of mass μ moves over the surface of a sphere of radius a.

If no external forces act on the rotator, the potential energy of the hypothetical particle can be set equal to zero. The Schrödinger equation then has the form

$$\Delta \Psi + \frac{8\pi^2 \mu E}{h^2} \Psi = 0 \qquad (3\text{-}80)$$

Passing from Cartesian coordinates to spherical coordinates (cf. Fig. 3-8) permits exploitation of the symmetry of the system and leads to a substantial simplification of Eq. (3-80). Expression of the Laplace operator Δ in spherical coordinates can be found in text-books on quantum mechanics (cf. e.g. Ref. 1 in Chapter 4):

$$\Delta \equiv \left[\frac{\partial}{\partial r} \left(r^2 \frac{\partial}{\partial r} \right) + \frac{1}{\sin^2 \Theta} \frac{\partial^2}{\partial \Phi^2} + \frac{1}{\sin \Theta} \frac{\partial}{\partial \Theta} \left(\sin \Theta \frac{\partial}{\partial \Theta} \right) \right] r^{-2}$$

As for a rigid rotator, coordinate r is equal to constant a, the first term differentiated with respect to r is omitted. Furthermore, in Eq. (3-80) the reduced mass μ can be expressed in terms of the moment of inertia I,

$$I = \mu a^2, \tag{3-81}$$

and $h/2\pi$ can be replaced by $\hbar$. The Schrödinger equation for a rigid rotator in spherical coordinates is thus obtained, as given in Table 3-1. The solution of differential equations of this type is well known (see Table 3-1).

The eigenfunctions satisfying the given characteristic equation are called spherical harmonics, $Y_{l,m}(\Theta, \Phi)$, and can be expressed in a separated form

$$Y_{l,m}(\Theta, \Phi) = Th_{l,m}(\Theta) \, F_m(\Phi), \tag{3-82}$$

where the indices indicate the dependence on integral values of quantum numbers l and m, where

$$l \geqq |m| \tag{3-83}$$

For the eigenvalues giving the allowed energies of the rotator it holds that

$$E_l = \frac{l(l+1)\hbar^2}{2I}, \tag{3-84}$$

where l is the rotational quantum number, which can assume values of positive integers including zero.

3.3.5 The hydrogen atom

Solution of the hydrogen atom problem is, for several reasons, of fundamental importance. First, it is one of the few important systems in chemistry that — as a two-body problem — is still exactly solvable. Further, it provides a natural starting point for discussion and solution of problems concerning many-electron atoms. Finally, atomic wave functions of the hydrogen type or related types represent elementary building units for constructing molecular wave functions. For all these reasons, atomic wave functions of the hydrogen type will be discussed in detail in this section. The problem of the hydrogen atom is similar to that of the rigid rotator. The boundary condition of a constant distance between the components of the rigid rotator is replaced in the hydrogen atom by the existence of Coulomb interaction between the nucleus and the electron. Because of the large difference between the masses of the nucleus of the hydrogen atom (M) and of the electron (m), the nucleus (i.e. the proton) can be considered as the centre of mass of the system and assumed to be at rest. Hydrogen-type atoms, which are systems consisting of a nucleus of Z protons and one electron [i.e. cations with a charge of $+ (Z - 1) e$ (Fig. 3-8)], are similar. The Schrödinger equation for these atoms appears far more complicated in spherical coordinates than in Cartesian coordinates (compare Table 3-1), but the use of spherical coordinates as for the rigid rotator permits easy separation of the variables, r, Θ, and Φ.

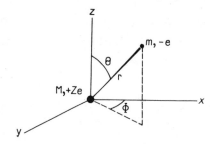

Fig. 3-8. Model of a hydrogen-like atom with nucleus of mass M and charge $+Ze$. The position of the electron is given in polar coordinates (r, Φ, Θ).

The solution can therefore begin with the Schrödinger equation for the hydrogen atom as given in Table 3-1, for which a solution in the form

$$\Psi = R(r) \, Y(\Theta, \Phi), \qquad (3\text{-}85)$$

will be sought, where R is a function of the radial coordinate r, alone, and Y is a function of the angular coordinates Θ and Φ. If Eq. (3-85) is substituted into the Schrödinger equation for the hydrogen atom, both sides of this equation are divided by function Ψ, and the expressions depending on angular variables are transferred to the right-hand side, then the equation

$$\frac{1}{R} \frac{\partial}{\partial r} \left(r^2 \frac{\partial R}{\partial r} \right) + \frac{2mr^2}{\hbar^2} \left(E + \frac{e^2}{4\pi\varepsilon_0 r} \right) =$$

$$= -\frac{1}{Y} \left[\frac{1}{\sin^2 \Theta} \frac{\partial^2 Y}{\partial \Phi^2} + \frac{1}{\sin \Theta} \frac{\partial}{\partial \Theta} \left(\sin \Theta \frac{\partial Y}{\partial \Theta} \right) \right] \qquad (3\text{-}86)$$

is obtained. Since either side depends on only one kind of independent variable, Eq. (3-86) can be satisfied only when

$$-\frac{1}{Y}\left[\frac{1}{\sin^2\Theta}\frac{\partial^2 Y}{\partial\Phi^2} + \frac{1}{\sin\Theta}\frac{\partial}{\partial\Theta}\left(\sin\Theta\frac{\partial Y}{\partial\Theta}\right)\right] = \lambda \qquad (3\text{-}87a)$$

and

$$\frac{1}{R}\frac{\partial}{\partial r}\left(r^2\frac{\partial R}{\partial r}\right) + \frac{2mr^2}{\hbar^2}\left(E + \frac{e^2}{4\pi\varepsilon_0 r}\right) = \lambda, \qquad (3\text{-}87b)$$

where λ is a constant. Comparison of Eq. (3-87a) with the equation for the rigid rotator in Table 3-1 demonstrates that they correspond to the same type of differential equation; the equations are identical when $\lambda = 2IE/\hbar^2$. Thus, the corresponding solutions for the eigenvalues and eigenfunctions for the rigid rotator can be used to express the quantities sought in Eq. (3-87a): function Y and the allowed values of constants λ are then given by

$$Y = Y_{l,m}(\Theta, \Phi) \qquad (3\text{-}88a)$$

$$\lambda = l(l + 1) \qquad (3\text{-}88b)$$

The symbols employed have the same meaning as for the rigid rotator. Due to the validity of Eq. (3-88b), function $R(r)$, as the solution of Eq. (3-87b), is obviously dependent on the value of quantum number l. As described in text-books on quantum mechanics, the approach to the solution of Eq. (3-87b) is similar to the calculation of the differential equation describing a harmonic oscillator (cf. Section 3.3.3). The final solution is, as in Eq. (3-57), sought in the form of a product of an approximate solution and a power series. The requirement of quadratic integrability of the wave function (that is, the requirement of normalizability) necessitates introduction of a further integral (positive) quantum number, n [similarly as in Eq.(3-73)]. As a final solution of the equation for the radial part of the wave function, the function $R_{n,l}(r)$ is obtained (see Tables 3-1 and 3-3).

Some Normalized Radial Wave Functions $R_{n,l}(r)$ for Hydrogen-like Atoms	Table 3-3

$$R_{1,0} = 2\left(\frac{Z}{a_0}\right)^{3/2}e^{-\varrho} \qquad\qquad R_{3,0} = \frac{2}{81\sqrt{3}}\left(\frac{Z}{a_0}\right)^{3/2}(27 - 18\varrho + 2\varrho^2)e^{-\varrho/3}$$

$$R_{2,0} = \frac{1}{2\sqrt{2}}\left(\frac{Z}{a_0}\right)^{3/2}(2 - \varrho)e^{-\varrho/2} \qquad\qquad R_{3,1} = \frac{4}{81\sqrt{6}}\left(\frac{Z}{a_0}\right)^{3/2}(6 - \varrho)\varrho\,e^{-\varrho/3}$$

$$R_{2,1} = \frac{1}{2\sqrt{6}}\left(\frac{Z}{a_0}\right)^{3/2}\varrho\,e^{-\varrho/2}$$

In Table 3-1 the Schrödinger equation for the hydrogen atom is given, together with its solution; conventional symbols are used. The symbol E_n denotes the energy of the n-th level of the hydrogen atom ($n = 1, 2, 3 \ldots$). The corresponding wave function Ψ (or $\Psi_{n,l,m}$), called the atomic orbital (AO), can be expressed using Eq. (3-85) as the product of the radial and angular parts (Fig. 3-8):

$$\Psi_{n,l,m} = R_{n,l}(r)\, Y_{l,m}(\Theta, \Phi), \tag{3-89}$$

where n denotes the principal quantum number, l is the azimuthal quantum number and m is the magnetic quantum number. Constant a_0 corresponds to the radius of the first Bohr orbit ($a_0 = 0.0529$ nm) and dimensionless parameter ϱ is defined by the relation

$$\varrho = \frac{Zr}{a_0}, \tag{3-90}$$

where Z is the number of protons in the nucleus.

The wave function $\Psi_{n,l,m}$ (3-89) is an eigenfunction not only of the Hamiltonian but also of the angular momentum operators (cf. Section 4.4). It holds

$$\mathscr{H}\Psi_{n,l,m} = E_n\Psi_{n,l,m} \tag{3-91}$$

$$\mathscr{L}^2\Psi_{n,l,m} = l(l+1)\hbar^2\Psi_{n,l,m} \tag{3-92}$$

$$\mathscr{L}_z\Psi_{n,l,m} = m\hbar\Psi_{n,l,m} \tag{3-93}$$

The principal quantum number n determines the energy of the electron, the quantum number l determines the orbital angular momentum of the electron, and the quantum number m determines its z component.

The quantum number n, can assume an arbitrary positive integral value (for large values of n quantization of the energy is less important). Quantum number l can assume any positive integral value from 0 to ($n - 1$), so that to each value of n corresponds a total of n different values of quantum number l, for which

$$0 \leqq l \leqq n - 1 \tag{3-94}$$

Quantum number m can assume any integral value in the interval $-l$ to $+l$ (including zero), i.e. to each value of l correspond $(2l + 1)$ different values of m, for which

$$|m| \leqq l \tag{3-95}$$

Numbers l and m (by analogy with the quantum numbers in the Bohr theory) are called *the azimuthal quantum number* and *the magnetic quantum number*, respectively; some authors call them *the angular momentum quantum numbers*. For lucidity, the individual values of the azimuthal quantum number are, as a rule, expressed by letters according to the following convention:

$$\text{value of } l: \quad 0 \ 1 \ 2 \ 3 \ 4$$

$$\text{designation:} \quad \text{s} \ \text{p} \ \text{d} \ \text{f} \ \text{g}$$

For elucidation of the shapes of the wave functions, some specific examples of radial, angular and total wave functions for hydrogen-like atoms are given in Tables 3-3 to 3-5. The partial and total functions are independently normalized to unity (cf. Section 3.1).

A disadvantage of the angular functions given in Table 3-4 lies in the fact that they are generally complex functions that cannot be represented in real space. However, equally good and real wave functions (atomic orbitals) are obtained through linear combination of spherical harmonics

Some Normalized Spherical Harmonics $Y_{l,m}(\Theta, \Phi)$ for Hydrogen-like Atoms   Table 3-4

s orbital	p orbitals	d orbitals
$Y_{0,0} = \dfrac{1}{2\sqrt{\pi}}$	$Y_{1,0} = \dfrac{\sqrt{3}}{2\sqrt{\pi}} \cos\Theta$	$Y_{2,0} = \dfrac{\sqrt{5}}{4\sqrt{\pi}}(3\cos^2\Theta - 1)$
	$Y_{1,\pm 1} = \dfrac{\sqrt{3}}{2\sqrt{(2\pi)}} \sin\Theta \, e^{\pm i\Phi}$	$Y_{2,\pm 1} = \dfrac{\sqrt{15}}{4\sqrt{(2\pi)}} \sin 2\Theta \, e^{\pm i\Phi}$
		$Y_{2,\pm 2} = \dfrac{\sqrt{15}}{4\sqrt{(2\pi)}} \sin^2\Theta \, e^{\pm 2i\Phi}$

Some Normalized Wave Functions $\Psi_{n,l,m}$ of Hydrogen-like Atoms   Table 3-5

n	l	m	
1	0	0	$\Psi_{1s} = \dfrac{1}{\sqrt{\pi}}\left(\dfrac{Z}{a_0}\right)^{3/2} e^{-\varrho}$
2	0	0	$\Psi_{2s} = \dfrac{1}{4\sqrt{(2\pi)}}\left(\dfrac{Z}{a_0}\right)^{3/2}(2-\varrho)e^{-\varrho/2}$
2	1	0	$\Psi_{2p_z} = \dfrac{1}{4\sqrt{(2\pi)}}\left(\dfrac{Z}{a_0}\right)^{3/2}\varrho \, e^{-\varrho/2}\cos\Theta$
2	1	± 1	$\begin{cases} \Psi_{2p_x} = \dfrac{1}{4\sqrt{(2\pi)}}\left(\dfrac{Z}{a_0}\right)^{3/2}\varrho \, e^{-\varrho/2}\sin\Theta\cos\Phi \\[2ex] \Psi_{2p_y} = \dfrac{1}{4\sqrt{(2\pi)}}\left(\dfrac{Z}{a_0}\right)^{3/2}\varrho \, e^{-\varrho/2}\sin\Theta\sin\Phi \end{cases}$

with equal values of l, and the same value of $|m|$, for instance:

$$\left(\sqrt{\tfrac{1}{2}}\right)(Y_{1,1} + Y_{1,-1}) = \frac{\sqrt{3}}{2\sqrt{\pi}}\sin\Theta\cos\Phi \qquad (3\text{-}96)$$

(It should be noted that the actual atomic orbitals are obtained by multiplying the angular part of the wave function by the radial part [cf. Eq. (3-89)]. This circumstance, of course, does not affect the symmetry considerations in any way.) This combination will be denoted p_x, since the expression on the right-hand side of Eq. (3-96) exhibits the same angular dependence as the expression for transformation of the x-coordinate when passing from Cartesian to spherical coordinates. Two more real atomic orbitals with $l = 1$ can be obtained in a similar way, their designation being apparent from the survey given below (symbol $Y_{1,1}$ is replaced by p_1, etc.):

$$\frac{p_1 + p_{-1}}{\sqrt{2}} \sim \sin\Theta\cos\Phi \sim x \qquad\qquad p_x$$

$$p_0 \sim \cos\Theta \sim z \qquad\qquad p_z$$

$$-i\,\frac{p_1 - p_{-1}}{\sqrt{2}} \sim \sin\Theta\sin\Phi \sim y \qquad\qquad p_y$$

Table 3-6

Survey of Atomic Orbitals for the Principal Quantum Numbers $n = 1$, $n = 2$, and $n = 3$

Quantum numbers			Symbol of atomic orbital $\Psi_{n,l,m}$	Conventional notation	
n	l	m		angular function (complex)	resulting real function (atomic orbital)
1	0	0	Ψ_{100}	1s	1s
2	0	0	Ψ_{200}	2s	2s
	1	-1	Ψ_{21-1}	$2p_{-1}$	$2p_x$
	1	0	Ψ_{210}	$2p_0$	$2p_y$
	1	1	Ψ_{211}	$2p_1$	$2p_z$
3	0	0	Ψ_{300}	3s	3s
	1	-1	Ψ_{31-1}	$3p_{-1}$	$3p_x$
	1	0	Ψ_{310}	$3p_0$	$3p_y$
	1	1	Ψ_{311}	$3p_1$	$3p_z$
	2	-2	Ψ_{32-2}	$3d_{-2}$	$3d_{xy}$
	2	-1	Ψ_{32-1}	$3d_{-1}$	$3d_{xz}$
	2	0	Ψ_{320}	$3d_0$	$3d_{yz}$
	2	1	Ψ_{321}	$3d_1$	$3d_{x^2-y^2}$
	2	2	Ψ_{322}	$3d_2$	$3d_{z^2}$

In a similar way, real atomic orbitals denoted d_{xy}, d_{xz}, d_{yz}, $d_{x^2-y^2}$, d_{z^2} are obtained from complex angular functions d_0, $d_{\pm 1}$, and $d_{\pm 2}$. A survey of atomic orbitals with principal quantum numbers $n = 1$, $n = 2$, and $n = 3$ is given in Table 3-6.

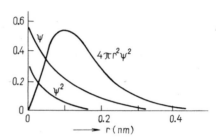

Fig. 3-9. The shape of the $\Psi(1s)$ atomic orbital and of quantities related to it in terms of dependence on the distance from the nucleus.

In chemical applications, graphic representation of wave functions is frequently used; however, their radial and angular parts are generally depicted separately. From this point of view, the 1s atomic orbital (i.e. the wave function Ψ_{1s}), is instructive; here the relationships are particularly simple as the function is spherically symmetrical and decreases exponentially with increasing distance from the nucleus. (Spherical symmetry is a characteristic property of all s-type atomic orbitals, whatever the value of the principal quantum number.) Fig. 3-9 represents the dependence of Ψ, Ψ^2 and $4\pi r^2\Psi^2$ on the value of r for the 1s orbital (Ψ designates the Ψ_{1s} function and Ψ^2 gives the probability density of an electron at any point at a distance r from the nucleus).

In Fig. 3-10 various modes of graphical representation of the 1s orbital are given: a) a section through a series of concentric spheres with the value of Ψ given for each ("contour" representation); b) designation of the envelope corresponding to the space where the probability of electron occurence anywhere inside this envelope is, for instance, 90 per cent (or, for example, 99 per cent); for all points on the envelope Ψ ($\equiv \Psi_{1s}$) has the same value; c) indication of the electron probability density at different points by the corresponding density of dots (for the 1s orbital the density of dots decreases exponentially with the distance from the nucleus).

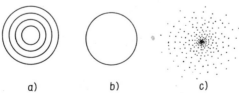

a)   b)   c)

Fig. 3-10. Various modes of graphical representation of the 1s atomic orbital: (a) by contours, (b) by a region with certain probability of electron occurrence, (c) by an electron cloud.

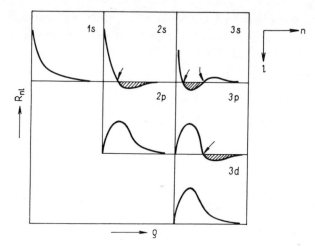

Fig. 3-11. Dependence of radial functions R_{nl} on ϱ (Table 3-3). The region corresponding to negative R_{nl} values is shaded, nodal points are indicated by arrows.

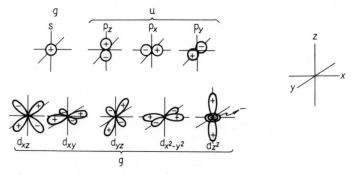

Fig. 3-12. Angular part of the s, p, d orbitals. The signs of the wave functions are indicated $(+, -)$ and symmetry with respect to inversion is given (g, u).

For atomic orbitals with higher values of n and l, the graphical representation becomes more complicated. In Figs. 3-11 and 3-12, radial and angular functions are pictured for several different orbitals. Whereas a similar mode was used for depicting the radial function, as for the 1s orbital in Fig. 3-9, the angular part of the orbital can most conveniently be depicted by mode b (cf. Fig. 3-10). While this representation encounters no difficulties for s-type orbitals, the other types of orbitals are treated in a way that can be illustrated, for instance, on the $2p_x$ orbital. For purposes of graphical representation, that part of the function which depends on variables Θ and Φ is separated from the expression for this

orbital and is denoted B; the resultant relationship then formally corresponds to the expression for this orbital where the radial part is by
convention set equal to unity $[f(r) = 1]$:

$$\Psi_{2p_x} = f(r)\underbrace{|\sin\Theta\cos\Phi|}_{B}$$

A representation using the square of the wave function can also be
employed:

$$\Psi^2_{2p_x} = f^2(r)\underbrace{|\sin\Theta\cos\Phi|^2}_{B^2}$$

Graphical representations are usually limited to two dimensions; orbital
$2p_x$ can be conveniently characterized by a section lying in the $x - y$
plane for which $\Theta = 90°$. Graphical representation of the dependence
of B or B^2 on Φ is called a polar plot (Fig. 3-13).

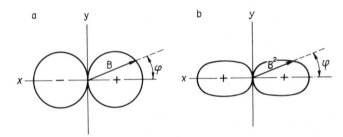

Fig. 3-13. (a) Graphical representation of values of the $2p_x[f(r) = 1]$ orbital denoted by B,
(b) A similar dependence as in the part (a) but for values of B^2.

To the graphical representation of the angular function can be
added that each value of l corresponds to a certain characteristic shape
independent of the principal quantum number. The s orbitals have, as
has already been mentioned, spherical symmetry, whereas the other
orbitals are directed (i.e. they have relatively large values in certain
directions from the nucleus); these facts are important for the theory
of the chemical bond. According to mode b in Fig. 3-10, the p orbitals
can be represented by a pair of ellipsoids, touching each other in one
point, which are symmetrical with respect to the axis designated by the
index of symbol p (i.e. the p_x, p_y and p_z orbitals are oriented in three
mutually perpendicular directions). For the sake of lucidity, in graphical
representation of these orbitals, cigar-like shapes are drawn instead of
rather voluminous ellipsoids. This is particularly advantageous in more
complicated representations encountered, for instance, with d orbitals,

each of which (except the d_{z^2} orbital) corresponds to four such cigar-like configurations. A complete wave function (the atomic orbital) is rarely treated graphically, since the form of the angular function is usually quite satisfactory. It is necessary, of course, in the individual parts of the representation (that is, in the individual cigar-like areas) to indicate the sign of the wave function Ψ in the respective part of space. This is, in fact, very important for considerations on orbital overlap in connection with the formation of chemical bonds.

The results of the study of atomic orbital properties in the hydrogen atom can be characterized as follows:

a) The larger the value of n, the larger the spatial area in which the wave function assumes non-vanishing values. This is not readily apparent from Fig. 3-11, however, as parameter ϱ is plotted on the x-axis instead of distance r, where

$$\varrho = \frac{2Zr}{na_0}$$

b) The probability density of an electron in the nucleus is zero, the only exception being s orbitals. This fact is very important for spectroscopy in the radiofrequency region.

c) The number of nodal surfaces of the atomic orbitals (that is, surfaces in which the wave function has zero value) depends on the values of numbers n and l. Spherical nodal surfaces occur in the radial part of the wave function; in the angular part nodal planes are involved. The number of these planes in the angular part equals number l. Since the total number of nodal surfaces equals $n - 1$, $(n - l - 1)$ nodal surfaces will remain for the radial part (after subtracting the l nodal planes of the angular part).

d) In atomic orbitals with identical values of n, the electron density close to the nucleus is smaller for larger values of number l. From the solution of the Schrödinger equation for the hydrogen atom (Table 3-1) it follows that the energy term appears only in its radial part, so that obviously the energy is independent of quantum number m (which occurs in the angular function, but not in the radial). As the radial function involves both n and l, it could be expected that the orbital energy will depend on both of these quantum numbers. It appears, however, that, owing to the special position of the Coulomb potential of the point charge among potentials of spherical symmetry, the orbital energy in hydrogen-like atoms depends on the principal quantum number n alone:

$$E_n = - \frac{Z^2 e^2}{2 . 4\pi\varepsilon_0 a_0 n^2} = - \frac{Z^2}{n^2} 13.60 \text{ eV} \qquad (3\text{-}97a)$$

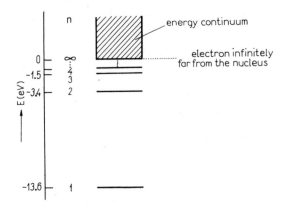

Fig. 3-14. Dependence of the orbital energy of the hydrogen atom [Eq. (3-97a); $Z = 1$] on quantum number n.

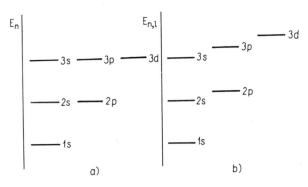

Fig. 3-15. Dependence of orbital energies on quantum numbers n and l for (a) an atom with one electron, (b) an atom with more electrons.

Negative values are obtained for orbital energies due to the fact that zero energy was, by definition, assigned to a system composed of a proton and of an infinitely distant electron (Fig. 3-14); all the remaining systems, where the electron is closer to the proton, have energy values below zero, i.e. negative.

It is worth noting that the energy of a particle exposed to a constant potential field in a "box" is directly proportional to n^2, whereas the energy of a charged particle in the central electrostatic field varies inversely with n^2. It is apparent from expression (3-97a) that, in hydrogen-like atoms, the 3s, 3p and 3d orbitals have the same energy (that is, they are many-fold degenerate states). It will be shown later that this degeneracy is considerably reduced in systems with more electrons (at least two) where repulsion among these electrons is considered (Fig. 3-15). In these systems, the energy is dependent not only on n, but also on l, so that only those orbitals that have the same combinations of n and l are degenerate; the order of degeneracy is equal to $2l + 1$ (hence, the order of degeneracy of p, d and f orbitals is 3, 5 and 7, respectively).

Transition of an electron from an orbital with principal quantum number n_1 into an orbital with this number equal to n_2 is accompanied by emission or absorption of energy in the form of electromagnetic radiation. For the hydrogen-like atom the following relationship (cf. Table 3-1) is valid:

$$E_{n_2} - E_{n_1} = \frac{Z^2 e^2}{8 \pi \varepsilon_0 a_0} \left(\frac{1}{n_1^2} - \frac{1}{n_2^2} \right) = h\nu = \frac{hc}{\lambda} \qquad (3\text{-}97b)$$

REFERENCES

References given in Chapter 4 can be also used in this chapter.

4. MATHEMATICS AND LOGIC OF QUANTUM MECHANICS

4.1 Linear operators and their properties

In the preceding chapter the concept of an operator was introduced, which will now be made more precise and specified in greater detail. The theory of linear operators is one of the fundamental mathematical tools of quantum mechanics[1−12].

Definition 1. The term *operator* $\mathcal{O}$ denotes an instruction according to which to function $f(x, y, ...)$ of coordinates $x, y, ...$ is assigned another function $F(x, y, ...)$ in the same variables:

$$\mathcal{O}f = F \tag{4-1}$$

(denotation of the dependence on the coordinates is omitted).

Definition 2. Operator $\mathcal{O}$ is linear if

$$\mathcal{O}(f_1 + f_2) = \mathcal{O}f_1 + \mathcal{O}f_2 \tag{4-2}$$

and

$$\mathcal{O}cf = c\mathcal{O}f, \tag{4-3}$$

f_1 and f_2 being functions and c an arbitrary constant.

From relations (4-2) and (4-3) it follows that

$$\mathcal{O}(c_1 f_1 + c_2 f_2) = c_1 \mathcal{O}f_1 + c_2 \mathcal{O}f_2, \tag{4-4}$$

where c_1 and c_2 are arbitrary constants. In all the following considerations it will be assumed that general operators satisfy properties (4-2) and (4-3). It can easily be verified that the operators introduced in the preceding chapter are linear.

We have already encountered operators requiring differentiation of the function or multiplication by one of the coordinates − such operators or combinations thereof will appear here most frequently. The following symbols

$$\mathcal{P}_x = \frac{\partial}{\partial x} \tag{4-5}$$

$$\mathcal{Q}_x = x \qquad (4\text{-}6)$$

can be introduced. By definition,

$$\mathcal{P}_x f = \frac{\partial f}{\partial x} \qquad (4\text{-}7)$$

$$\mathcal{Q}_x f = x \cdot f \qquad (4\text{-}8)$$

The action of operator $\mathcal{Q}_x$ on Eq. (4-7) gives

$$\mathcal{Q}_x \mathcal{P}_x f = x \frac{\partial f}{\partial x} \qquad (4\text{-}9)$$

and $\mathcal{P}_x$ acting on Eq. (4-8) gives

$$\mathcal{P}_x \mathcal{Q}_x f = \frac{\partial}{\partial x}(x \cdot f) = f + x \frac{\partial f}{\partial x}, \qquad (4\text{-}10)$$

from which it follows that the result depends on the sequence of operations. In other words, operators $\mathcal{P}_x$ and $\mathcal{Q}_x$ are not commutative. Subtracting Eq. (4-9) from Eq. (4-10) gives

$$[\mathcal{P}_x \mathcal{Q}_x - \mathcal{Q}_x \mathcal{P}_x] f = f, \qquad (4\text{-}11)$$

and since f is an arbitrary function of its coordinates, Eq. (4-11) can be changed into a form in which only operators occur:

$$\mathcal{P}_x \mathcal{Q}_x - \mathcal{Q}_x \mathcal{P}_x = 1, \qquad (4\text{-}12)$$

where 1 is the identity operator.

Definition 3. Operators satisfying the equation

$$[\mathcal{P}, \mathcal{Q}] \equiv \mathcal{P}\mathcal{Q} - \mathcal{Q}\mathcal{P} = 0 \qquad (4\text{-}13)$$

are termed *commutative* operators. The symbol $[\mathcal{P}, \mathcal{Q}]$ is called *the commutator* of $\mathcal{P}$ and $\mathcal{Q}$.

For example, operators $\mathcal{Q}_x$ and $\mathcal{P}_y$ are a pair of commutative operators.

Definition 4. Operator $\mathcal{O}(x)$ is a Hermitian operator if

$$\int f_1^*(x)\, \mathcal{O}(x)\, f_2(x)\, dx = \int f_2(x)\, \mathcal{O}^*(x)\, f_1^*(x)\, dx, \qquad (4\text{-}14)$$

where an asterisk denotes complex conjugate quantities and integration is performed over all possible values of variable x. If more variables are involved, dx is replaced by the volume element $d\tau = dx\, dy \ldots$ and integration is performed over variables $x, y, \ldots$.

Definition 5. To each operator $\mathcal{O}$ can be assigned a linear equation of the type

$$\mathcal{O}f = of, \qquad (4\text{-}15)$$

where o is a constant which, in general, can be a complex number. This equation is called *an eigenvalue* (or *characteristic*) *equation*, function f and constant o are called, respectively, *an eigenfunction* (or *characteristic function*) and an *eigenvalue* (or *characteristic value*) of operator $\mathcal{O}$.

Equation (4-15) can be satisfied by a number of functions with various properties; among them is the trivial solution, $f = 0$. In order to ensure that the solution obtained is reasonable from a physical point of view (cf. Eq. (4-22)) the eigenfunctions must fulfil the following requirements:

a) function f exists in the whole region in which the variables lie,

b) function f must be continuous and finite everywhere in this region (with the exception of singular points),

c) function f must be single-valued.

Other functions can, of course, satisfy Eq. (4-15) so that

$$\mathcal{O}f_k = o_k f_k, \tag{4-16}$$

where the subscript indicates that there can be further eigenvalues whose spectrum can pass through discrete or continuous values.

An interesting consequence for its eigenvalues follows from the definition of the hermicity of operator $\mathcal{O}$. If Eq. (4-15) is multiplied from the left by the function f^* and integration is carried out over the entire space, then

$$o = \frac{\int f^* \mathcal{O} f \, d\tau}{\int |f|^2 \, d\tau} \tag{4-17}$$

Similarly, if we start with the complex conjugate of Eq. (4-15) and multiply it by function f, after integration we obtain

$$o^* = \frac{\int f \mathcal{O}^* f^* \, d\tau}{\int |f|^2 \, d\tau} \tag{4-18}$$

From comparison of Eqs. (4-17) and (4-18) and from the condition of hermicity of operator (4-14) it can be seen that

$$o = o^* \tag{4-19}$$

Thus, it obviously holds that:

Theorem 1. The eigenvalues of Hermitian operators are real numbers.

4.2 Axiomatic foundation of quantum mechanics

Scientific disciplines, deductive by character, depend on axioms or postulates which are considered to be fundamental and non-deducible. It is necessary to realize that the justification of postulates depends on the ability of the theory based on them to interpret observations (to correlate data) and to predict experimental facts.

There is, as a rule, a number of ways of forming the axiomatic basis of a certain branch of science. Individual modes can differ in character, generality and the number of postulates required. Classical mechanics is a good example: it can begin from Newton's laws or from the principle of least action and the properties of an inertial frame of reference. It is evident that, while the first approach allows simple formulation of "normal" problems from the field of mechanics up to the motion of celestial bodies, the second is, because of its generality, also useful for investigating problems in which electrical, magnetic and relativistic phenomena appear. The example given also shows that the system of postulates can be chosen so that, within a specific application, it may be possible to use the respective laws directly without further derivation.

For our purpose, i.e. for the application of quantum mechanics to the problems of chemical bonding, it will be satisfactory to axiomatically introduce the Schrödinger equation, interpretation of the wave function and the requirements imposed on it. Even so, it is not surprising that the postulates of quantum mechanics are not immediately clear as they concern the properties of particles of the microcosmos, with which we have no direct experience.

To investigate a system composed of n particles, the classical description requires knowledge of $3n$ coordinates and $3n$ momenta at a given instant in time. Description of this system in quantum mechanics can be performed after introduction of the following postulates.

Postulate 1. To every physical quantity M corresponds a linear, Hermitian operator $\mathcal{M}$ (observable), which can be obtained by the following steps: in the classical expression for the corresponding physical quantity expressed in terms of Cartesian coordinates and momenta

  a) time and coordinates will remain unchanged

  b) linear momentum p_x will be replaced by the operator

$$\not\!p_x = \frac{h}{2\pi \mathrm{i}} \frac{\partial}{\partial x}$$

Postulate 2. Every dynamic state of the particle system is fully described by a function of the coordinates and time, *wave function* Φ. The wave function (in the form normalized to unity) must satisfy conditions *a)* to *c)* in Section 4.1, and has the following physical interpretation: the expression $\Phi^*\Phi \, \mathrm{d}\tau$ gives the probability that at time t the variables lie in the intervals x_1 to $x_1 + \mathrm{d}x_1$, y_1 to $y_1 + \mathrm{d}y_1$, z_1 to $z_1 + \mathrm{d}z_1$, x_2 to $x_2 + \mathrm{d}x_2$, ... z_n to $z_n + \mathrm{d}z_n$, where $\mathrm{d}\tau = \mathrm{d}x_1 \, \mathrm{d}y_1 \, \mathrm{d}z_1 \ldots \mathrm{d}z_n$. Each of the n

particles occurs somewhere in space and therefore the integrated probability density must equal 1:

$$\int \Phi^* \Phi \, d\tau = 1 \qquad (4\text{-}20)$$

Equation (4-20) must be understood as a condition which is satisfied by function Φ at a certain instant t. The time dependence of the wave function then need not be considered; it is necessary, however, to require that the norm be maintained throughout the time development of the system.

Postulate 3. In *the time-dependent Schrödinger equation* the wave function Φ satisfies the relationship

$$\mathcal{H}\Phi = i\hbar \frac{\partial \Phi}{\partial t}, \qquad (4\text{-}21)$$

where $\mathcal{H}$ is the Hamiltonian operator of the given system and t is the time.

Postulate 4. The only possible values obtained by measuring the physical quantity M are the eigenvalues, m_k, of the equation

$$\mathcal{M}\Psi_k = m_k \Psi_k, \qquad (4\text{-}22)$$

where Ψ_k fulfils conditions a) to c) in Section 4.1 and $\mathcal{M}$ is the corresponding observable.

Postulate 5. If the state of the given system is described by wave function Φ, the mean (or expectation) value $\bar{m}$ of physical quantity M is given by the expression

$$\bar{m} = \frac{\int \Phi^* \mathcal{M}\Phi \, d\tau}{\int \Phi^* \Phi \, d\tau}. \qquad (4\text{-}23)$$

In the next section, the relationship between the mean value defined in this way and the mean value of an experimental quantity obtained by a series of measurements will be described.

4.3 Consequences of the axiomatic system

To facilitate analysis of the properties of the wave function, some important concepts must be introduced:

Definition 6. The set of functions $\varphi_1(x)$, $\varphi_2(x)$, ..., $\varphi_k(x)$, ..., defined for variable x in the interval (a, b), is orthonormal if the scalar product satisfies the relationship

$$\int_a^b \varphi_i^*(x)\, \varphi_j(x)\, dx = \langle \varphi_i \,|\, \varphi_j \rangle = \delta_{ij}, \qquad (4\text{-}24)$$

where δ_{ij} is the Kronecker symbol ($\delta_{ii} = 1$; $\delta_{ij} = 0$ for $i \neq j$), and is complete if the arbitrary function $f(x)$ satisfying properties a) to c) introduced in Section 4.1 can be approximated by

$$f(x) \approx f_m(x) = \sum_{i=1}^{m} c_i \varphi_i(x), \tag{4-25}$$

where

$$\int_a^b |f(x) - f_m(x)|^2 \, dx \tag{4-26}$$

approaches zero provided m tends to infinity.

Similar sets of functions exist for functions of more variables and also for infinite intervals.

Expansions of type (4-25) are useful for representing the wave function of a system in terms of a linear combination of functions $\varphi_i(x)$. If the analytical form of the function f is known, the expansion coefficients, c_i, can easily be calculated. Multiplying Eq. (4-25) from the left by the function $\varphi_k^*(x)$, performing integration over the given interval and considering the orthonormality of the functions leads to the expression

$$c_k = \int_a^b \varphi_k^*(x) \, f(x) \, dx \tag{4-27}$$

The form of expansion (4-25) is also useful for finding the wave function as a solution of an eigenvalue problem. The solution leads, as a rule, to a linear problem in variables c_i that is relatively easy to solve.

Theorem 2. The eigenfunctions of the Hermitian operator form a complete orthonormal set.

Proof of the completeness of a set of functions is a difficult problem, so that only an outline of the proof of the orthonormality of eigenfunctions of the Hermitian operator will be given here.

Let $\mathcal{M}$ be the Hermitian operator, Ψ_1 and Ψ_2 its two eigenfunctions and m_1 and m_2 the respective eigenvalues. Therefore

$$\mathcal{M}\Psi_1 = m_1 \Psi_1 \tag{4-28}$$

$$\mathcal{M}\Psi_2 = m_2 \Psi_2 \tag{4-29}$$

Multiplication of Eq. (4-28) from the left by function Ψ_2^* and integration gives

$$\int \Psi_2^* \mathcal{M}\Psi_1 \, d\tau = m_1 \int \Psi_2^* \varphi_1 \, d\tau, \tag{4-30}$$

and, because of the hermiticity of $\mathcal{M}$

$$\int \Psi_1 \mathcal{M}^* \Psi_2^* \, d\tau = m_1 \int \Psi_2^* \Psi_1 \, d\tau \tag{4-31}$$

The complex conjugate of Eq. (4-29) is then multiplied from the left by Ψ_1 and integrated to give

$$\int \Psi_1 \mathcal{M}^* \Psi_2^* \, d\tau = m_2^* \int \Psi_1 \Psi_2^* \, d\tau = m_2 \int \Psi_1 \Psi_2^* \, d\tau \qquad (4\text{-}32)$$

which is due to the fact that the eigenvalues of Hermitian operators are real numbers. The left-hand sides of Eqs. (4-31) and (4-32) are identical, so that

$$m_1 \int \Psi_2^* \Psi_1 \, d\tau = m_2 \int \Psi_1 \Psi_2^* \, d\tau, \qquad (4\text{-}33)$$

which for $m_1 \pm m_2$ can be valid only if

$$\int \Psi_2^* \Psi_1 \, d\tau = 0 \qquad (4\text{-}34)$$

For $m_1 = m_2$, the proof is insufficient; however, it is still possible to find a set of orthogonal eigenfunctions by orthogonalization, employed in linear algebra.

Theorem 2 guarantees the possibility of expanding wave function Φ in a certain instant in terms of eigenfunctions Ψ_i, corresponding to a certain observable. Substituting the expansion

$$\Phi = \sum_i c_i \Psi_i, \qquad (4\text{-}35)$$

into Eq. (4-20), where Ψ_i satisfies Eq. (4-22), and considering the orthonormality of functions Ψ_i leads to the relationship

$$\sum_i |c_i|^2 = 1 \qquad (4\text{-}36)$$

Substituting Eqs. (4-20) and (4-35) into expression (4-23) for the mean value $\bar{m}$ of the physical quantity M gives

$$\bar{m} = \sum_i |c_i|^2 m_i \qquad (4\text{-}37)$$

Relationship (4-37) is analogous to the definition of the mean experimental value of quantity M, for which value m_1 was determined p_1 times, value m_2 determined p_2 times, etc. It then holds that

$$\bar{m} = \frac{1}{N} \sum_i p_i m_i,$$

where N is the total number of measurements ($N = \sum_i p_i$) and p_i/N is the probability of finding value m_i during the measurements. The sum of all the probabilities fulfils the necessary condition that

$$\sum_i \frac{p_i}{N} = 1$$

In condition (4-36) the square of the absolute value of coefficient c_i can be interpreted as follows:

Theorem 3. If a system is in a state described by wave function Φ expressible in terms of the expansion $\Phi = \sum_i c_i \Psi_i$, the value $|c_i|^2$ is the probability that value m_i, corresponding to the eigenfunction Ψ_i, is found as the measured value of quantity M.

The definition of commutative operators was given in Section 4.1. Here it will be shown that the commutation of two operators reflects significant physical properties of the system.

Theorem 4. The necessary and sufficient condition for two physical quantities K and M to simultaneously assume the precise values k_i and m_i during the measurement is the commutability of their operators $\mathcal{K}$ and $\mathcal{M}$.

To verify the validity of this theorem, it is necessary to show that, if there is a complete set of orthogonal functions which are simultaneously eigenfunctions of both operators $\mathcal{K}$ and $\mathcal{M}$, then these operators commute. Conversely, it also holds that, if the two operators $\mathcal{K}$ and $\mathcal{M}$ commute, they have a set of common eigenfunctions.

The first part of the theorem can easily be proved. It suffices to examine, according to Eqs. (4-11) to (4-13), the action of the respective commutator on an arbitrary function f. According to the above assumption it holds that

$$\mathcal{K}\Psi_i = k_i\Psi_i \tag{4-38}$$

$$\mathcal{M}\Psi_i = m_i\Psi_i \tag{4-39}$$

and

$$[\mathcal{K}, \mathcal{M}]f = [\mathcal{K}, \mathcal{M}]\sum_i c_i\Psi_i, \tag{4-40}$$

where Theorem 2 was used for expansion of function f in a series of functions Ψ_i. Using Eqs. (4-38) and (4-39) for modification of Eq. (4-40) leads to the result

$$\sum_i c_i(k_i m_i - m_i k_i)\Psi_i = 0, \tag{4-41}$$

thus proving the commutability of operators $\mathcal{K}$ and $\mathcal{M}$ in the sense of Definition 3. The second part of the theorem will be proved for a special case assuming that the eigenvalues of one of the operators (say $\mathcal{K}$) are *not degenerate*, in other words, that in Eq. (4-38) only one function Ψ_i corresponds to value k_i. Generalization of the proof for the degenerate case renders no difficulty; it is, however, too lengthy to be given in detail here. Equation (4-38) is multiplied from the left by operator $\mathcal{M}$ and $\mathcal{M}\mathcal{K}$ is substituted for $\mathcal{K}\mathcal{M}$ as implied in the assumption; the relation

$$\mathcal{K}(\mathcal{M}\Psi_i) = k_i(\mathcal{M}\Psi_i), \tag{4-42}$$

is then obtained, from which it follows that $\Lambda_i = \mathcal{M}\Psi_i$ is also an eigenfunction of $\mathcal{K}$ with the eigenvalue k_i. However, since k_i corresponds to a non-degenerate state, Λ_i can differ from Ψ_i only by a constant factor, i.e. $\Lambda_i = m_i\Psi_i$, and

$$\mathcal{M}\Psi_i = m_i\Psi_i$$

Thus it has been shown that Ψ_i is also an eigenfunction of operator $\mathcal{M}$.

The final theorem is of no less importance and its usefulness for facilitating some calculations will be appreciated later.

Theorem 5. Let $\mathcal{K}$ and $\mathcal{M}$ be Hermitean operators that commute. Let Ψ_1 and Ψ_2 be eigenfunctions of operator $\mathcal{M}$ and m_1 and m_2 be the respective eigenvalues. If $m_1 \neq m_2$, then the integral $\int \Psi_1^* \mathcal{K} \Psi_2 \, d\tau$ equals zero.

It obviously holds that

$$\int \Psi_1^* \mathcal{K} \mathcal{M} \Psi_2 \, d\tau = m_2 \int \Psi_1^* \mathcal{K} \Psi_2 \, d\tau \qquad (4\text{-}43)$$

and, moreover,

$$\int \Psi_1^* \mathcal{K} \mathcal{M} \Psi_2 \, d\tau = \int \Psi_1^* \mathcal{M} f \, d\tau; \qquad f = \mathcal{K}\Psi_2, \qquad (4\text{-}44)$$

where the commutation property of the operators was employed. Hermicity of operator $\mathcal{M}$ permits rewriting the right-hand side of Eq. (4-44) in the form

$$\int f \mathcal{M}^* \Psi_1^* \, d\tau = m_1 \int f \Psi_1^* \, d\tau = m_1 \int \Psi_1^* \mathcal{K} \Psi_2 \, d\tau, \qquad (4\text{-}45)$$

taking into account Definition 4 and Theorem 1. The left-hand sides of Eqs. (4-43) and (4-44) are identical; therefore

$$(m_2 - m_1) \int \Psi_1^* \mathcal{K} \Psi_2 \, d\tau = 0, \qquad (4\text{-}46)$$

whence it follows that, for $m_1 \neq m_2$, the corresponding integral must vanish.

To conclude, it is necessary to state that, for the sake of simplicity, we have assumed (and shall continue to assume) a discrete spectrum of eigenvalues and that generalization to a continuous spectrum is possible.

4.4 Constants of motion.
The Pauli principle

Among quantum mechanical operators, the Hamiltonian is undoubtedly the most important, not only because of its relation to the total energy of the system but also for its role in the time-dependent

Schrödinger equation (see Postulate 3). If $\mathcal{H}$ does not depend explicitly upon the time, $\mathcal{H}$ is identical with the total energy operator. Then the Schrödinger equation (4-21) also has a solution that can be found by separation of the spatial coordinates and time.

It can be assumed that wave function Φ can be written as the product of two functions,

$$\Phi(r, t) = f(t)\, \Psi(r), \tag{4-47}$$

where f depends only on time, t, and $\Psi(r)$ only on the spatial coordinates, symbolically denoted by r. Substitution of assumption (4-47) into Eq. (4-21) permits rewriting the Schrödinger equation in the form

$$i\hbar \frac{1}{f(t)} \frac{\partial f(t)}{\partial t} = \frac{\mathcal{H}(r)\, \Psi(r)}{\Psi(r)}, \tag{4-48}$$

where each side of the equation depends on a different type of variable, which is possible only if both the left-hand and right-hand sides equal a common constant, denoted by E. Thus two equations result:

$$\mathcal{H}(r)\, \Psi(r) = E\Psi(r) \tag{4-49}$$

and

$$i\hbar \frac{df(t)}{f(t)} = E\, dt, \tag{4-50}$$

where the first is the eigenvalue equation for the energy, that is, *the time-independent Schrödinger equation*. States represented by wave function $\Psi_k(r)$ and with a precise value of energy E_k [the solution of Eq. (4-49)] are called *stationary* states.

If, in addition, the form of function $\Phi(r, t)$ is to be found, Eq. (4-50) must be integrated, giving

$$f(t) = A\, e^{-iEt/\hbar}, \tag{4-51}$$

where A is the integration constant. Hence the particular solution of Eq. (4-47) is obtained:

$$\Phi_k(r, t) = \Psi_k(r)\, e^{-iE_k t/\hbar} \tag{4-52}$$

In general the initial state of a system is described by a wave function which can be expressed as an expansion in a series of eigenfunctions of Eq. (4-49):

$$\Phi(r, 0) = \sum_k c_k \Psi_k(r) \tag{4-53}$$

According to the theory of differential equations, the following solution satisfies Eq. (4-21):

$$\Phi(r, t) = \sum_k c_k \Psi_k(r)\, e^{-iE_k t/\hbar} \qquad (4\text{-}54)$$

In Eqs. (4-53) and (4-54), the c_k denote the expansion coefficients. By substituting expansion (4-54) into expression (4-23) for calculation of the mean energy value, it follows that not only the total energy of a system but also the statistical energy distribution is constant in time:

$$\bar{E} = \int \Phi^* \mathcal{H} \Phi \, d\tau = \sum_k |c_k|^2 \int \Psi_k^* \mathcal{H} \Psi_k \, d\tau = \sum_k |c_k|^2 E_k \qquad (4\text{-}55)$$

Similarly, there are other physical quantities whose statistical distribution does not vary in time. This property is possessed by any observable $\mathcal{M}$ that does not depend explicitly on time and commutes with the Hamiltonian:

$$[\mathcal{M}, \mathcal{H}] = 0 \qquad (4\text{-}56)$$

From the rule of differentiation of the product of functions, it follows for the derivative of the mean value of M with respect to time that

$$\frac{d}{dt} \int \Phi^* \mathcal{M} \Phi \, d\tau = \int \frac{\partial \Phi^*}{\partial t} \mathcal{M} \Phi \, d\tau + \int \Phi^* \mathcal{M} \frac{\partial \Phi}{\partial t} \, d\tau$$

$$+ \int \Phi^* \frac{\partial \mathcal{M}}{\partial t} \Phi \, d\tau \qquad (4\text{-}57)$$

In Eq. (4-57) Postulate 1, which stated that operators can be treated as functions of time, was used. The time variation of the wave function is given by the Schrödinger equation (4-21) and the derivative $\partial \Phi^*/\partial t$ can therefore be expressed by the complex conjugate of Eq. (4-21):

$$i\hbar \frac{\partial \Phi^*}{\partial t} = -\mathcal{H}^* \Phi^* \qquad (4\text{-}58)$$

Substituting these expressions into Eq. (4-57) gives

$$\frac{1}{i\hbar} \left\{ -\int (\mathcal{H}^* \Phi^*)\, \mathcal{M} \Phi \, d\tau + \int \Phi^* \mathcal{M} \mathcal{H} \Phi \, d\tau \right\} + \int \Phi^* \frac{\partial \mathcal{M}}{\partial t} \Phi \, d\tau =$$

$$= \frac{1}{i\hbar} \int \Phi^* [\mathcal{M}, \mathcal{H}]\, \Phi \, d\tau + \int \Phi^* \frac{\partial \mathcal{M}}{\partial t} \Phi \, d\tau, \qquad (4\text{-}59)$$

where the fact that $\mathcal{H}$ is a Hermitian operator was employed. Equation (4-59) expresses the general time dependence of the mean value of physical quantity M. If $\partial \mathcal{M}/\partial t = 0$ and $\mathcal{M}$ satisfies Eq. (4-56), the right-hand side of Eq. (4-59) is equal to zero. The observable $\mathcal{M}$ is then, by analogy with classical mechanics, called *a constant of motion*. It is apparent that, if the wave function in the initial instant ($t = 0$) is identical with the eigenfunction Ψ_k of $\mathcal{M}$ [cf. Eqs. (4-52) and (4-53)] with eigenvalue m_k, then, during the evolution of the system, physical quantity M retains the value m_k. Then m_k is said to be a "valid quantum number". It can be used for classification of stationary states, since, employing Theorem 4, conditions are obviously fulfilled for the energy of the system to assume precise values of E_k.

As a constant of motion, *the angular momentum* is particularly important for systems with spherical symmetry, represented, for example by atoms, where the electrons move in the electrostatic field of the point charge of the atomic nucleus. According to the classical definition, the angular momentum L of a point particle of mass m with linear momentum $p = mv$, whose position is specified by vector r, is given by the relationship

$$|L| = |r \times p| = |r||p|\sin\Theta \tag{4-60}$$

(cf. Fig. 4-1).

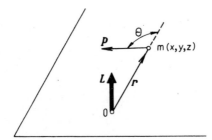

Fig. 4-1. Angular momentum vector **L**.

In three-dimensional space the expression for L can be written in the form of a determinant

$$L = \begin{vmatrix} i, & j, & k \\ x, & y, & z \\ p_x, & p_y, & p_z \end{vmatrix} \begin{array}{l} \dots \text{ unit vectors} \\ \dots \text{ components of } r \\ \dots \text{ components of } p \end{array} \tag{4-61}$$

Considering Postulate 1, the corresponding operator can be written as

$$\mathcal{L} = -i\hbar \begin{vmatrix} i, & j, & k \\ x, & y, & z \\ \dfrac{\partial}{\partial x}, & \dfrac{\partial}{\partial y}, & \dfrac{\partial}{\partial z} \end{vmatrix} \tag{4-62}$$

Expressions for components of this operator can be obtained by expanding the determinant

$$\mathscr{L}_x = -i\hbar\left(y\frac{\partial}{\partial z} - z\frac{\partial}{\partial y}\right) \tag{4-63a}$$

$$\mathscr{L}_y = -i\hbar\left(z\frac{\partial}{\partial x} - x\frac{\partial}{\partial z}\right) \tag{4-63b}$$

$$\mathscr{L}_z = -i\hbar\left(x\frac{\partial}{\partial y} - y\frac{\partial}{\partial x}\right) \tag{4-63c}$$

For the operators of components of the position vector and the momentum, the commutation relations

$$[x, p_x] = [y, p_y] = [z, p_z] = i\hbar \tag{4-64}$$

are valid [called *Heisenberg's commutation rules*, cf. Postulate 1 and Eqs. (4-5) to (4-13)]; the commutation relations

$$[\mathscr{L}_x, \mathscr{L}_y] = i\hbar\mathscr{L}_z \tag{4-65a}$$

$$[\mathscr{L}_y, \mathscr{L}_z] = i\hbar\mathscr{L}_x \tag{4-65b}$$

$$[\mathscr{L}_z, \mathscr{L}_x] = i\hbar\mathscr{L}_y \tag{4-65c}$$

are then obtained for the components of the angular momentum, indicating that simultaneous measurement of the components of the angular momentum is precluded (cf. Theorem 4).

The operator of the square of the angular momentum can be defined as

$$\mathscr{L}^2 = \mathscr{L}_x^2 + \mathscr{L}_y^2 + \mathscr{L}_z^2 =$$
$$= -\hbar^2\left\{\left(y\frac{\partial}{\partial z} - z\frac{\partial}{\partial y}\right)^2 + \left(z\frac{\partial}{\partial x} - x\frac{\partial}{\partial z}\right)^2 + \left(x\frac{\partial}{\partial y} - y\frac{\partial}{\partial x}\right)^2\right\} \tag{4-66}$$

Using relations (4-64) and (4-65), it can be shown that each component of the angular momentum commutes with $\mathscr{L}^2$, so that

$$[\mathscr{L}_x, \mathscr{L}^2] = [\mathscr{L}_y, \mathscr{L}^2] = [\mathscr{L}_z, \mathscr{L}^2] = 0 \tag{4-67}$$

On the basis of relationships (4-65) and (4-67), $\mathscr{L}^2$ and one of the operators of the angular momentum components that mutually commute, usually $\mathscr{L}_z$, can be selected from the four operators defined by Eqs. (4-63) and (4-66). This ensures that both these physical quantities are, in principle, simultaneously measurable. It is now necessary to determine the eigenvalues for these observables, on the basis of Postulates 1 and 4, which can be carried out, similarly as with the rigid rotator and the hydrogen atom, by introducing spherical coordinates (cf. Fig. 3-8). A rather lengthy transformation of Eqs. (4-66) and (4-63c) results in the expressions

$$\mathscr{L}^2 = -\frac{\hbar^2}{\sin\Theta}\frac{\partial}{\partial\Theta}\left(\sin\Theta\frac{\partial}{\partial\Theta}\right) + \frac{1}{\sin^2\Theta}\frac{\partial^2}{\partial\Phi^2} \tag{4-68}$$

$$\mathscr{L}_z = -i\hbar\frac{\partial}{\partial\Phi} \tag{4-69}$$

Comparison of Eq. (4-68) with the formulation of the Schrödinger equation given for the rigid rotator in Table 3-1 permits prediction of eigenvalues and eigenfunctions for the equation

$$\mathscr{L}^2\Psi_l = L_l^2\Psi_l, \tag{4-70}$$

namely

$$L_l^2 = \hbar^2 l(l+1) \tag{4-71}$$

and

$$\Psi_l \equiv Y_{l,m}(\Theta,\Phi) = \mathrm{Th}_{l,m}(\Theta)\,\mathrm{F}_m(\Phi), \tag{4-72}$$

where

$$\mathrm{F}_m(\Phi) = \frac{1}{\sqrt{(2\pi)}}\,e^{im\Phi} \tag{4-73}$$

The other functions are given in Table 3-1. l and m are integral quantum numbers for which condition (3-83),

$$l \geq |m|,$$

is valid. From the separated form of function Ψ_l in Eq. (4-72) it follows that Ψ_l is also an eigenfunction of $\mathscr{L}_z$, since, from Eq. (4-69),

$$\mathscr{L}_z\mathrm{F}_m \equiv -i\hbar\frac{d\mathrm{F}_m(\Phi)}{d\Phi} = -i\hbar(im)\,\mathrm{F}_m(\Phi) = \hbar m\mathrm{F}_m(\Phi), \tag{4-74}$$

and hence, considering Eq. (3-83), the eigenvalues of L_z are

$$L_z = \hbar m; \quad m = 0, \pm 1, \ldots, \pm l \tag{4-75}$$

Thus, the angular momentum is a vector whose length equals

$$\hbar\sqrt{[l(l+1)]}$$

This vector is oriented so that the component in a chosen direction, z, is an integral multiple of $\hbar$. The behaviour of the vector of the angular momentum, as follows from Eqs. (4-70), (4-71), (4-74) and (4-75), can be understood as its spatial quantization.

Comparing the eigenfunctions of $\mathscr{L}^2$, i.e. Eq. (4-72), with the eigenfunctions for the hydrogen atom (see Table 3-1) shows that the angular parts (spherical harmonics) are identical for the two cases. Since the radial part of the hydrogen wave function is constant with respect to operators $\mathscr{L}^2$ and $\mathscr{L}_z$ it can, according to Theorem 4, be assumed that operators $\mathscr{H}$, $\mathscr{L}^2$ and $\mathscr{L}_z$ commute. The same result could be derived using commutation relations for the corresponding operators expressed in analytical form. Thus it follows that, for the hydrogen atom, all three observables, $\mathscr{H}$, $\mathscr{L}^2$ and $\mathscr{L}_z$, are constants of motion.

In the many-electron atom the quantities derived from the total angular momentum act as constants of motion. For a system of n electrons the operator of the z-component, $\mathscr{L}_z$, of the total (orbital) angular momentum, L, is defined as the sum of all the z-components

$$\mathscr{L}_z = \sum_{k=1}^{n} \mathscr{L}_{kz} \tag{4-76}$$

of the orbital angular momenta of the individual electrons; a similar definition is also valid for the two remaining components, $\mathscr{L}_x$ and $\mathscr{L}_y$. By using commutation relations (4-65) for the operators of the individual electrons, formally identical commutation relations can be derived for operators of the components of the total momentum. On this basis, the angular momentum can be defined in general as any vector satisfying commutation relations (4-65). For the sake of completeness it can be added that the operator of the square of the total angular momentum is again defined by Eq. (4-66), that is, as the sum of the squares of three components of type (4-76), and that, for the many-electron atom, neglecting spin-orbit interaction, operators $\mathscr{H}$, $\mathscr{L}^2$ and $\mathscr{L}_z$ commute. In connection with the eigenvalues of the total angular momentum, it is particularly interesting to determine these values from the eigenvalues of the constituent quantities. The procedure used for addition of two momenta represents a general algorithm that is also applicable to a many-component system; the process of vector addition is then repeated stepwise for all pairs of vectors.

Assuming that there is a system of two commuting angular momenta, corresponding either to two independent particles, such as a pair of electrons, or to two independent coordinates of the same particle, such as the spatial and spin coordinates of an electron, and if the squares of the momenta [cf. Eq. (4-71)] are given by the expressions

$$L_{l1}^2 = \hbar^2 l_1(l_1 + 1) \tag{4-77a}$$

and

$$L_{l2}^2 = \hbar^2 l_2(l_2 + 1) \tag{4-77b}$$

then the square of the total angular momentum has the value

$$L_l^2 = \hbar^2 l(l + 1),\tag{4-78}$$

where

$$l = l_1 + l_2, l_1 + l_2 - 1, ..., |l_1 - l_2|\tag{4-79}$$

The result has a simple vectorial interpretation if vectors of length l_1 and l_2 are added vectorially; the largest value of the resulting vector corresponds to parallel alignment; further values vary by 1 owing to spatial quantization and the lowest value corresponds to antiparallel alignment of vectors l_1 and l_2. The spatial quantization of the resulting vector, **L**, is also subject to a rule of type (4-75).

In addition to orbital angular momentum, an electron has an intrinsic angular momentum — *the spin*. It has been experimentally determined that its components in a particular direction can have values of $\pm\frac{1}{2}\hbar$. Dirac demonstrated that the existence of spin naturally follows from the relativistic description of an electron moving in an electromagnetic field. For practical purposes, however, it is preferable to introduce spin by the Pauli procedure, where the spin of an electron is treated as a physical quantity corresponding to the angular momentum with the quantum number $l \equiv s = \frac{1}{2}$. The existence of spin momentum **S**, independent of the orbital momentum, **L**, is thus postulated. In connection with the spin, similar operators $\mathscr{S}^2$ and $\mathscr{S}_z$ can be defined. These operators satisfy commutation rules analogous to those valid for operators $\mathscr{L}^2$, $\mathscr{L}_z$, etc., for example

$$[\mathscr{S}^2, \mathscr{S}_z] = 0\tag{4-80}$$

$$[\mathscr{S}_z, \mathscr{S}_x] = i\hbar\mathscr{S}_y\tag{4-81}$$

These operators are applied to spin functions α and β, which have the properties of orthonormal functions:

$$\int \alpha^*(\sigma)\,\alpha(\sigma)\,d\sigma = 1; \quad \int \beta^*(\sigma)\,\beta(\sigma)\,d\sigma = 1\tag{4-82}$$

$$\int \alpha^*(\sigma)\,\beta(\sigma)\,d\sigma = 0,\tag{4-83}$$

where σ represents the discrete spin variable corresponding to the eigenvalue of $\mathscr{S}_z$ and assumes the two values $\pm\frac{1}{2}$. The characteristic equations for operators $\mathscr{S}^2$ and $\mathscr{S}_z$ have the form

$$\mathscr{S}^2\alpha = \frac{1}{2}\left(1 + \frac{1}{2}\right)\hbar^2\alpha, \quad \mathscr{S}_z\alpha = \frac{\hbar}{2}\alpha\tag{4-84a}$$

$$\mathscr{S}^2\beta = \frac{1}{2}\left(1 + \frac{1}{2}\right)\hbar^2\beta, \quad \mathscr{S}_z\beta = -\frac{\hbar}{2}\beta\tag{4-84b}$$

In contrast to all the cases discussed so far, the analytical form of functions α and β cannot be established. In practice, the spin-dependent part of a one-electron function is defined by specifying the value of the quantum number, $m_s = \pm\hbar/2$.

Similarly as with the orbital angular momentum, the total spin angular momentum can be introduced; for a system of n electrons,

$$S = \sum_{i=1}^{n} S_i \tag{4-85}$$

From the procedure of vector addition it is apparent that quantum number S [by analogy with l in Eq. (4-78)] assumes integral values if n is even and half-integral values if n is odd. The largest possible eigenvalue of $\mathscr{S}_z$ is evidently $\frac{1}{2}n\hbar$.

The total angular momentum of a system of electrons, considering the spin, is defined as

$$J = L + S = \sum_{i=1}^{n} J_i = \sum_{i=1}^{n} L_i + \sum_{i=1}^{n} S_i \tag{4-86}$$

and

$$\mathscr{J}^2 = \mathscr{J}_x^2 + \mathscr{J}_y^2 + \mathscr{J}_z^2 \tag{4-87}$$

The following quantum numbers correspond to the given six operators related to the many-electron atom:

$$L, \; M_L$$
$$S, \; M_S$$
$$J, \; M_J$$

For the total operators the commutation rules given in Table 4-1 are valid.

The computation of physical properties of atomic systems is carried out by setting up stationary wave functions that are simultaneously eigenfunctions of a maximum number of commuting operators. Physical quantities are generally expressed in terms of matrix elements of type (4-97); according to Theorem 5, the matrix elements of the total energy operator of a system are non-zero only if the Hamiltonian in the matrix element is surrounded by functions with eigenvalues identical of the other commuting operators. The extent of the calculation can then be reduced considerably. It follows from Table 4-1 that the group of operators $\mathscr{L}^2$, $\mathscr{L}_z$, $\mathscr{S}^2$ and $\mathscr{S}_z$ is suitable for the Hamiltonian, exhibiting no explicit dependence on the electronic spin. If subtler types of interaction are also included in the Hamiltonian, such as, for instance, spin-orbit coupling, which contributes the term

$$\mathscr{H}_{(SL)} = \sum_{i=1}^{n} \xi(r_i)\,(\mathscr{L}_{ix}\mathscr{S}_{ix} + \mathscr{L}_{iy}\mathscr{S}_{iy} + \mathscr{L}_{iz}\mathscr{S}_{iz}) \qquad (4\text{-}88)$$

to the non-relativistic Hamiltonian, then the four given operators are not constants of motion and their role is assumed by operators $\mathscr{J}^2$ and $\mathscr{J}_z$. $\xi(r_i)$ is a function of the radial coordinate of the i-th electron alone, and depends on the type of spherically symmetrical potential field in which the electron is moving.

So far only those constants of motion that have classical analogues have been discussed. There are also other types of operators commuting with the Hamiltonian. Let us consider a system of identical particles; this can be an atom, a molecule, or a solid substance, where it is assumed that n electrons move in the electrostatic field of the rigidly fixed nuclei. Since the electrons are indistinguishable, the Hamiltonian is invariant under any transposition of the electrons. Mathematically this property can be expressed by the relationship

$$\mathscr{P}_{jk}\mathscr{H} = \mathscr{H}\mathscr{P}_{jk}, \qquad (4\text{-}89)$$

where $\mathscr{P}_{jk}$ denotes *the transposition operator* of the k-th and j-th electrons. It follows from relationship (4-89) that operator $\mathscr{P}_{jk}$ is a constant of motion and that $\mathscr{P}_{jk}$ and $\mathscr{H}$ have common eigenfunctions.

Table 4-1

Commutation Properties of Operators Corresponding to Quantities L, S, J, and H (Hamiltonian where $L-S$ coupling is not considered). 0 denotes that the operators in the respective row and column commute, the dash denotes that they do not commute [Ref. 5]

	$\mathscr{L}_x$	$\mathscr{L}_y$	$\mathscr{L}_z$	$\mathscr{L}^2$	$\mathscr{S}_x$	$\mathscr{S}_y$	$\mathscr{S}_z$	$\mathscr{S}^2$	$\mathscr{J}_x$	$\mathscr{J}_y$	$\mathscr{J}_z$	$\mathscr{J}^2$	$\mathscr{H}$
$\mathscr{L}_x$	0	–	–	0	0	0	0	0	0	–	–	–	0
$\mathscr{L}_y$	–	0	–	0	0	0	0	0	–	0	–	–	0
$\mathscr{L}_z$	–	–	0	0	0	0	0	0	–	–	0	–	0
$\mathscr{L}^2$	0	0	0	0	0	0	0	0	0	0	0	0	0
$\mathscr{S}_x$	0	0	0	0	0	–	–	0	0	–	–	–	0
$\mathscr{S}_y$	0	0	0	0	–	0	–	0	–	0	–	–	0
$\mathscr{S}_z$	0	0	0	0	–	–	0	0	–	–	0	–	0
$\mathscr{S}^2$	0	0	0	0	0	0	0	0	0	0	0	0	0
$\mathscr{J}_x$	0	–	–	0	0	–	–	0	0	–	–	0	0
$\mathscr{J}_y$	–	0	–	0	–	0	–	0	–	0	–	0	0
$\mathscr{J}_z$	–	–	0	0	–	–	0	0	–	–	0	0	0
$\mathscr{J}^2$	–	–	–	0	–	–	–	0	0	0	0	0	0
$\mathscr{H}$	0	0	0	0	0	0	0	0	0	0	0	0	0

If the wave function

$$\Psi(1, 2, ..., j, ..., k, ..., n) = \Psi(j, k) \tag{4-90}$$

is a function of the space and spin coordinates of a system of electrons, then

$$\mathscr{P}_{jk}\Psi(j, k) = \lambda\Psi(j, k) = \Psi(k, j), \tag{4-91}$$

where λ is a real eigenvalue, and $\mathscr{P}_{jk}$ is a Hermitian operator (the right-hand side of the second equation expresses the result of the operation on the function). If the operator $\mathscr{P}_{jk}$ is repeatedly applied to Eq. (4-91), then

$$\mathscr{P}_{jk}^2\Psi(j, k) = \lambda^2\Psi(j, k) = \Psi(j, k) \tag{4-92}$$

whence it follows that

$$\lambda^2 = 1, \qquad \lambda = \pm 1 \tag{4-93}$$

Thus it has been shown that the solutions Ψ of the stationary Schrödinger equation may or may not change sign on transposition of two identical particles. States of the first type are referred to as symmetric, those of the second type as antisymmetric states. The functions that can be considered to be actual solutions are given by *the Pauli exclusion principle*. according to which, of all possible solutions of the Schrödinger equation for electrons, only those which are antisymmetric are to be considered. The Pauli principle, originally only a hypothesis, was later shown experimentally to be valid.

4.5 Matrix representation of operators and operations with matrices

It has been shown that the mean values of physical quantities for a system which is in a stationary state are constant in time [cf. Eq. (4-59)]. In quantum chemistry these states are generally most interesting and therefore the dependence of operators and of the wave function on time will be excluded from further discussion.

The expectation value of physical quantity K can be written according to Postulate 5 as follows:

$$\bar{k} = \int \Phi^*\mathscr{K}\Phi \, d\tau: \qquad \int \Phi^*\Phi \, d\tau = 1 \tag{4-94}$$

It is sometimes expedient to expand wave function Φ in Eq. (4-94) in terms of a set of eigenfunctions (see Theorem 2) of observable $\mathscr{M}$,

$$\Phi = \sum_i c_i\Psi_i, \tag{4-95}$$

so that

$$\bar{k} = \sum_{i,j} c_i^* c_j K_{ij}, \tag{4-96}$$

where

$$K_{ij} = \int \Psi_i^* \mathcal{K} \Psi_j \, d\tau \tag{4-97}$$

By introducing a certain sequence for functions Ψ_i, expressions (4-97) can be systematically arranged in an array

$$\left\|\begin{array}{llll} K_{11}, & K_{12}, & ..., & K_{1j}, & ... \\ K_{21}, & K_{22}, & ..., & K_{2j}, & ... \\ \vdots & \vdots & & \vdots \\ K_{i1}, & K_{i2}, & ..., & K_{ij}, & ... \\ \vdots & \vdots & & \vdots \end{array}\right\| = \mathbf{K} \equiv \| K_{ij} \| \tag{4-98}$$

This scheme is usually referred to as *a matrix*. Matrix element K_{ij} is located in matrix $\mathbf{K}$ on the intersection of the i-th row and the j-th column. Matrices can assume finite as well as "infinite" dimensions. When the matrix consists of m rows and n columns it is called an m, n-type matrix. The matrix of the m, m type is termed *square* matrix.

Matrix $\mathbf{K}$ is said to form *the matrix representation of the operator* $\mathcal{K}$. The respective functions Ψ_i, $i = 1, 2, ...,$ are the basis vectors of the representation. If functions Ψ_i, $i = 1, 2, ...,$ are eigenfunctions of operator $\mathcal{K}$, it follows from Theorem 2, Section 4.3, that matrix $\mathbf{K}$ has non-zero elements only on the main diagonal.

Since the operators considered here are linear, it is useful to introduce some matrix operations.

Let us assume that operator $\mathcal{K}$ is defined as the sum of two operators:

$$\mathcal{K} = \mathcal{A} + \mathcal{B} \tag{4-99}$$

By substituting Eq. (4-99) into Eq. (4-97), it follows for the matrix elements that

$$K_{ij} = A_{ij} + B_{ij} \tag{4-100}$$

and, consequently, that the sum of two matrices,

$$\mathbf{K} = \mathbf{A} + \mathbf{B}, \tag{4-101}$$

can be defined so that the matrix elements satisfy Eq. (4-100).

To derive the algorithm for matrix multiplication it is necessary to utilize the fact that the function $f = \mathcal{K} \Psi_j$ can be expanded using Theorem 2, as

$$\mathcal{K} \Psi_j = \sum_i K_{ij} \Psi_i \tag{4-102}$$

Operator $\mathcal{K}$ can be defined as

$$\mathcal{K} = \mathcal{A}\mathcal{B} \tag{4-103}$$

Then successive action of the operators on Ψ_j gives

$$\mathcal{A}(\mathcal{B}\Psi_j) = \mathcal{A}\sum_k B_{kj}\Psi_k = \sum_k B_{kj}\mathcal{A}\Psi_k = \sum_{ki} B_{kj}A_{ik}\Psi_i \tag{4-104}$$

Comparison of coefficients of the same function in Eqs. (4-102) and (4-104) yields an expression for the general element K_{ij} of matrix **K** in terms of matrix elements of **A** and **B**:

$$K_{ij} = \sum_k A_{ik}B_{kj} \tag{4-105}$$

For the sake of completeness, multiplication of matrix **K** by a constant, k, can be defined as

$$\mathbf{A} = k\mathbf{K}, \tag{4-106}$$

where

$$A_{ij} = kK_{ij} \tag{4-107}$$

Some special kinds of matrices can be given:
Null matrix **0**, definition: $K_{ij} = 0$ for all i and j.
Unit matrix **1**, definition: $K_{ii} = 1$ for all i; $K_{ij} = 0$, for $i = j$.
Diagonal matrix, definition: non-zero elements on the main diagonal alone.
Inverse $\mathbf{K}^{-1}$ of matrix **K**, definition: $\mathbf{K}\mathbf{K}^{-1} = \mathbf{K}^{-1}\mathbf{K} = \mathbf{1}$.
Transposed matrix $\mathbf{K}^{\mathrm{T}}$ to matrix **K**, definition: $(\mathbf{K}^{\mathrm{T}})_{ij} = K_{ji}$.
Complex conjugate matrix $\overline{\mathbf{K}}$ to matrix **K**, definition: $(\overline{\mathbf{K}})_{ij} = K_{ij}^*$.
Hermitean conjugate matrix $\mathbf{K}^{\mathrm{H}}$ to matrix **K**, definition: $(\mathbf{K}^{\mathrm{H}})_{ij} = K_{ji}^*$.
If **K** is *a Hermitean matrix*, then

$$\mathbf{K} = \mathbf{K}^{\mathrm{H}}, \quad \text{or} \quad K_{ij} = K_{ji}^* \tag{4-108}$$

and it is evident that, for real matrix elements, the matrix is *symmetrical* and

$$\mathbf{K} = \mathbf{K}^{\mathrm{T}}, \quad \text{or} \quad K_{ij} = K_{ji} \tag{4-109}$$

It can easily be verified that a matrix representing a Hermitian operator is also Hermitian.
A unitary matrix is defined as

$$\mathbf{K}^{\mathrm{H}}\mathbf{K} = \mathbf{K}\mathbf{K}^{\mathrm{H}} = \mathbf{1} \tag{4-110}$$

For real matrices definition (4-110) reduces to give

$$\mathbf{K}^{\mathrm{T}}\mathbf{K} = \mathbf{K}\mathbf{K}^{\mathrm{T}} = \mathbf{1} \tag{4-111}$$

and matrix **K**, satisfying relation (4-111), is called *an orthogonal matrix*.

The matrix formulation of some problems permits more lucid and more compact recording of complex expressions, as is demonstrated in several examples given below. Let us define c as a one-column matrix and arrange functions Ψ_i in the corresponding order in one-row matrix Ψ. Expression (4-95) can then be written in the form of the matrix product

$$\Phi = \| \Psi_1, \Psi_2, \dots \| \begin{Vmatrix} c_1 \\ c_2 \\ \vdots \end{Vmatrix} \equiv \Psi c \qquad (4\text{-}112)$$

Let us suppose that the set of functions Ψ_i, $i = 1, 2, \dots$, is normalized but not orthogonal. Basis sets of this type are frequently used in quantum chemical calculations. Therefore,

$$\int \Psi_i^* \Psi_j \, d\tau = S_{ij} \qquad (4\text{-}113)$$

with $S_{ii} = 1$ for all i's. S is called *the metric or overlap matrix* of the corresponding basis set. The norm of wave function Φ is then given, instead of by Eq. (4-36), by the more general expression

$$\int \Phi^* \Phi \, d\tau = \sum_{ij} c_i^* c_j \int \Psi_i^* \Psi_j \, d\tau \qquad (4\text{-}114)$$

In order that expressions of type (4-114) be recordable by the matrix formalism, a two-dimensional matrix to the basis Ψ_i, $i = 1, 2, \dots$, is formally assigned, where the column index specifies functions Ψ_i arranged in a certain order, and the row index (continuous) lies in the region of the integration variable. In accordance with this notation, the chosen function Ψ_i is a one-column matrix, and the integral in Eq. (4-114) can be formally written as

$$\int \Psi_i^* \Psi_j \, d\tau = \Psi_i^H \Psi_j, \qquad (4\text{-}115)$$

where the integration is expressed by the summation involved in the matrix multiplication. This notation also implies that

$$S = \Psi^H \Psi \qquad (4\text{-}116)$$

and

$$\int \Phi^* \Phi \, d\tau = \Phi^H \Phi = c^H \Psi^H \Psi c = c^H S c, \qquad (4\text{-}117)$$

where the property of matrix multiplication has been employed:

$$[AB]^H = B^H A^H \qquad (4\text{-}118)$$

The problem of transformation of the operator from one representation to another merits particular attention. Let us assume that the representation of operator $\mathscr{K}$ for the basis formed of eigenfunctions Ψ_i, $i = 1, 2, \ldots$, of operator $\mathscr{M}$ [i.e. matrix $\mathbf{K}$ defined by Eqs. (4-97) and (4-98)] and the following transformation are known:

$$\Psi_1 = \sum_i U_{i1}\varphi_i \qquad (4\text{-}119a)$$

$$\Psi_2 = \sum_i U_{i2}\varphi_i \qquad (4\text{-}119b)$$

These equations relate the set of functions Ψ_i, $i = 1, 2, \ldots$, to a new set φ_j, $j = 1, 2, \ldots$, which is also assumed to be complete and orthonormal. Now the relationship between matrices $\mathbf{K}$ and $\mathbf{K}'$ can be investigated, where

$$K'_{ij} = \int \varphi_i^* \mathscr{K} \varphi_j \, d\tau \qquad (4\text{-}120)$$

First, system of linear equations (4-119) can be written in matrix form:

$$\boldsymbol{\Psi} = \boldsymbol{\varphi}\mathbf{U} \qquad (4\text{-}121)$$

Forming the product according to Eq. (4-116)

$$\boldsymbol{\Psi}^H\boldsymbol{\Psi} = \mathbf{U}^H\boldsymbol{\varphi}^H\boldsymbol{\varphi}\mathbf{U}, \qquad (4\text{-}122)$$

then, because

$$\boldsymbol{\Psi}^H\boldsymbol{\Psi} = \boldsymbol{\varphi}^H\boldsymbol{\varphi} = \mathbf{1} \quad \text{(the bases are orthonormal)} \qquad (4\text{-}123)$$

the expression [cf. Eq. (4-110)]

$$\mathbf{U}^H\mathbf{U} = 1 \qquad (4\text{-}124)$$

is obtained. Therefore, transformation between two orthonormal basis sets is evidently mediated by a unitary matrix. Further, by substituting expansions for Ψ_i and Ψ_j from expressions (4-119) into Eq. (4-97), it follows that

$$K_{ij} = \sum_{k,l} U_{ki}^* U_{lj} K'_{kl}, \qquad (4\text{-}125)$$

so that the transformation of the whole matrix can be written as follows:

$$\mathbf{K} = \mathbf{U}^H\mathbf{K}'\mathbf{U} \qquad (4\text{-}126a)$$

By matrix multiplication of Eq. (4-126a) from the left by $\mathbf{U}$ and from the right by $\mathbf{U}^H$ and using property (4-110), the inverse transformation

$$\mathbf{U}\mathbf{K}\mathbf{U}^H = \mathbf{K}' \qquad (4\text{-}126b)$$

is obtained. One more important property of a unitary transformation deserves mentioning. A unitary transformation retains the sum of the diagonal elements; this sum is called the *trace of matrix* $\mathbf{K}$ and is designated by

$$\text{Tr } \mathbf{K} = \sum_i K_{ii} \tag{4-127}$$

From Eq. (4-125) it follows that

$$\text{Tr } \mathbf{K} = \sum_{i,k,l} U_{ki}^* U_{li} K_{kl}' = \sum_{k,l} K_{kl}' \sum_i U_{ki}^* U_{li} \tag{4-128}$$

According to the definition of a unitary matrix (4-110), its rows and columns form orthonormal vectors in the sense

$$\sum_i U_{ki}^* U_{li} = \sum_i U_{ik}^* U_{il} = \delta_{kl}, \tag{4-129}$$

where δ_{kl} is the Kronecker symbol defined in Eq. (4-24). Hence, from Eq. (4-128),

$$\text{Tr } \mathbf{K} = \sum_{k,l} K_{kl}' \delta_{kl} = \text{Tr } \mathbf{K}', \tag{4-130}$$

i.e. the trace of the matrix is invariant under unitary transformation.

4.6 Approximate solution of the Schrödinger equation: variation and perturbation methods

With the exception of quite simple problems (whose significance for chemistry is limited), the Schrödinger equation is not exactly solvable. Thus it is often necessary to use approximate solutions. This is undoubtedly undesirable from a mathematical point of view, but in physics and chemistry approximate solutions can be quite useful. In fact, it is often possible to find approximate solutions that are very close to the exact solution.

In this connection two methods are particularly important: the variation and perturbation methods. The first is more important for the applications discussed here and so will be discussed first.

The variation method is based on *the variation principle*: if f is an arbitrary function satisfying the condition

$$\int f^* f \, d\tau = 1,$$

then

$$\int f^* \mathscr{H} f \, d\tau \geq E_0, \tag{4-131}$$

where E_0 denotes the ground state energy of the system (that is, the lowest eigenvalue of Hamiltonian $\mathscr{H}$).

The proof of the theorem is simple. It is assumed that the solutions Ψ_i of the stationary Schrödinger equation

$$\mathscr{H}\Psi_i = E_i\Psi_i$$

are known. According to Theorem 2, function f can be expressed in the form of expansion (4-25)

$$f = \sum_i c_i\Psi_i \tag{4-132}$$

with expansion coefficients c_i satisfying condition (4-36). Substituting this expansion into the integral in Eq. (4-131) yields

$$\int f^*\mathscr{H}f\,d\tau = \sum_i |c_i|^2 \int \Psi_i^*\mathscr{H}\Psi_i\,d\tau = \sum_i |c_i|^2 E_i \geq E_0 \sum_i |c_i|^2, \tag{4-133}$$

where the fact that the energies of the excited states E_i ($i = 1, 2, \ldots$) of the system are higher than (or at least equal to) the energy of the ground state E_0 was employed; finally, using Eq. (4-36), relationship (4-131) is obtained.

The energy calculated using the approximate function f is therefore, according to the variation principle, higher than energy E_0 (or, in the limiting case, equal to it). When the exact wave function cannot be determined because of the mathematical complexity of the problem, then a formulation is used where f is to be determined in such a way that the integral, $\int f^*\, \mathscr{H}f\,d\tau$, has a minimum value. The exact energy value of the ground state E_0 is then approached most closely. It is worth noting that the variation method can be modified so that it can be used for the calculation of excited states of the system.

In practical calculations based on the variation method the following procedure is employed. A trial wave function is proposed (on the basis of experience) for the problem under study, containing initially variable parameters. The energy corresponding to this wave function is expressed in terms of these parameters; their numerical values are then determined so that the calculated energy of the ground state (the state with the minimum allowed energy) is as low as possible. The proposed approximate function f can have an arbitrary form (as long as it satisfies the usual requirements imposed on wave functions). Usually it has the form of a linear combination of other functions (which do not form a complete system):

$$f = \sum_{i=1}^n c_i\Phi_i, \tag{4-134}$$

where coefficients c_i are the parameters to be optimized.

It can be mentioned, for illustration, that the Φ_i's can be atomic orbitals when forming molecular orbitals, or that the Φ_i's can denote valence bond (canonical) structures in the valence bond method or, finally, that the Φ_i's can be Slater determinants (or their linear combinations) corresponding to distinct configurations in the configuration interaction method.

In the mathematical formulation of the problem, the minimum value of the expression

$$\min \int f^* \mathcal{H} f \, d\tau \tag{4-135}$$

should be found, which, in terms of variation calculus, can be expressed by the condition

$$\delta \int f^* \mathcal{H} f \, d\tau = 0 \tag{4-136}$$

It is necessary to realize that the variation parameters, c_i, are not independent and are subject to the condition

$$\int f^* f \, d\tau = \sum_{i,j=1}^{n} c_i^* c_j \int \Phi_i^* \Phi_j \, d\tau = \mathbf{c}^{\mathbf{H}} \mathbf{S} \mathbf{c} = 1, \tag{4-137}$$

where, for the elements of the overlap matrix, the denotation

$$S_{ij} = \int \Phi_i^* \Phi_j \, d\tau$$

was employed, using the matrix notation from Eq. (4-117). In accordance with the method of Lagrange multipliers, instead of Eq. (4-136), the equation

$$\delta \left[\int f^* \mathcal{H} f \, d\tau - E \left(\int f^* f \, d\tau - 1 \right) \right] = 0, \tag{4-138}$$

will be solved, in which E is the Lagrange multiplier, so that all parameters c_i can formally be considered independent. In this calculation stage it is necessary to be aware that, in general, a complex number is given by both its real and imaginary parts, and that, consequently, c_i and its conjugate quantity c_i^* are two independent parameters. Substituting expansion (4-134) into Eq. (4-138) gives

$$\sum_{i=1}^{n} \delta c_i^* \sum_{j=1}^{n} c_j [H_{ij} - E S_{ij}] + \sum_{j=1}^{n} \delta c_j \sum_{i=1}^{n} c_i^* [H_{ij} - E S_{ij}] = 0, \tag{4-139}$$

where $\| H_{ij} \|$ is the matrix representing the operator $\mathcal{H}$ within the basis $\Phi_i, i = 1, 2, \ldots,$

$$H_{ij} = \int \Phi_i^* \mathcal{H} \Phi_j \, d\tau$$

Since all the variations in Eq. (4-139) are now independent, it can be satisfied only if the coefficients of all the variations are zero, so that

$$\sum_{j=1}^{n} c_j [H_{ij} - ES_{ij}] = 0, \qquad \text{for } i = 1, 2, ..., n \qquad (4\text{-}140\text{a})$$

$$\sum_{i=1}^{n} c_i^* [H_{ij} - ES_{ij}] = 0, \qquad \text{for } j = 1, 2, ..., n \qquad (4\text{-}140\text{b})$$

Systems of equations (4-140a) and (4-140b) determine coefficients c_i and their complex conjugate values c_i^*. As both $\mathbf{H}$ and $\mathbf{S}$ are Hermitian matrices, that is, $H_{ji} = H_{ij}^*$ and $S_{ji} = S_{ij}^*$, it follows that system of equations (4-140b) is the complex conjugate of system (4-140a). Then system of equations (4-140a) can be considered as the final result which represents the condition for the optimum values of coefficients c_i. For the sake of lucidity, the linear system of equations with unknown c_j's can be written as

$$c_1 [H_{11} - E] \quad + c_2 [H_{12} - ES_{12}] + ... + c_n [H_{1n} - ES_{1n}] = 0$$
$$c_1 [H_{21} - ES_{21}] + c_2 [H_{22} - E] \quad + ... + c_n [H_{2n} - ES_{2n}] = 0$$
$$\vdots$$
$$c_1 [H_{n1} - ES_{n1}] + c_2 [H_{n2} - ES_{n2}] + ... + c_n [H_{nn} - E] \quad = 0 \quad (4\text{-}141)$$

In Eq. (4-141) the normality of functions Φ_i, $\int \Phi_i^* \Phi_i \, d\tau = 1$, for all i's, was taken into account. It is obvious that this system of equations is satisfied for values $c_1 = c_2 ... c_n = 0$. However, this is not a physically significant solution. A non-trivial solution can be found if

$$\det \| H_{ij} - ES_{ij} \| =$$

$$= \begin{vmatrix} H_{11} - E, & H_{12} - ES_{12}, & ..., & H_{1n} - ES_{1n} \\ H_{21} - ES_{21}, & H_{22} - E, & ..., & H_{2n} - ES_{2n} \\ \vdots & & & \\ H_{n1} - ES_{n1}, & H_{n2} - ES_{n2}, & ..., & H_{nn} - E \end{vmatrix} = 0 \quad (4\text{-}142)$$

Equation (4-142), called *the secular equation*, represents an n-th order algebraic equation in E and has, consequently, n real roots: $E_1, E_2 ... E_n$. The reality of the roots is a consequence of the hermicity of matrices $\mathbf{H}$ and $\mathbf{S}$. Provided function f represents a state of the system, then, according to the variation principle, the smallest root is the best approximation to the ground state energy. If good judgment is applied in choosing the variation function, the approximate energy value may be very close to the true energy. The other roots can be interpreted as approximate values of the energies of the excited states. The entire derivation could be carried out in such a compact form only because of use of the matrix formalism. System of equations (4-141) can be rewritten in the form

$$(\mathbf{H} - E\mathbf{S})\mathbf{c} = \mathbf{0}, \tag{4-143}$$

where the notation used in Eqs. (4-112) and (4-117) is employed. By multiplying from the left by matrix $\mathbf{c}^H$ it can easily be proven that E is an energy value (cf. Eq. (4-23)) since, employing Eq. (4-137), it follows that

$$\mathbf{c}^H\mathbf{H}\mathbf{c} = E\mathbf{c}^H\mathbf{S}\mathbf{c} = E \tag{4-144}$$

When solving a system of equations of type (4-141), the procedure leading to the determination of optimum values for variable parameters (called MO expansion coefficients, expansion coefficients of the VB or CI wave function) is as follows:

1. The matrix elements of the Hamiltonian (H_{ij}) and the elements of the overlap matrix (S_{ij}) are calculated.

2. The determinant in Eq. (4-142) is expanded and the algebraic equation of the n-th degree is solved. In this way, n real energy values are obtained (where multiple roots corresponding to degenerate levels can appear). These values correspond to n wave functions, each of which is expressed as a linear combination of n functions (Φ_i).

3. For each of the energy values, a system of equations is solved for the unknown expansion coefficients. For each energy level, E_i, n values of c_{ji} ($j = 1, 2, ..., n$) are obtained; since there are n energy values (E_i, $i = 1, 2, ..., n$), a total of n^2 expansion coefficients is found; these coefficients are not independent because of the condition of orthonormality of the individual solutions.

From a broader point of view, *the perturbation method* has a particularly important position in quantum mechanics. It is, however, less important here than the variation method and will therefore only be outlined in general. The perturbation method is suitable for the solution of the Schrödinger equation (E_i, Ψ_i) for a problem which differs only slightly from another, related problem, for which the solution (E_i°, Ψ_i°) is known. The required solution, i.e. E_i and Ψ_i, is then expressed in terms of known values, E_i° and Ψ_i°. The formation of the investigated system from the initial system is considered to be the result of a particular perturbation. The situation can be visualized as follows:

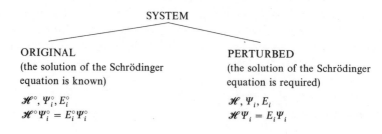

SYSTEM

ORIGINAL
(the solution of the Schrödinger
equation is known)

$\mathcal{H}^\circ, \Psi_i^\circ, E_i^\circ$
$\mathcal{H}^\circ \Psi_i^\circ = E_i^\circ \Psi_i^\circ$

PERTURBED
(the solution of the Schrödinger
equation is required)

$\mathcal{H}, \Psi_i, E_i$
$\mathcal{H} \Psi_i = E_i \Psi_i$

Among various forms of the perturbation theory[13], the Rayleigh-Schrödinger theory is of main importance here. The Hamiltonian $\mathscr{H}$ of the studied system is considered to be resolvable into two parts,

$$\mathscr{H} = \mathscr{H}^\circ + \lambda\mathscr{V}, \qquad (4\text{-}145)$$

where $\mathscr{H}^\circ$ is the Hamiltonian of the unperturbed system and $\lambda\mathscr{V}$ is a perturbation term. λ in Eq. (4-145) is a parameter which may (but need not) have physical significance and whose main purpose is to distinguish different orders of perturbation contributions. It is evident that, if λ approaches zero, the solution of the perturbed Hamiltonian, $\mathscr{H}$, converges to the solution of the unperturbed Hamiltonian $\mathscr{H}^\circ$; below we will consider first the case when, in the process $\lambda \to 0$ (the value of λ tends to zero), it holds that $\Psi_i \to \Psi_i^\circ$ and $E_i \to E_i^\circ$, where E_i° is a non-degenerate eigenvalue.

The equation to be solved can be written as

$$(\mathscr{H}^\circ + \lambda\mathscr{V} - E_i)\,\Psi_i = 0 \qquad (4\text{-}146)$$

According to the basic assumption of the Rayleigh-Schrödinger perturbation theory, Ψ_i and E_i can be expanded into a power series of parameter λ:

$$\Psi_i = \Psi_i^\circ + \lambda\Psi_i^{(1)} + \lambda^2\Psi_i^{(2)} + \ldots \qquad (4\text{-}147\text{a})$$

$$E_i = E_i^\circ + \lambda E_i^{(1)} + \lambda^2 E_i^{(2)} + \ldots \qquad (4\text{-}147\text{b})$$

Expansions (4-147) are substituted into Eq. (4-146) and, because the resulting equation must be satisfied for all values of λ, the coefficient of each power of λ must be equal to zero. As a result a system of equations is obtained:

0-th order: $\qquad (\mathscr{H}^\circ - E_i^\circ)\,\Psi_i^\circ = 0 \qquad\qquad\qquad (4\text{-}148\text{a})$

1st order: $\qquad (\mathscr{H}^\circ - E_i^\circ)\,\Psi_i^{(1)} + (\mathscr{V} - E_i^{(1)})\,\Psi_i^\circ = 0 \qquad (4\text{-}148\text{b})$

2nd order: $\qquad (\mathscr{H}^\circ - E_i^\circ)\,\Psi_i^{(2)} + (\mathscr{V} - E_i^{(1)})\,\Psi_i^{(1)} - E_i^{(2)}\Psi_i^\circ = 0 \quad (4\text{-}148\text{c})$

Solved successively, these equations yield the correction contributions of the individual orders. The function Ψ_i° as a solution of the unperturbed problem is assumed to be normalized, $\int (\Psi_i^\circ)^* \, \Psi_i^\circ \, d\tau = 1$. The properties of function Ψ_i must be also defined. It seems natural to choose Ψ_i normalized, which is the usual approach. For certain reasons, however, it is more advantageous to choose

$$\int (\Psi_i^\circ)^* \, \Psi_i \, d\tau = 1 \qquad (4\text{-}149)$$

With this type of normalization condition, Ψ_i approaches Ψ_i° as λ

approaches zero and, in addition, multiplying Eq. (4-147a) by function $(\Psi_i^\circ)^*$ from the left and integrating yields

$$\int (\Psi_i^\circ)^* \, \Psi_i^{(1)} \, d\tau = \int (\Psi_i^\circ)^* \, \Psi_i^{(2)} \, d\tau = \ldots =$$
$$= \int (\Psi_i^\circ)^* \, \Psi_i^{(n)} \, d\tau = \ldots = 0 \qquad (4\text{-}150)$$

Equations (4-150) can clearly be interpreted geometrically; an integral of type $\int \Psi_1^* \Psi_2 \, d\tau$ has the properties of a scalar product and, from elementary vector calculus, the scalar product of two vectors is zero if these vectors lie perpendicular to each other, so that the projection of the one on the other is zero. Equation (4-150) can be interpreted analogously, so that the wave function corrections have no "component" in common with the unperturbed function Ψ_i°.

Under these conditions it is easy to obtain explicit expressions for the energy corrections $E_i^{(n)}$. Multiplying Eq. (4-148) from the left by function $(\Psi_i^\circ)^*$ and integrating, and using both the hermicity of $\mathscr{H}^\circ$ and orthogonality relations (4-150), it follows that

$$E_i^{(1)} = \int (\Psi_i^\circ)^* \, \mathscr{V} \Psi_i^\circ \, d\tau \qquad (4\text{-}151a)$$
$$E_i^{(2)} = \int (\Psi_i^\circ)^* \, \mathscr{V} \Psi_i^{(1)} \, d\tau \qquad (4\text{-}151b)$$
$$\vdots$$

In order to obtain the expressions for the perturbation corrections to the wave function, the completeness of the orthonormal set of functions Ψ_i°, $i = 1, 2, \ldots$, will be employed. Function $\Psi_i^{(1)}$ in Eq. (148b) can then be expressed in the form of an expansion

$$\Psi_i^{(1)} = \sum_j c_j \Psi_j^\circ \qquad (4\text{-}152)$$

and substituted into Eq. (4-148b). Multiplying the resulting expression from the left by function $(\Psi_k^\circ)^*$ for $k \neq i$ and integrating yields

$$c_k(E_k^\circ - E_i^\circ) + \int (\Psi_k^\circ)^* \mathscr{V} \Psi_i^\circ \, d\tau = 0 \qquad (4\text{-}153)$$

and hence

$$c_k = \frac{-V_{ki}}{E_k^\circ - E_i^\circ}, \qquad \text{for } k \neq i, \qquad (4\text{-}154)$$

where $V_{ki} = \int (\Psi_k^\circ)^* \mathscr{V} \Psi_i^\circ \, d\tau$ is a matrix element of the perturbation operator within the basis of unperturbed functions. Coefficient c_i equals zero due to condition (4-150), so that, for the first-order correction,

$$\Psi_i^{(1)} = -\sum_j' \frac{V_{ji}}{E_j^\circ - E_i^\circ} \, \Psi_j^\circ, \qquad (4\text{-}155)$$

where the prime on the summation sign denotes that the term for $j = i$ is removed from the summation.

Perturbation contributions of higher orders can, similarly as for the first order, be obtained by solution of the corresponding equation of system of equations (4-148). The complexity of the expressions increases very rapidly with the order of the perturbation contribution and it is therefore evident that the usefulness of employing the perturbation treatment for the solution of a problem is related to the rate of convergence of the perturbation series. In practice, the expression for the energy contribution of the second order is still useful (and is used, for example, in the interpretation of NMR spectra); this expression is obtained by substitution of Eq. (4-155) into Eq. (4-151b):

$$E_i^{(2)} = \sum_j{}' \frac{|V_{ji}|^2}{E_i^\circ - E_j^\circ}$$

Once again it must be emphasized that the perturbation treatment can be used only when the perturbation is relatively small; the condition

$$|V_{ji}| \ll |E_j^\circ - E_i^\circ|$$

can be regarded as an approximate criterion of the validity of the use of this method.

In previous considerations we started from the assumption that unperturbed energy level E_i° corresponds to a single eigenfunction. It has been shown that, on perturbation, this level is simply shifted by the value of the correction contributions. Often it is necessary, however, to solve problems where two or more orthonormal eigenfunctions correspond to one unperturbed eigenvalue. If they are g in number, $\Psi_1^\circ, \Psi_2^\circ, ..., \Psi_g^\circ$, then *the level* corresponding to eigenvalue E_i° is said to be *g-fold degenerate*. As a rule, the introduction of a perturbation removes this degeneracy, so that, in practice, the original unperturbed level splits into several levels with different energy values. In contrast to the non-degenerate case where the perturbed function of the zeroth-approximation is already known [cf. Eq. (4-147a)], the situation is more complicated, because under continuous removal of the perturbation, function Ψ_i can, in general, tend to function f in the form of a linear combination of functions

$$f = \sum_{k=1}^{g} a_k \Psi_k^\circ, \qquad (4\text{-}156)$$

which is also an eigenfunction of the unperturbed operator, $\mathscr{H}^\circ$, with the eigenvalue E_i°. It then follows that it is necessary to replace Eq. (4-147a) by the assumption that

$$\Psi_i = f + \lambda \Psi_i^{(1)} + \lambda^2 \Psi_i^{(2)} + ... \qquad (4\text{-}157)$$

with the consequence that even system of equations (4-148) is changed as follows:

0-th order $\quad (\mathscr{H}^\circ - E_i^\circ)f = 0 \qquad\qquad\qquad$ (4-158a)

1st order $\quad (\mathscr{H}^\circ - E_i^\circ)\,\Psi_i^{(1)} + (\mathscr{V} - E_i^{(1)})f = 0 \quad$ (4-158b)

$\vdots$

While Eq. (4-158a) is automatically satisfied, Eq. (4-158b) will be used for the determination of both expansion coefficients a_k and the correction term for energy $E_i^{(1)}$. Similarly as in the normalization condition expressed by Eqs. (4-149) and (4-150), it will be necessary that corrections of the first and higher orders be orthogonal to the space which is spanned by functions $\Psi_1^\circ,\ \Psi_2^\circ,\ \dots,\ \Psi_g^\circ$. Using this condition in Eq. (4-158b) and multiplying it from the left by function $(\Psi_1^\circ)^*$, after integration and successive application of further functions $\Psi_2^\circ,\ \dots,\ \Psi_g^\circ$, a system of g equations for unknown coefficients a_k is obtained:

$$\sum_{k=1}^{g} a_k(V_{jk} - E_i^{(1)}\delta_{jk}) = 0, \qquad\qquad (4\text{-}159)$$

where $j = 1, 2, \dots, g$.

Matrix designation of the integral

$$\int (\Psi_j^\circ)^* \mathscr{V} \Psi_k^\circ \, d\tau = V_{jk}$$

has been introduced, and the orthonormality of functions

$$\Psi_1^\circ, \Psi_2^\circ, \dots, \Psi_g^\circ, \qquad \int (\Psi_j^\circ)^* \, \Psi_k^\circ \, d\tau = \delta_{jk},$$

for $j, k = 1, \dots, g$, has been taken into account. Solution of system of equations (4-159) is mathematically equivalent to problem (4-141), which was solved when dealing with the linear variation problem. Moreover, there is also a simplifying circumstance here in that the overlap matrix $\mathbf{S}$ is a unit matrix. Therefore, all the possible values of $E_i^{(1)}[(E_i^{(1)})_1, (E_i^{(1)})_2, \dots, (E_i^{(1)})_g]$ can be obtained by solving the secular determinant

$$\det \| V_{jk} - E_i^{(1)}\delta_{jk} \| = 0 \qquad\qquad (4\text{-}160)$$

Successive substitution of the energy values $(E_i^{(1)})_1, \dots$ in system of equations (4-159) yields the corresponding series of expansion coefficients bound by the normalization condition

$$\sum_{k=1}^{g} |a_k|^2 = 1$$

In conclusion it is necessary to add that not all of the g values $(E_i^{(1)})_1, (E_i^{(1)})_2, \dots, (E_i^{(1)})_g$ need be different. Supposing that there are m

different energy values ($m \leqq g$), then up to first-order accuracy the original energy level E_i° splits, owing to the perturbation, into m levels with the following energies:

$$E_i = E_i^\circ + (E_i^{(1)})_j, \qquad (4\text{-}161)$$

where j indicates the different correction values.

REFERENCES

1. Blokhintsev D. I.: *Osnovy kvantovoj mekhaniki*. Gos. izd. techn. teoret. lit., Moscow 1949. *Quantum Mechanics*. D. Reidel Publ. Company, Dordrecht 1964. *Mécanique quantique*. Edit. Masson et Cie, Paris 1967.
2. Davtyan O. K.: *Kvantovaya chimiya*. Gos. izd. "Visshaya shkola", Moscow 1962.
3. Eyring H., Walter J., Kimball G. E.: *Quantum Chemistry*. Wiley, New York 1944.
4. Daudel R., Lefebvre R., Moser C.: *Quantum Chemistry*. Interscience Publishers, New York 1959.
5. Murrell J. N., Kettle S. F. A., Tedder J. M.: *Valence Theory*. Wiley, London 1965.
6. Hanna M. W.: *Quantum Mechanics in Chemistry*. Benjamin, New York 1966.
7. Phillips L. F.: *Basic Quantum Chemistry*. Wiley, New York 1965.
8. Coulson C. A.: *Valence*. Clarendon Press, Oxford 1952.
9. Veselov M. G.: *Elementarnaya kvantovaya teoria atomov i molekul*. Gos. izd. fiz.-mat. lit., Moscow 1962.
10. Messiah A.: *Quantum Mechanics*. North-Holland Publishing Company, Amsterdam 1965.
11. Thouless D. J.: *The Quantum Mechanics of Many-Body Systems*. Academic Press, New York 1961.
12. Fermi E.: *Notes on Quantum Mechanics*. The University of Chicago Press, Chicago 1960.
13. Hirschfelder J. O., Brown W. B., Epstein S. T.: *Advan. Quantum Chem.* **1**, 255 (1964).

5. BASIC APPROXIMATIONS IN THE THEORY OF THE CHEMICAL BOND

5.1 Introductory comments

The modern theory of the chemical bond is based on the quantum mechanics of systems composed of electrons and atomic nuclei, assuming that solution of the fundamental quantum mechanical equation leads to a complete description of the system. As follows from the two preceding chapters, difficulties lie not in the formulation of the Schrödinger equation, but in its solution. As even three-particle systems are not exactly solvable, problems interesting for chemists must be simplified by conversion into *model systems*. The mere fact that, as a rule, an isolated system is treated (an atom, a molecule, a solid or a system of several partial subsystems) is a kind of abstraction, as the influence of the surrounding medium, for example the influence of a solvent, is frequently ignored.

Let us start from the general formulation of the atom, molecule or solid matter as a system composed of N atomic nuclei and n electrons. This system is described by the time-dependent Schrödinger equation introduced by Postulate 3. If only stationary cases are studied (see Section 4.4), a description of the system can be obtained by solving the equation

$$\mathcal{H}_t \Phi_i(y) = W_i \Phi_i(y), \tag{5-1}$$

where $\mathcal{H}_t$ is the total Hamiltonian of the system, $\Phi_i(y)$ are the wave functions of the stationary states (the variable y expresses the dependence on the spatial $[R]$ and spin $[\Sigma]$ coordinates of the nuclei and on the spatial $[r]$ and spin $[\sigma]$ coordinates of the electrons, subscript i indicates different solutions) and W_i is the corresponding eigenvalue of the total energy. The total Hamiltonian can be expressed as the sum of three contributions:

$$\mathcal{H}_t = \mathcal{H}_{en} + \mathcal{H}_{ext} + \mathcal{H}_{int}, \tag{5-2}$$

where $\mathcal{H}_{en}$ contains the operators of the kinetic energy and the electrostatic interactions of all the participating particles, $\mathcal{H}_{ext}$ expresses all types of interactions of the system with the external magnetic and electric fields and $\mathcal{H}_{int}$ contains all types of non-electrostatic interactions between particles in the given system, related to the spins of the electrons and nuclei.

Equation (5-1) represents an exact formulation for an isolated system. Approximations enabling the solution of a particular problem can, in principle, be divided into two groups. The first contains approximations that simplify the Hamiltonian, limit the form of the wave function, and, as a rule, have a more general character so that they can be used in a number of various methods. In the second group belong approximations related to determination of the values of the integrals appearing in the calculation schemes. This type of approximation is frequently specific for a particular method and will therefore be discussed when describing the individual methods. This chapter contains a brief outline of approximations employed in the first group. It should be added that the effect of approximations in individual methods on the ability to predict the physical and chemical properties of systems can be judged only on the basis of comparison of theoretical results with experimental data. This is not always true of non-empirical methods: sometimes the quality of approximations is studied by comparing the results of less precise procedures with those obtained using more precise ones.

5.2 Neglecting of non-electrostatic interactions

The approximation of neglecting non-electrostatic interactions consists of ignoring terms $\mathcal{H}_{ext}$ and $\mathcal{H}_{int}$ in the total Hamiltonian (5-2), i.e. contributions following from the spins of the electrons and atomic nuclei and the influence of the external field. This approximation is used in nearly all quantum-chemical methods. Investigation of interactions of the spin-spin coupling type (interaction of two charged particles via magnetic dipoles corresponding to their spins) and spin-orbit coupling type (interaction of charged particles via magnetic dipoles corresponding to spin and orbital motions) is important for the study of atomic fine structure; the magnitude of the effect increases with increasing atomic number. Part of the Hamiltonian $\mathcal{H}_{ext}$ is taken into account in studying the effect of external fields on molecular systems, for example when interpreting NMR and ESR spectra.

The independence of the Hamiltonian of the spin coordinates of the electrons has an important theoretical consequence: the Hamiltonian commutes with the spin operators and the total spin quantum numbers are thus "valid" quantum numbers for characterizing the electronic states.

5.3 The Born-Oppenheimer and adiabatic approximations

Within the framework of the approximation neglecting all non-electrostatic interactions, the Hamiltonian of the system can be expressed by the relationship

$$\mathscr{H}_{en} = \mathscr{T}_n + \mathscr{T}_e + \mathscr{V}_{en} + \mathscr{V}_{ee} + \mathscr{V}_{nn}, \qquad (5\text{-}3)$$

where the individual terms correspond to the following contributions to the total energy of the system:

$$\mathscr{T}_n = -\frac{\hbar^2}{2} \sum_{I=1}^{N} \frac{1}{M_I} \Delta_I$$

($\mathscr{T}_n$ is the kinetic energy operator of the atomic nuclei, and M_I is the mass of the nucleus I),

$$\mathscr{T}_e = -\frac{\hbar^2}{2m} \sum_{i=1}^{n} \Delta_i$$

($\mathscr{T}_e$ is the kinetic energy operator of the electrons, and m is the mass of an electron),

$$\mathscr{V}_{en} = -\sum_{i=1}^{n} \sum_{I=1}^{N} \frac{Z_I e^2}{4\pi\varepsilon_0 \left| \mathbf{r}_i - \mathbf{R}_I \right|}$$

($\mathscr{V}_{en}$ is the potential energy operator of the electrostatic interaction between the electrons and the nuclei, Z_I is the charge of nucleus I in units of the elementary charge e, $\mathbf{r}_i$ is the positional vector describing the position of electron i, and $\mathbf{R}_I$ is the positional vector describing the position of nucleus I with respect to the origin of the coordinate system),

$$\mathscr{V}_{ee} = \sum_{i<j}^{n} \frac{e^2}{4\pi\varepsilon_0 \left| \mathbf{r}_i - \mathbf{r}_j \right|}$$

($\mathscr{V}_{ee}$ is the potential energy operator of electrostatic interactions between the electrons),

$$\mathscr{V}_{nn} = \sum_{I<J}^{N} \frac{Z_I Z_J e^2}{4\pi\varepsilon_0 \left| \mathbf{R}_I - \mathbf{R}_J \right|}$$

($\mathscr{V}_{nn}$ is the potential energy operator of electrostatic interactions between the nuclei).

A further simplification of the Hamiltonian (5-3) can be based on the fact that electrons have a substantially smaller mass than nuclei; for the hydrogen atom the mass ratio is $5 : 10^4$.

First, the interaction of the electron — atomic nucleus pair will be considered. Since the same force acts on both particles, the lighter of the two, the electron, acquires a much greater acceleration according to Newton's second law, so that the average velocity of the electron is substantially higher than that of the nucleus. On this basis, for an initial system of n electrons and N nuclei the following model[1] can be formed: slowly moving atomic nuclei create an electrostatic field in which the electrons move with a much greater velocity; the ratio between the mean velocity of the electrons and the mean velocity of the nuclei is so large that the motion of the electrons is almost instantly adaptable to changes in the configuration of the nuclei, and, conversely, the nuclei are exposed to such rapid fluctuations of electrostatic potential from the electrons that they "see" only its mean value.

In the zeroth approximation, the mass of the nuclei can be considered to be infinitely large. This assumption leads to the conclusion that both the acceleration imparted to the nuclei and their velocity become zero, and consequently, the kinetic energy of the nuclei can be considered equal to zero. An attempt to find the stationary states of the system leads to the Schrödinger equation

$$[\mathscr{H}_e - E_i(R)] \, \Psi_i(R, r) = 0, \tag{5-4}$$

where the nuclei are assumed to be fixed in space, so that the coordinates of the atomic nuclei in an arbitrary configuration can then be considered as parameters (denoted by symbol R). The Hamiltonian, $\mathscr{H}_e$, is defined by the established notation as

$$\mathscr{H}_e = \mathscr{T}_e + \mathscr{V}_{en} + \mathscr{V}_{ee} + \mathscr{V}_{nn} \tag{5-5}$$

$E_i(R)$ is the total energy of a system in a state described by wave function $\Psi_i(R, r)$, where for simplicity the spin part of the coordinates is omitted, as it has no direct relation to this problem.

This simplification is called *the Born-Oppenheimer approximation* and leads, as has been shown, to complete separation of the motion of the electrons from that of the nuclei. By solving Eq. (5-4) a system of eigenvalues $E_i(R)$ and eigenfunctions $\Psi_i(r, R)$ can be obtained for any nuclear configuration R, where subscript i characterizes the set of quantum numbers determining the corresponding stationary state.

It is, of course, necessary to determine the magnitude of the error introduced into the calculation when using the Born-Oppenheimer approximation. If the concept of fixed nuclei is abandoned, then the problem is expressed by the Schrödinger equation in the form

$$(\mathscr{H}_{en} - W) \, \Phi(R, r) = 0 \tag{5-6}$$

The solution of the Born-Oppenheimer approximation [represented by Eq. (5-4)] will be assumed to be known, enabling wave function $\Phi(R, r)$

to be sought in the form of an expansion in a series whose terms are the orthonormalized functions $\Psi_i(R, r)$:

$$\Phi(R, r) = \sum_i \Xi_i(R)\,\Psi_i(R, r) \tag{5-7}$$

For the sake of simplicity, it is assumed that only a discrete spectrum of eigenvalues of operator $\mathscr{H}_e$ exists. Functions $\Xi_i(R)$ act as expansion coefficients in Eq. (5-7) and are dependent only on the coordinates of the nuclei.

Before substituting Eq. (5-7) into Eq. (5-6), the action of the operator of the total kinetic energy of the nuclei $\mathscr{T}_n$ on the product of functions $\Xi_i(R)\,\Psi_i(R, r)$ will be investigated. The Laplace symbol Δ_I is equal to the sum of three expressions of the type $\partial^2/\partial X_I^2$, where X_I represents one of the three rectangular coordinates describing the position of nucleus I. According to the rule of differentiation of the product of two functions the expression

$$\frac{\partial^2(\Xi\Psi)}{\partial X^2} = \Psi\,\frac{\partial^2\Xi}{\partial X^2} + 2\,\frac{\partial\Xi}{\partial X}\,\frac{\partial\Psi}{\partial X} + \Xi\,\frac{\partial^2\Psi}{\partial X^2} \tag{5-8}$$

is obtained. The required expression then follows

$$\mathscr{T}_n[\Xi_i(R)\,\Psi_i(R, r)] = \Psi_i(R, r)\,\mathscr{T}_n\Xi_i(R) - \hbar^2\sum_I\frac{1}{M_I}[\nabla_I\Xi_i(R)]\,[\nabla_I\Psi_i(R, r)] +$$

$$+ \Xi_i(R)\,\mathscr{T}_n\Psi_i(R, r) \tag{5-9}$$

From Eq. (5-9) and the fact that $\mathscr{H}_{en} = \mathscr{H}_e + \mathscr{T}_n$ and after substituting series (5-7) into Eq. (5-6) the expression

$$\sum_i \{(\mathscr{H}_e - W)\,\Xi_i(R)\,\Psi_i(R, r) + \Psi_i(R, r)\,\mathscr{T}_n\Xi_i(R) -$$

$$- \hbar^2\sum_I\frac{1}{M_I}[\nabla_I\Xi_i(R)]\,[\nabla_I\Psi_i(R, r)] + \Xi_i(R)\,\mathscr{T}_n\Psi_i(R, r)\} = 0 \tag{5-10}$$

is obtained. Functions $\Psi_i(R, r)$ in Eq. (5-10) are assumed to be orthonormal,

$$\int \Psi_j^*(R, r)\,\Psi_i(R, r)\,\mathrm{d}r = \delta_{ij}, \tag{5-11}$$

where symbol $\mathrm{d}r$ indicates integration over the coordinates of all the electrons. Of the operators depending on the variables denoted by R, operator $\mathscr{H}_e$ contains only those which have a simple dependence on R, so that, by multiplying Eq. (5-10) from the left by function $\Psi_j^*(R, r)$, integrating over the coordinates of electrons r, and employing Eq. (5-4), the expression

$$[\mathscr{T}_n + E_j(R) - W]\,\Xi_j(R) = \sum_i \Lambda_{ji}\Xi_i(R) \tag{5-12}$$

is obtained, where operator Λ_{ji} is defined as follows:

$$\Lambda_{ji} = \hbar^2 \sum_I \frac{1}{M_I} \int \Psi_j^*(R, r) \left[\nabla_I \Psi_i(R, r)\right] dr \, \nabla_I -$$

$$- \int \Psi_j^*(R, r) \, \mathscr{T}_n \Psi_i(R, r) \, dr \qquad (5-13)$$

Expression (5-12) is actually a system of differential equations because Eq. (5-10) can be multiplied not only by function $\Psi_j^*(R, r)$ but by any other function which is also a solution of Eq. (5-4).

No approximations were used, so that system of equations (5-12) gives a true picture of the relationship between the electron and the nuclear motion in the form of the square matrix, Λ_{ji}. If only the diagonal elements are considered in this matrix, then the original system of equations (5-12) is separated into a system of independent equations

$$\left[\mathscr{T}_n + \varepsilon_j(R)\right] \Xi_j(R) = W\Xi_j(R), \qquad (5-14)$$

where $\varepsilon_j(R)$ $\left[\varepsilon_j(R) = E_j(R) - \Lambda_{jj}\right]$ is the corrected electronic energy obtained within the framework of the Born-Oppenheimer approximation, where the correction Λ_{jj} can change the original value in either the positive or negative sense. This approximation allowing separation of the electron and the nuclear motion and, in addition, taking into account the weak interaction of both motions is called *the adiabatic approximation*[2].

Similar to expression $E_j(R)$ in the Born-Oppenheimer approximation, the term $\varepsilon_j(R)$ in the adiabatic approximation can be interpreted in the eigenvalue equation for the nuclear motion (5-14) as a potential in whose field the atomic nuclei move. The parallel between the two types of approximation is retained for the total molecular wave function, which is, in both cases, equal to the simple product

$$\Phi_{j,v} = \Xi_{j,v}(R) \, \Psi_j(R), \qquad (5-15)$$

where a new index v was introduced to distinguish the different solutions of Eq. (5-14).

Generally, the adiabatic approximation is justified when the solution of Eq. (5-14) differs only slightly from the solution of system of equations (5-12). In the Schrödinger perturbation theory, where the Born-Oppenheimer approximation is considered to be a zero approximation, the adiabatic correction, Λ_{jj}, corresponds to the first-order perturbation term; similarly, the off-diagonal elements, which correspond to the interaction of various electronic levels, can be interpreted as representing contributions of the second order (cf. Section 4.6). From the convergence condition of the perturbation expansion it follows that the adiabatic approximation is a reasonable approximation if

$$\int \Xi_{i,v}^*(R)\,\Lambda_{ij}\Xi_{j,v'}(R)\,\mathrm{d}R \ll |W_{i,v} - W_{j,v'}| \qquad (5\text{-}16)$$

for $i \neq j$ and arbitrary v, v', where $W_{i,v}$ is the eigenvalue corresponding to the eigenfunction $\Xi_{i,v}$. Criterion (5-16) is satisfied for a majority of molecules; it is, however, not fulfilled for some large systems where the energy levels are very close together and explains the failure of the adiabatic approximation for these cases. The relationship between the electron and the nuclear motion should then be examined so that the largest interaction elements of operator Λ_{ij} are included in the calculation. Terms $E_j(R)$ or $\varepsilon_j(R)$, of course, can no longer be interpreted as the potential field in which the nuclei are moving.

Table 5-1

Experimental and Calculated Dissociation Energies (in cm^{-1}) of Some Molecules in the Ground State

	H_2	HD	D_2
experimental	36 113.6	36 400.5	36 744.2
Born – Oppenheimer approximation	36 112.2	36 401.5	36 745.6
adiabatic approximation	36 118.0	36 405.7	36 748.3
non-adiabatic approach	36 114.7	36 402.9	36 746.2

The Born-Oppenheimer approximation generally represents a very good approximation to real systems. Errors arising from its use are much smaller than those encountered using other approximations. Reliable numerical verification is possible solely in the smallest systems where the same accuracy or greater as that of the measured quantities can be achieved. Since the ratio of the masses of an electron and the participating nuclei is minimal in these systems, the deviations from validity of the Born-Oppenheimer or the adiabatic approximation are, consequently, maximal, assuming that criterion (5-16) is not considered. Table 5-1, from Wolniewicz's paper[3], gives the dissociation energies of the H_2, HD and D_2 molecules in the ground state. Included in the table are experimental values and values calculated within the framework of the Born-Oppenheimer approximation together with the adiabatic correction and the correction for non-adiabaticity. It is obvious that the agreement with experiment is very good in all cases and that the adiabatic correction to the Born-Oppenheimer approximation decreases with increasing masses of the participating nuclei, beginning with a correction of 0.016 % for the hydrogen molecule and ending with a value of 0.007 % for D_2. In addition, Table 5-1 shows that, of the two types of approximation, only the adiabatic approximation fulfils the variation principle, and thus the corresponding calculated energy values lie above the actual experimentally determined values.

In conclusion it is necessary to state that the Born-Oppenheimer approximation is used in practically all quantum chemical calculations of the electronic structure of molecules and solid substances and therefore the solution of Eq. (5-4) will receive a great deal of attention here. It will be useful to simplify the notation in this equation: first, designation of the dependence of the nuclear coordinates, which has been shown to be of only parametric character, will be omitted. Furthermore, the electronic Hamiltonian $\mathscr{H}_e$ will be designated $\mathscr{H}$ and the term $\mathscr{V}_{nn}$, which contributes a constant amount to the total energy of a system with a given nuclear configuration, irrespective of the state of the system, will be omitted. Wave function Ψ depends on the spin and spatial coordinates of all the electrons. In this modified notation, the Schrödinger equation assumes the form

$$(\mathscr{H} - E)\,\Psi(r, \sigma) = 0, \tag{5-17}$$

where it is convenient to separate $\mathscr{H}$ into one-electron and two-electron contributions:

$$\mathscr{H} = \mathscr{T}_e + \mathscr{V}_{en} + \mathscr{V}_{ee} = \sum_{i=1}^{n} \hbar(i) + \sum_{i<j}^{n} g(i, j), \tag{5-18}$$

where

$$\hbar(i) = -\frac{\hbar^2}{2m}\Delta_i - \sum_{I=1}^{N} \frac{Z_I e^2}{4\pi\varepsilon_0 \,|\,r_i - R_I\,|} \tag{5-19a}$$

and

$$g(i, j) = \frac{e^2}{4\pi\varepsilon_0 \,|\,r_i - r_j\,|} \tag{5-19b}$$

5.4 The method of configuration interaction

In Section 4.3 the concept of a complete orthonormal set of functions was introduced. Let us assume, in accordance with this definition, that there is a complete orthonormal set of one-electron functions, $\lambda_k(x)$, $k = 1, 2, 3, \ldots$, where each of the functions depends on the spatial coordinates (r) and on the spin coordinate (σ). Functions of this type are generally referred to as *spin orbitals*. Since set of functions λ_k is complete, wave function Φ, which describes any stationary state of the electron, can be expressed in the form of the expansion

$$\Phi(x) = \sum_{i} c_i \lambda_i(x) \tag{5-20}$$

(It should again be noted that, for the sake of simplicity, only functions corresponding to discrete spectra are considered.)

The set of spin orbitals λ_k, $k = 1, 2, \ldots$, can be used to express the n-electron wave function Ψ, which is the solution of Eq. (5-17). To this end it will be convenient to investigate first a system composed of two electrons.

The wave function $\Psi(x_1, x_2)$ of a two-electron system, where x_1 and x_2 denote the space-spin coordinates of the first and second electrons, can be expressed in terms of functions λ_k as follows: assuming that electron 2 is fixed in space and that, consequently, its coordinates can be considered to be a set of constants, it follows that

$$\Psi(x_1, x_2 \equiv \text{const}) = \sum_i c_i \lambda_i(x_1) \tag{5-21a}$$

Release of electron 2 from its fixed position can be expressed in the following manner: all quantities c_i (initially constant) in Eq. (5-21a) will become a function of the instantaneous coordinate of electron 2; therefore it follows that

$$c_i(x_2) = \sum_j c_{i,j} \lambda_j(x_2) \tag{5-21b}$$

and, after substituting Eq. (5-21b) into Eq. (5-21a)

$$\Psi(x_1, x_2) = \sum_{i,j} c_{i,j} \lambda_i(x_1) \lambda_j(x_2), \tag{5-22}$$

where $c_{i,j}$ are the corresponding expansion coefficients that are not, however, independent. As mentioned previously (cf. Section 4.4), a wave function describing a system of electrons must obey the Pauli principle, i.e. its sign changes on exchanging the coordinates of an arbitrary pair of electrons:

$$\mathscr{P}_{12} \Psi(x_1, x_2) = -\Psi(x_2, x_1), \tag{5-23}$$

where $\mathscr{P}_{12}$ is the transposition operator of electrons 1 and 2. Condition (5-23) results in the expression

$$c_{i,j} = -c_{j,i} \tag{5-24}$$

and permits Eq. (5-22) to be written in the form

$$\Psi(x_1, x_2) = \sum_{i<j} c_{i,j} [\lambda_i(x_1) \lambda_j(x_2) - \lambda_i(x_2) \lambda_j(x_1)] \tag{5-25}$$

Equation (5-25) embodies the property of diagonal expansion coefficients following from Eq. (5-24),

$$c_{i,i} = 0 \qquad \text{for all values of } i,$$

and includes the assumption that the spin orbitals are ordered in a fixed sequence. The expression in square brackets in Eq. (5-25) is obviously an expanded second-order determinant

$$\det \| \lambda_i(1), \ \lambda_j(2) \| = \begin{vmatrix} \lambda_i(x_1), \ \lambda_i(x_2) \\ \lambda_j(x_1), \ \lambda_j(x_2) \end{vmatrix} \tag{5-26}$$

which can be taken as an element of the total set of functions used for expanding the two-electron function. In order to conform with the standard manner of expanding functions, expressed by Eqs. (4-35) and (4-36), the elements must be both normalized and mutually orthogonal functions. First, from the condition

$$1 = \frac{1}{k^2} \int \begin{vmatrix} \lambda_i(x_1), \ \lambda_i(x_2) \\ \lambda_j(x_1), \ \lambda_j(x_2) \end{vmatrix}^* \begin{vmatrix} \lambda_i(x_1), \ \lambda_i(x_2) \\ \lambda_j(x_1), \ \lambda_j(x_2) \end{vmatrix} dx_1 \, dx_2 =$$

$$= \frac{1}{k^2} \{ \int \lambda_i^*(x_1) \, \lambda_i(x_1) \, dx_1 \int \lambda_j^*(x_2) \, \lambda_j(x_2) \, dx_2 +$$

$$+ \int \lambda_i^*(x_2) \, \lambda_i(x_2) \, dx_2 \int \lambda_j^*(x_1) \, \lambda_j(x_1) \, dx_1 \} = \frac{1}{k^2} \{ 1 + 1 \}, \tag{5-27}$$

where the orthonormality of spin orbitals has been taken into account, it follows that the two-electron determinant function

$$\frac{1}{k} \begin{vmatrix} \lambda_i(x_1), \ \lambda_i(x_2) \\ \lambda_j(x_1), \ \lambda_j(x_2) \end{vmatrix}$$

is normalized if $k = \sqrt{2}$. It similarly follows that the determinants are orthogonal if they differ in at least one spin orbital.

On generalization of the above considerations, it follows that any n-electron wave function can be expanded as a linear combination of determinants in the form[4]

$$\Psi(1, 2, \ldots, n) = \sum_K C_K \Delta_K(1, 2, \ldots, n), \tag{5-28}$$

where the summation is carried out over all the ordered configurations K of the spin orbitals. An ordered configuration K is a certain selection of n indices fulfilling the condition $i < j < \ldots < k$. C_K is the expansion coefficient; its significance is such that $|C_K|^2$ gives the weight of the function

$$\Delta_K(1, 2, \ldots, n) = \frac{1}{\sqrt{n!}} \begin{vmatrix} \lambda_i(x_1), & \lambda_i(x_2), & \ldots, & \lambda_i(x_n) \\ \lambda_j(x_1), & \lambda_j(x_2), & \ldots, & \lambda_j(x_n) \\ \vdots & & & \\ \lambda_k(x_1), & \lambda_k(x_2), & \ldots, & \lambda_k(x_n) \end{vmatrix} = |\lambda_i, \lambda_j, \ldots, \lambda_k| \tag{5-29}$$

in expansion (5-28). The cofactor $1/\sqrt{n!}$ of the determinant function ensures its normalization. The function defined by Eq. (5-29) is termed *the Slater determinant.*

Expansion of the wave function in form (5-28) is, as a rule, not practically feasible, as the complete sets of one-electron functions are not usually of finite dimension. Numerical applications must be limited to bases of finite dimentions, where the solutions assume an approximate character. It was noted in Section 4.6 that, for calculations of this type, the variation method is useful.

Now let us assume that, in contrast to the preceding instance, a finite incomplete set of m spin orbitals $\lambda_k(x)$ is considered. This will permit a wave function in form (5-28) to be sought, that is, in the form of a linear combination of Slater determinants, with the restriction, however, that the summation includes a finite number of configurations. If the summation is carried out over all possible configurations of the given spin-orbital space we shall speak again of *"complete configuration interaction"*; if function Ψ is to be found in the form of a linear combination of specially selected configurations, this is termed *"limited configuration interaction"*. This involves a certain nomenclatural inaccuracy, since both complete and incomplete sets of one-electron functions are assigned a wave function in the form of a "complete configuration interaction", although in the first instance it represents a precise function, in the second, that is, for an incomplete set, it is only an approximate wave function, which can be at best optimized. The best solution in the chosen extent of configuration interaction is obtained by solving equations of types (4-141) and (4-142), where the corresponding matrix elements are now defined with respect to the (5-18) type of Hamiltonian:

$$H_{IJ} = \int \Delta_I^* \mathcal{H} \Delta_J \, d\tau = I_{IJ} + G_{IJ}, \qquad (5\text{-}30a)$$

$$S_{IJ} = \int \Delta_I^* \Delta_J \, d\tau, \qquad (5\text{-}30b)$$

where

$$I_{IJ} = \int \Delta_I^* \left[\sum_{i=1}^{n} h(i) \right] \Delta_J \, d\tau$$

and

$$G_{IJ} = \int \Delta_I^* \left[\sum_{i<j} g(i,j) \right] \Delta_J \, d\tau$$

The solution of the secular problem is a standard task, but the calculation of the matrix elements of operator $\mathcal{H}$ enclosed by Slater determinants is a specific problem occurring in calculations using the configuration interaction method. It is summarized in *Slater's rules*[5]. In calculating integrals (5-30) it is convenient to classify cases according to the number of spin orbitals in which the Δ_I and Δ_J functions differ from each other. The number of differing spin orbitals can be determined so that, using the rule for the interchange of rows in the determinant, the Δ_I and Δ_J determinant functions are converted to a mutual maximum coincidence and then compared row by row. For illustration, $\Delta_1 \equiv \ \equiv |\lambda_i, \lambda_k, \lambda_j|$ is not compared with $\Delta_2 \equiv |\lambda_i, \lambda_j, \lambda_l|$ but $-\Delta_1 \equiv \ \equiv |\lambda_i, \lambda_j, \lambda_k|$ with Δ_2, where $(-\Delta_1)$ differs from Δ_2 in one spin orbital. In calculation of the corresponding matrix elements, a number of cases will be distinguished:

    1. Δ_I and Δ_J do not differ from each other

    2. Δ_I and Δ_J differ in one spin orbital; λ_a in Δ_I is replaced in Δ_J by spin orbital λ_c

    3. Δ_I and Δ_J differ in two spin orbitals; λ_a and λ_b in Δ_I are replaced in Δ_J by spin orbitals λ_c and λ_d

    4. Δ_I and Δ_J differ from each other in more than two spin orbitals.

Table 5-2

Matrix Elements in the Secular Equation Employed in the CI Method

Cases	1	2	3	4
	$\Delta_I \equiv \Delta_J$	$\Delta_I(\lambda_a); \Delta_J(\lambda_c)$	$\Delta_I(\lambda_a, \lambda_b); \Delta_J(\lambda_c, \lambda_d)$	
S_{IJ}	1	0	0	0
I_{IJ}	$\sum_{i=1}^{n} \langle \lambda_i | \hbar | \lambda_i \rangle$	$\langle \lambda_a | \hbar | \lambda_c \rangle$	0	0
G_{IJ}	$\sum_{i<j} [\langle \lambda_i\lambda_j | g | \lambda_i\lambda_j \rangle -$ $- \langle \lambda_i\lambda_j | g | \lambda_j\lambda_i \rangle]$	$\sum_{i(\neq a,c)} [\langle \lambda_i\lambda_a | g | \lambda_i\lambda_c \rangle -$ $- \langle \lambda_i\lambda_a | g | \lambda_c\lambda_i \rangle]$	$\langle \lambda_a\lambda_b | g | \lambda_c\lambda_d \rangle -$ $- \langle \lambda_a\lambda_b | g | \lambda_d\lambda_c \rangle$	0

The matrix elements of the overlap matrix S_{IJ}, the one-electron part of Hamiltonian I_{IJ}, and the two-electron part of Hamiltonian G_{IJ} are given in Table 5-2. The matrix elements were written using the Dirac notation, according to which

$$\langle \lambda_a | \hbar | \lambda_b \rangle = \int \lambda_a^*(x_1)\, \hbar(1)\, \lambda_b(x_1)\, dx_1 \qquad (5\text{-}31a)$$

$$\langle \lambda_a\lambda_b | g | \lambda_c\lambda_d \rangle = $$
$$= \iint \lambda_a^*(x_1)\, \lambda_b^*(x_2)\, g(1, 2)\, \lambda_c(x_1)\, \lambda_d(x_2)\, dx_1\, dx_2 \qquad (5\text{-}31b)$$

The expressions can be somewhat simplified—as will be justified later—by noting that spin orbitals have the form of a simple product of the spatial, orbital depending on the space coordinates and of spin function α or β; thus

$$\lambda_i(x) = \varphi_i(\mathbf{r})\,\eta(\sigma), \tag{5-32}$$

where η is either function α or function β.

If spin orbitals are expressed in the form (5-32), integration can easily be performed in matrix elements (5-31) over the spin variable [cf. Eqs. (4-82) and (4-83)], since operators h and g do not depend on the spin coordinates. We then obtain

$$\langle \lambda_a | h | \lambda_b \rangle = \begin{cases} \langle \varphi_a | h | \varphi_b \rangle & \text{(if } \lambda_a \text{ and } \lambda_b \text{ have the same spin)} \\ 0 & \text{(if } \lambda_a \text{ and } \lambda_b \text{ have a different spin)} \end{cases} \tag{5-33}$$

$$\langle \lambda_a \lambda_b | g | \lambda_c \lambda_d \rangle = \begin{cases} \langle \varphi_a \varphi_b | g | \varphi_c \varphi_d \rangle & \text{(if } \lambda_a \text{ has the same spin} \\ & \text{as } \lambda_c, \text{ and } \lambda_b \text{ has the} \\ & \text{same spin as } \lambda_d) \\ 0 & \text{(if the above condition} \\ & \text{is not satisfied)} \end{cases} \tag{5-34}$$

Relations (5-33) and (5-34) enable us to express the matrix element types given in Table 5-2 in terms of integrals over space coordinates alone, provided the specific way in which the Δ_I and Δ_J Slater determinants are occupied by spin orbitals is known.

The choice of suitable one-electron functions, $\lambda_i(x)$, forming an incomplete set has not yet been discussed. The suitability of use of the configuration interaction method and the convergence properties of an expansion of type (5-28) depend on this choice. A frequently used procedure begins with a one-particle approximation to the given problem as the first calculation step, giving the requisite one-electron functions.

5.5 The independent electron model (one-electron approximation)

The term $\sum_{i<j} g(i,j)$ in Hamiltonian (5-18) expressing the electrostatic interaction between electrons renders the differential equations describing many-electron systems exactly unsolvable. The Schrödinger equation can be converted into a solvable problem by separating the sum of the two-electron operators into contributions which can be summed up over the individual electrons:

$$\sum_{i<j} g(i,j) \approx \sum_i \mathcal{V}(i) \tag{5-35}$$

Assumption (5-35) can be physically interpreted so that each of the electrons moves in the electrostatic field of the nuclei and in the space- and time-averaged potential of the remaining electrons. Substitution of Eq. (5-35) into Eq. (5-18) yields an approximate Hamiltonian for the n-electron system

$$\mathscr{H} \approx \sum_{i=1}^{n} [\hbar(i) + \mathscr{V}(i)] = \sum_{i=1}^{n} \mathscr{H}'(i), \qquad (5\text{-}36)$$

which allows formulation of a simple solution $\Psi(x)$ of the Schrödinger equation for our model case,

$$[\sum_{i=1}^{n} \mathscr{H}'(i)] \, \Psi(x) = E\Psi(x), \qquad (5\text{-}37)$$

that is, in the form of the product of one-electron functions:

$$\Psi(x) = \lambda_1(x_1) \, \lambda_2(x_2) \dots \lambda_n(x_n) \qquad (5\text{-}38)$$

If Eq. (5-38) is substituted into Eq. (5-37), bearing in mind that each of the $\mathscr{H}'(j)$ operators acts on only one function, $\lambda_j(x_j)$, and if Eq. (5-37) is multiplied by the expression $1/\Psi(x)$, then the relationship

$$\sum_{i=1}^{n} \frac{\mathscr{H}'(i) \, \lambda_i(x_i)}{\lambda_i(x_i)} = E \qquad (5\text{-}39)$$

is obtained. Since individual terms in the summation are independent, Eq. (5-39) can be satisfied only when each of the terms equals a constant:

$$\frac{\mathscr{H}'(i) \, \lambda_i(x_i)}{\lambda_i(x_i)} = E_i, \qquad (5\text{-}40\text{a})$$

where

$$\sum_i E_i = E \qquad (5\text{-}40\text{b})$$

Equation (5-40a) is the one-particle Schrödinger equation for the i-th electron; however, since it has the same form for all electrons, all possible one-electron states represented by functions λ_i must satisfy Eqs. (5-40a) and (5-40b). The one-electron Hamiltonian, $\mathscr{H}'(i)$, is assumed not to contain spin variables (interactions of a non-electrostatic nature are neglected), so that the assumption expressed by Eq. (5-32) can be applied to function $\lambda_i(x_i)$, and the one-electron Schrödinger equation can be written in the form

$$\mathscr{H}'(i) \, \varphi_i(\mathbf{r}) = E_i \varphi_i(\mathbf{r}) \qquad (5\text{-}41)$$

Since integration was carried out over the spin variable, only space coordinates occur in the equation. Function $\varphi_i(\mathbf{r})$ is, according to the circumstances, called either *an atomic* or *a molecular orbital*. It is now necessary to arrange the form of the product function (5-38) so that $\Psi(x)$ satisfies the Pauli principle. It follows from the preceding section that the Slater determinant represents such a function, and thus the final equation for a system described within a one-particle approximation assumes the form

$$\Psi(x) = \left| \lambda_1(x_1), \lambda_2(x_2), \ldots, \lambda_n(x_n) \right|, \qquad (5-42)$$

where the notation introduced in Eq. (5-29) is used, according to which symbolically only the diagonal elements of the determinant are written out. It is worth noting that function (5-42) expresses an older formulation of the Pauli principle, stating that two electrons cannot occupy the same one-electron function as, owing to the equality of two rows, the determinant vanishes.

It has been shown experimentally that the majority of molecules contain an even number of electrons and that, in its energy-lowest state, the total spin equals zero, so that the electrons occupy the same number of α and β spin states. Wave function (5-42) then assumes the form

$$\Psi(x) = \left| \varphi_1(\mathbf{r}_1)\,\alpha(\sigma_1), \varphi_1(\mathbf{r}_2)\,\beta(\sigma_2), \ldots, \varphi_{n/2}(\mathbf{r}_{n-1})\,\alpha(\sigma_{n-1}), \varphi_{n/2}(\mathbf{r}_n)\,\beta(\sigma_n) \right|, \quad (5-43)$$

where every φ_i orbital is occupied by two electrons, one in the α spin state and another in the β state. Such a system is usually termed a system with *closed shells*; if each orbital is not occupied by a pair of electrons, it is termed a system with *open shells*.

In conclusion, it is desirable to describe methods for setting up the approximate one-particle potential energy, $\mathscr{V}$, defined by Eq. (5-35). In principle, two procedures are possible:

    a) *the semiempirical method*, where a potential is produced, such that experimental data can be reproduced by computation,

    b) *the "self-consistent field" method*.

The Hückel method and the extended Hückel method, to be described in Chapter 10, are examples of the first category. The "self-consistent field" method is based on the requirement that the functional

$$E = \frac{\int \Psi^* \mathscr{H} \Psi \, d\tau}{\int \Psi^* \Psi \, d\tau} \qquad (5-44)$$

has a minimum, where the Hamiltonian $\mathscr{H}$ is defined by Eq. (5-18) and function Ψ by Eq. (5-42). The variation variables are represented by the one-electron functions $\lambda_k(x_k)$. In other words, the optimum spin orbitals which yield the best, i.e. the minimum, estimate of the total energy of the system, expressed by Eq. (5-44), are sought.

The problem can also be approached using the configuration inter-action method, which affords a more accurate estimate of the wave function and according to which the wave function is expressed in the form of a linear combination of Slater determinants [cf. Eq. (5-28)] corresponding to particular spin orbital configurations. Let us assume that the first n spin orbitals of an orthonormalized basis set minimize expression (5-44); function $\Psi \equiv \Delta_0$ is evidently one of the determinants in expansion (5-28). The corresponding configuration, $K_0 \equiv 1, 2, ..., n$, is termed *the ground state configuration*, and can obviously be expected to appear in the expansion with the maximum weight (represented by the value $|C_0|^2$) compared with the other configurations. In the configura-tion expansion of wave function Φ, of course, also appear configurations that differ from the ground state configuration in one, two, or more spin orbitals:

$$\Phi = C_0\Delta_0 + \sum_{i=1}^{n} \sum_{q(>n)} C_{iq}\Delta_{iq} + ..., \qquad (5\text{-}45)$$

where Δ_{iq} is the Slater determinant of the configuration in which the λ_i spin orbital from the ground state configuration is replaced by the λ_q spin orbital, which lies outside the ground state configuration.

It will be demonstrated that, when

$$\Delta_0 \equiv |\lambda_1, \lambda_2, ..., \lambda_i, ..., \lambda_n|$$

minimizes expression (5-44), the relation

$$\int \Delta_0^* \mathcal{H} \Delta_{iq} \, d\tau = 0, \quad \text{for all} \quad 1 \leq i \leq n \quad \text{and} \quad q > n, \qquad (5\text{-}46)$$

called *the Brillouin theorem*, is valid.

The proof of the theorem is relatively simple. Let us assume that, although Δ_0 is an "optimum" Slater determinant, the Brillouin theorem (5-46) does not hold; for example, let

$$H_{01} = \int \Delta_0^* \mathcal{H} \Delta_{iq} \, d\tau \neq 0 \qquad (5\text{-}47)$$

Assumption of the validity of Eq. (5-47) justifies considering a function of the form

$$\Psi = C_0\Delta_0 + C_1\Delta_{iq} \qquad (5\text{-}48)$$

and specifying coefficients C_0 and C_1 so that the normalized function Ψ yields an energy minimum. According to Eqs. (4-142) and (5-30) and due to the validity of the expression $S_{12} = \int \Delta_0^* \Delta_{iq} \, d\tau = 0$ (cf. Table 5-2), it is necessary to solve the secular equation

$$\begin{vmatrix} H_{00} - E & H_{01} \\ H_{01}^* & H_{11} - E \end{vmatrix} = 0, \qquad (5\text{-}49)$$

where an analogous notation is used for the matrix elements as that employed in Eq. (5-47) and the hermiticity of operator $\mathcal{H}$ is employed. The smaller of the two roots,

$$E_\pm = \frac{H_{00} + H_{11}}{2} \pm \sqrt{\left[\frac{(H_{00} - H_{11})^2}{4} + |H_{01}|^2\right]}, \qquad (5\text{-}50)$$

has a lower value than that corresponding to diagonal element H_{00}; rearranging the expression for E_-, the relationship

$$E_- = H_{00} + \frac{H_{11} - H_{00}}{2} - \frac{|H_{00} - H_{11}|}{2} \sqrt{\left[1 + \frac{4|H_{01}|^2}{(H_{00} - H_{11})^2}\right]}$$

$$= H_{00} - \frac{(H_{11} - H_{00})}{2}\left\{\sqrt{\left[1 + \frac{4|H_{01}|^2}{(H_{00} - H_{11})^2}\right]} - 1\right\} \qquad (5\text{-}51)$$

is obtained, where the facts that the value of H_{00} is negative and that the inequality $H_{00} < H_{11}$ holds, were utilized. Since wave function Ψ given by Eq. (5-48) corresponds to the sum of two determinants differing in only one row, function Ψ can, because of the basic properties of determinants, be expressed as a single determinant,

$$\Psi = |\lambda_1, \lambda_2, ..., C_0\lambda_i + C_1\lambda_q, ..., \lambda_n| \qquad (5\text{-}52)$$

This corresponds to $E_- < \int \Delta_0^* \mathcal{H} \Delta_0 \, d\tau$, which is, however, contradictory to the initial assumption that Δ_0 is the optimum determinant function, whereby the indirect proof is complete.

Brillouin's theorem can now be used to formulate *the Hartree-Fock equations*. Solving them leads to spin orbitals λ_k, $k = 1, 2, ..., i, ..., n$, which represent elements of the optimum Slater determinant (in the sense of the variation principle[6,7]).

Using the second column of Table 5-2, integral (5-46) can be expressed as follows:

$$\int \lambda_i^*(x_1) \, h(1) \, \lambda_q(x_1) \, dx_1 + \sum_{\substack{j=1 \\ (\neq i)}}^{n} \left[\int \lambda_j^*(x_1) \, \lambda_i^*(x_2) \, g(1, 2) \, \lambda_j(x_1) \, \lambda_q(x_2) \, dx_1 \, dx_2 \right.$$

$$\left. - \int \lambda_j^*(x_1) \, \lambda_i^*(x_2) \, g(1, 2) \, \lambda_q(x_1) \, \lambda_j(x_2) \, dx_1 \, dx_2\right] = 0 \qquad (5\text{-}53)$$

In contrast to Table 5-2, the Dirac notation is not used in Eq. (5-33) [cf. Eq. (5-31)]. If Eq. (5-53) is converted into its complex conjugate, if the hermiticity of operator $h(1)$ is employed and if the notation of variables x_1 and x_2 is interchanged in the first integral within the square brackets to facilitate further modifications (which is permissible), then the relationship

$$\int \lambda_q^*(x_1) \, \textit{h}(1) \, \lambda_i(x_1) \, dx_1 + \sum_{\substack{j=1 \\ (\neq i)}}^{n} [\int \lambda_q^*(x_1) \, \lambda_j^*(x_2) \, \textit{g}(1,2) \, \lambda_i(x_1) \, \lambda_j(x_2) \, dx_1 \, dx_2 -$$

$$- \int \lambda_q^*(x_1) \, \lambda_j^*(x_2) \, \textit{g}(1,2) \, \lambda_j(x_1) \, \lambda_i(x_2) \, dx_1 \, dx_2] = 0 \qquad (5\text{-}54)$$

is obtained.

The integration in Eq. (5-54) can be divided into two stages: integration over variables denoted by x_2 and then integration over the x_1 coordinates. The relationship is then converted to the form

$$\int \lambda_q^*(x_1) \{ \textit{h}(1) \, \lambda_i(x_1) + \sum_{\substack{j=1 \\ (\neq i)}}^{n} [\int | \lambda_j(x_2) |^2 \, \textit{g}(1,2) \, dx_2 \, \lambda_i(x_1) -$$

$$- \int \lambda_j^*(x_2) \, \lambda_i(x_2) \, \textit{g}(1,2) \, dx_2 \, \lambda_j(x_1)] \} \, dx_1 = 0 \qquad (5\text{-}55)$$

Equation (5-55) can be conceived as a condition to be fulfilled by the overlap integral

$$\int \lambda_q^*(x_1) f(x_1) \, dx_1,$$

where $f(x_1)$ represents the expression in the braces, which is no longer dependent on x_2 after integration over variables x_2. Let us now examine the general form of function $f(x_1)$.

The condition is satisfied trivially if λ_q and f (or λ_q and λ_i) have different spin functions. Generally, it is possible, however, to express function f in a form that satisfies Eq. (5-55) as a linear combination of spin orbitals occupied in the ground state configuration

$$f = \sum_{l=1}^{n} \varepsilon_{li} \lambda_l, \qquad (5\text{-}56)$$

as spin orbital λ_q lies outside this configuration, so that it is assumed to be orthogonal to all the λ_l's, $l = 1, 2, ..., n$. Substitution for f in Eq. (5-56) from Eq. (5-55) leads to the general Hartree-Fock equations. Condition (5-55) is also fulfilled for a particular case when

$$f(x_1) = \varepsilon_i \lambda_i(x_1) \qquad (5\text{-}57)$$

In order to avoid the above trivial case, the same spin functions must correspond to spin orbitals λ_i and λ_q. Thus the so-called canonical form of the Hartree-Fock equations is obtained:

$$\textit{h}(1) \, \lambda_i(x_1) + \sum_{\substack{j=1 \\ (\neq i)}}^{n} [\int | \lambda_j(x_2) |^2 \, \textit{g}(1,2) \, dx_2 \, \lambda_i(x_1) -$$

$$- \int \lambda_j^*(x_2) \, \lambda_i(x_2) \, \textit{g}(1,2) \, dx_2 \, \lambda_j(x_1)] = \varepsilon_i \lambda_i(x_1) \qquad (5\text{-}58)$$

$$\text{for} \qquad i = 1, 2, ..., n,$$

which must be satisfied by the spin orbitals if the one-determinant approximation is to be optimum for calculation of the energy of the system.

Equation (5-58) can be very simply expressed using the spatial orbitals φ_i, provided that the ground state configuration has the character of closed shells, so that Δ_0 corresponds to (5-43). Using Eq. (5-33) and (5-34) and the properties of the spin functions, the relationship

$$\mathscr{F}(1)\,\varphi_i(\mathbf{r}_1) = \varepsilon_i\varphi_i(\mathbf{r}_1), \qquad i = 1, 2, ..., n/2, \tag{5-59a}$$

is obtained, where

$$\mathscr{F}(1) = \hbar(1) + \sum_{j=1}^{n/2} [2\mathscr{J}_j(1) - \mathscr{K}_j(1)] \tag{5-59b}$$

and $\mathscr{J}_j$ and $\mathscr{K}_j$ denote *the Coulomb* and *exchange operators*:

$$\mathscr{J}_j(1)\,\varphi_i(\mathbf{r}_1) = \int |\varphi_j(\mathbf{r}_2)|^2\,g(1, 2)\,d\mathbf{r}_2\,\varphi_i(\mathbf{r}_1) \tag{5-59c}$$

$$\mathscr{K}_j(1)\,\varphi_i(\mathbf{r}_1) = \int \varphi_j^*(\mathbf{r}_2)\,\varphi_i(\mathbf{r}_2)\,g(1, 2)\,d\mathbf{r}_2\,\varphi_j(\mathbf{r}_1) \tag{5-59d}$$

In expression (5-59b), account is taken of the fact that, for $j = i$, the Coulomb and exchange operators are identical — thus the limitation $j \neq i$ used in Eq. (5-58) can be omitted in the summation index in Eq. (5-59b).

Operator $\mathscr{F}$ is referred to as *the Hartree-Fock operator* and from expressions (5-59) it is clear that it is a one-electron operator. It can be easily verified that all the operators $\mathscr{K}_j$, $\mathscr{J}_j$, and $\mathscr{F}$ are linear and Hermitian. Operator $\mathscr{J}_j$ can be interpreted physically: the matrix element $\langle \varphi_i | \mathscr{J}_j | \varphi_i \rangle$ [cf. notation in Eq. (5-31)] expresses the electrostatic interaction between two charge clouds whose spatial charge density is given by the expressions $|\varphi_j(\mathbf{r}_2)|^2$ and $|\varphi_i(\mathbf{r}_1)|^2$. Operator $\mathscr{K}_j$ cannot be interpreted according to classical conceptions; it represents the exchange interaction between two electrons which is a consequence of the Pauli exclusion principle.

The solution of the Hartree-Fock equations, (5-58) or (5-59), represents a non-linear problem for the required one-particle functions, as these functions which act as eigenfunctions are, moreover, included in the Coulomb and exchange operators. Because of this kind of non-linearity, the Hartree-Fock equations are, as a rule, solved in an iterative manner. In the first calculation stage an estimate of the form of the one-electron functions is made and then these approximate functions $\varphi_i^{(0)}$ ($i = 1, 2, ..., n/2$) are substituted into the expressions for the Coulomb and exchange integrals, which for a closed-shell consist of the terms in the summation in Eq. (5-59b). This step enables construction of the operator $\mathscr{F}(1)$ in the zeroth approximation and, after solving system of equations (5-59a), calculation of the partially corrected one-electron functions, $\varphi_i^{(1)}$. If $n/2$ functions corresponding to the $n/2$ lowest eigenvalues are selected, then the calculation can be repeated until the $\varphi_i^{(k)}$ functions calculated in the k-th step of the iterative process are sufficiently similar to functions $\varphi_i^{(k-1)}$, the criterion of convergence being chosen according to the requirements on the accuracy of the calculation. The $\varphi_i^{(k)}$ functions fulfilling this criterion are then considered to be the solution to the problem.

The method of solution of the Hartree-Fock equations leads to the term "*self-consistent field method*" (*SCF method*); however, the term Hartree-Fock method (HF method) is also employed.

According to Table 5-2 and Eqs. (5-33) and (5-34), the total energy of a system can be calculated within the framework of the SCF approximation (for a system with closed shells):

$$E = \int \varDelta_0 \mathscr{H} \varDelta_0 \, d\tau =$$

$$= \sum_{i=1}^{n} \langle \lambda_i | \hbar | \lambda_i \rangle + \sum_{i<j}^{n} \sum^{n} [\langle \lambda_i \lambda_j | g | \lambda_i \lambda_j \rangle - \langle \lambda_i \lambda_j | g | \lambda_j \lambda_i \rangle] =$$

$$= 2 \sum_{i=1}^{n/2} \langle \varphi_i | \hbar | \varphi_i \rangle + \sum_{i<j}^{n/2} \sum^{n/2} [4\langle \varphi_i \varphi_j | g | \varphi_i \varphi_j \rangle - 2\langle \varphi_i \varphi_j | g | \varphi_j \varphi_i \rangle] +$$

$$+ \sum_{i=1}^{n/2} \langle \varphi_i \varphi_i | g | \varphi_i \varphi_i \rangle =$$

$$= 2 \sum_{i=1}^{n/2} \langle \varphi_i | \hbar | \varphi_i \rangle + \sum_{i=1}^{n/2} \sum_{j=1}^{n/2} [2\langle \varphi_i \varphi_j | g | \varphi_i \varphi_j \rangle - \langle \varphi_i \varphi_j | g | \varphi_j \varphi_i \rangle]$$

$$(5-60)$$

Since according to Eq. (5-59) the relation

$$\varepsilon_i = \langle \varphi_i | \hbar | \varphi_i \rangle + \sum_{j=1}^{n/2} [2\langle \varphi_i \varphi_j | g | \varphi_i \varphi_j \rangle - \langle \varphi_i \varphi_j | g | \varphi_j \varphi_i \rangle] \quad (5-61)$$

is valid for ε_i, Eq. (5-60) can be modified to give

$$E = 2 \sum_{i=1}^{n/2} \varepsilon_i - \sum_{i=1}^{n/2} \sum_{j=1}^{n/2} [2\langle \varphi_i \varphi_j | g | \varphi_i \varphi_j \rangle - \langle \varphi_i \varphi_j | g | \varphi_j \varphi_i \rangle], \quad (5-62)$$

from which it follows that, in the HF method, relation (5-40b) derived for a model of independent particles based on assumption (5-35) does not apply for ε_i, since the physical basis for the HF model expressed by Eq. (5-59) [or (5-58)] takes into account the motion of each electron in the field of all the remaining electrons, and ε_i, consequently, denotes the energy of the i-th electron in the field of all the rest. The sum of the ε_i

values over all the electrons $2\sum_{i=1}^{n/2}\varepsilon_i$ (consider that each φ_i orbital is occupied by two electrons — by one in the α state and by another in the β state) necessarily includes an interelectron interaction twice, which must be taken into account when expressing the total energy of system (5-62).

5.6 The method of molecular orbitals as linear combinations of atomic orbitals

In connection with the determination of optimum one-electron functions it becomes necessary to choose the analytical form of these functions so that the variation procedure satisfies requirements imposed on the accuracy of the calculation and, at the same time, is mathematically manageable with relative ease.

Due to the spherical symmetry of one-electron potentials, it is reasonable when considering atoms to utilize various modifications of functions that are solutions of the Schrödinger equation for the hydrogen atom and are furnished with suitable variation parameters as variation functions.

One-particle potentials in molecules are not characterized by spherical symmetry and thus the form of the one-electron functions is not immediately apparent. It is then possible to tentatively write functions φ_i as an expansion in a series of a complete set of functions, for instance, using all the atomic eigenfunctions corresponding to one atom in the molecule. Such a procedure would, of course, be mathematically unmanageable.

Generally, however, it can be assumed that the molecular orbital φ will follow the shape of the molecule and that the electron close to the atomic nucleus will primarily "feel" the influence of the potential at this nucleus.

These conditions will be satisfied by one-electron functions, molecular orbitals, of the type

$$\varphi_i = \sum_\mu c_{\mu i}\chi_\mu \qquad (i = 1, 2, ...), \tag{5-63}$$

where functions χ_μ are atomic orbitals located on the atoms of a given molecule. Theoretical justification of this assumption encounters some difficulties and, moreover, the type and number of functions can hardly be anticipated. However, the quality of the atoms forming a molecule is taken into account in this manner. From experience it follows that the atomic orbitals which describe the properties of the most loosely bonded electrons, i.e. the valence orbitals, will contribute most substantially to the description of the bonding in a molecule.

Optimization of the expansion coefficients in Eq. (5-63) leads to the linear variation problem whose solution is expressed by Eqs. (4-141) and (4-142). It is necessary to be aware that the optimization of one-electron energies, $E_i = \langle \varphi_i | \mathscr{H}' | \varphi_i \rangle$, simultaneously leads to minimization of the total energy of the system, E, due to the validity of relationship (5-40b). If the one-electron Hamiltonian is given by Eqs. (5-35), (5-36) and (5-41), then the coefficients of Eq. (5-63) must be found as a solution of a system of homogeneous linear algebraic equations in unknown c_{vi}'s:

$$\sum_v c_{vi}[\langle \chi_\mu | \mathscr{H}' | \chi_v \rangle - E_i \langle \chi_\mu | \chi_v \rangle] = 0, \qquad \mu = 1, 2, ..., \qquad (5\text{-}64)$$

where E_i is the one of the roots of the secular determinant [cf. Eq. (4-142)],

$$|\langle \chi_\mu | \mathscr{H}' | \chi_v \rangle - E \langle \chi_\mu | \chi_v \rangle| = 0, \qquad (5\text{-}65)$$

and where again the notation

$$\int \chi_\mu^*(1) \, \mathscr{H}'(1) \, \chi_v(1) \, d\mathbf{r}_1 = \langle \chi_\mu | \mathscr{H}' | \chi_v \rangle \qquad (5\text{-}66a)$$

$$\int \chi_\mu^*(1) \, \chi_v(1) \, d\mathbf{r}_1 = \langle \chi_\mu | \chi_v \rangle \qquad (5\text{-}66b)$$

is employed.

Analogously, within the framework of the Hartree-Fock scheme, determination of the optimum linear combination of atomic orbitals of the (5-63) type leads to equations formally similar to Eqs. (5-64) and (5-65), except that, instead of $\mathscr{H}'$ and E_i, the Hartree-Fock operator $\mathscr{F}$ and the eigenvalues ε_i appear. The Hartree-Fock equations in the LCAO approximation are sometimes referred to as Roothaan equations[8]. The principal difference between them and equations of the (5-64) type is related to the well-known fact that the set of equations

$$\sum_v c_{vi}[\langle \chi_\mu | \mathscr{F} | \chi_v \rangle - \varepsilon_i \langle \chi_\mu | \chi_v \rangle] = 0, \qquad \mu = 1, 2, ..., \qquad (5\text{-}67)$$

is not linear in the coefficients $c_{\mu i}$, these coefficients also appearing in operator $\mathscr{F}$ [cf. Eq. (5-59)].

The comments on the Hartree-Fock equations in the conclusion to the previous section also hold for the numerical solution to Eqs. (5-67). It should also be noted that the description of open shell systems can be analogous, though more complex.

REFERENCES

1. Davydov A. S.: *Kvantovaya mekhanika*. Gos. izd. fiz.-mat. lit., Moscow 1963, § 116.
2. Kołos W.: *Advan. Quantum Chem.* **5**, 99 (1970).
3. Wolniewicz L.: *J. Chem. Phys.* **45**, 515 (1966).
4. Löwdin P. O.: *Phys. Rev.* **97**, 1474 (1955).
5. Slater J. C.: *Phys. Rev.* **38**, 1109 (1931).
6. Lefebvre R.: *Cahiers Phys.* (Paris) **13**, 369 (1959).
7. Mayer I.: *Acta Phys. Hung.* **30**, 373 (1971).
8. Roothaan C. C. J.: *Rev. Mod. Phys.* **23**, 69 (1951).

6. SYMMETRY IN QUANTUM CHEMISTRY

6.1 Introduction

Planary and spatial configurations are sometimes characterized by a property which is usually referred to as the geometrical regularity or symmetry. For description of this symmetry, *symmetry operations* are introduced that transform the original configuration into a physically identical configuration, while the individual points need not return to their original positions. Typical symmetry operations are:

    a) *rotation* by an angle φ, about axis ξ, designated by $\mathscr{R}(\xi, \varphi)$,

    b) *reflection* in the plane $\xi\eta$, designated by $\sigma_{\xi\eta}$,

    c) *inversion* through a point (a centre of symmetry), designated by i,

or various combinations of these operations. However, with symmetry operations a "leaving the configuration at rest" operation must also be considered, *the identity operation*, denoted $\mathscr{E}$.

In Table 6-1 all the symmetry operations of a rectangle, whose apices are designated by numbers, are given. Symmetry operations are denoted by the previously introduced symbols, where, for example, $\mathscr{R}(z, \pi)$ is a rotation from the original position by $180°$ about the z-axis and σ_{yz} is a reflection in the yz-plane. It is evident that two symmetry operations performed successively give another symmetry operation of the particular configuration — for such cases we use a multiplicative notation. For instance, if operations σ_{xz} ($\equiv A$) and σ_{yz} ($\equiv B$) are carried out successively, then

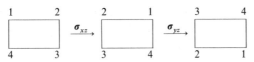

It can readily be seen that the final configuration corresponds to the last figure in Table 6-1 and that it can be obtained from the original configuration by the single operation $\mathscr{R}(z, \pi)$, ($\equiv C$). This relation can

Table 6-1

Symmetry Operations of a Rectangle

Original configuration	Symmetry operation	Symbol	Final configuration	Further operations of the D_{2h} group leading to the same final configuration
	$\mathscr{E}$	E	(1 2 / 4 3)	$\sigma_{xy}; \sigma_h$
	σ_{xz}	A	(2 1 / 3 4)	$\mathscr{R}(x, \pi); \mathscr{C}'_2$
	σ_{yz}	B	(4 3 / 1 2)	$\mathscr{R}(y, \pi); \mathscr{C}''_2$
	$\mathscr{R}(z, \pi)$	C	(3 4 / 2 1)	$i; \mathscr{C}_2$

be expressed symbolically by the equation

$$C = BA, \tag{6-1}$$

where the first operation performed is always written on the right in the product. This convention is important because the sequence of partial operations can, in general, affect the result of the overall operation. Further, for each symmetry operation an inverse operation exists which returns the configuration to its original position, so that the product of these operations is equal to the identity operation. It can happen, of course, that a particular operation is, in itself, an inverse operation. The result of multiplication of the four operators in Table 6-1 can be expressed in the form of a table (Table 6-2). Using this scheme it can easily be verified that symmetry operations of a rectangle, forming a set of elements E, A, B, C, satisfy the following conditions:

a) A "multiplication" operation is defined so that element Z is assigned to any ordered pair of elements X, Y of the same set,

$$XY = Z \tag{6-2}$$

b) The associative law holds for the multiplication of three elements, written as

$$(XY)Z = X(YZ) \tag{6-3}$$

c) An identity element E exists where, for any element X of the set, it holds that

$$EX = XE = X \tag{6-4}$$

d) An element Y is assigned to every element X so that

$$XY = YX = E \tag{6-5}$$

Symbol Y denotes the inverse of element X so that

$$Y = X^{-1} \tag{6-6}$$

Table 6-2

Multiplication Table for Symmetry Operations of a Rectangle (operations are defined in Table 6-1)

		second factor		
	E	*A*	*B*	*C*
E	*E*	*A*	*B*	*C*
A	*A*	*E*	*C*	*B*
B	*B*	*C*	*E*	*A*
C	*C*	*B*	*A*	*E*

first factor

A set of elements (of any kind) satisfying the four above conditions is called *a group*. The number of elements is *the order of the group*. Showing that symmetry operations of a certain configuration satisfy the four given axioms, i.e. that they constitute a group, bring symmetry considerations into a well-studied field of mathematics — the theory of abstract groups — as will be utilized below. It is also necessary to mention that if the symmetry operations are such that one point of the configuration (e.g., the centre of a rectangle) stays fixed in space, then we speak of *point groups*.

6.2 Symmetry transformations
of the Hamiltonian

It is evident that the operations presented under points a) to c) in Section 6.1 leading to positional changes in configurations can be realized in two physically equivalent ways: either by movement of the configuration in a fixed coordinate system, or by a change in the coordinate system with unchanged position of the configuration. As functions and operators depending on space coordinates will also be discussed below, an unambiguous definition of the transformation of Cartesian coordinates will be introduced.

First, let us imagine that a rectangular Cartesian coordinate system is rigidly connected with the configuration under study in such a way that all the rotation axes pass through the origin of the coordinate system. If the configuration is rotated, it is possible to consider the initial and final positions of the coordinate system as defining two coordinate systems. Therefore, to the same point in space can be assigned either rectangular coordinates x, y, z with respect to the original system or coordinates x', y', z' with respect to the new coordinate system and their relationship is expressed by an orthogonal matrix [see Eqs. (4-111) and (4-124)]:

$$\left\| \begin{array}{c} x \\ y \\ z \end{array} \right\| = \left\| \begin{array}{ccc} a_{11} & a_{12} & a_{13} \\ a_{21} & a_{22} & a_{23} \\ a_{31} & a_{32} & a_{33} \end{array} \right\| \left\| \begin{array}{c} x' \\ y' \\ z' \end{array} \right\| \tag{6-7}$$

The matrix elements a_{ij} belong to the transformation matrix a. Equation (6-7) represents three equations, the first of which is

$$x = a_{11}x' + a_{12}y' + a_{13}z'$$

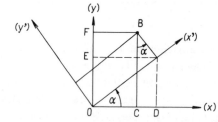

Fig. 6-1. Transformation of the coordinates on rotation of the Cartesian coordinate system about the z-axis through angle α.

From Fig. 6-1 the form of the transformation matrix for the rotation of the configuration by an angle α about the z-axis is apparent. For the calculation of coordinates of point B in the primed and unprimed systems, it is necessary to bear in mind that

$$x = \overline{OD} - \overline{CD} = x' \cos \alpha - y' \sin \alpha \qquad \text{(6-8a)}$$
$$y = \overline{OE} + \overline{EF} = x' \sin \alpha + y' \cos \alpha \qquad \text{(6-8b)}$$
$$z = z' \qquad \text{(6-8c)}$$

and thus the corresponding transformation matrix takes the form

$$\mathbf{a}_1 = \begin{Vmatrix} \cos \alpha & -\sin \alpha & 0 \\ \sin \alpha & \cos \alpha & 0 \\ 0 & 0 & 1 \end{Vmatrix} \qquad \text{(6-9)}$$

Transformation matrices corresponding to reflections in the planes determined by the coordinate axes or to reflections through the origin have a particularly simple form. For instance, reflection in the zy-plane leads to matrix $\mathbf{a}_2$:

$$\mathbf{a}_2 = \begin{Vmatrix} -1 & 0 & 0 \\ 0 & 1 & 0 \\ 0 & 0 & 1 \end{Vmatrix} \qquad \text{(6-10)}$$

The following transformation matrix corresponds to the reflection operation through the origin (inversion):

$$\mathbf{a}_3 = \begin{Vmatrix} -1 & 0 & 0 \\ 0 & -1 & 0 \\ 0 & 0 & -1 \end{Vmatrix} \qquad \text{(6-11)}$$

An inverse matrix can easily be constructed [see Eq. (4-111)] for an orthogonal matrix so that primed coordinates can also easily be expressed as a linear combination of unprimed coordinates:

$$\begin{Vmatrix} x' \\ y' \\ z' \end{Vmatrix} = \mathbf{a}^{\mathrm{T}} \begin{Vmatrix} x \\ y \\ z \end{Vmatrix} \qquad \text{(6-12)}$$

Thus, for example, for coordinate x',

$$x' = a_{11}x + a_{21}y + a_{31}z$$

Operation of a certain transformation on a general function f or on a general operator $\mathcal{O}$ can best be expressed in operator form: symbol $\mathcal{T}$ will denote the operator that represents the rotation, reflection, or inversion. The operations $\mathcal{T}f(x, y, z)$ or $\mathcal{T}\mathcal{O}(x, y, z)$ then simply mean that substitution in the corresponding expressions must be performed according to (6-7), where matrix $\mathbf{a}$ expresses the transformation of coordinates which occurred as a result of the corresponding operation. Therefore,

$$\mathscr{T} f(x, y, z) = f(x', y', z') \tag{6-13a}$$

and

$$\mathscr{T} \mathcal{O}(x, y, z)\, \varphi = \mathcal{O}(x', y', z')\, \mathscr{T}\varphi, \tag{6-13b}$$

where φ is an arbitrary function (cf. Section 4.1). The function $f(x', y', z')$ generally has a different analytical form than the original function $f(x, y, z)$. If, however, the analytical form is preserved after the transformation, a symmetry transformation of the function $f(x, y, z)$ has occurred. The same considerations, of course, also hold for Eq. (6-13b); if operator $\mathcal{O}$ remains invariant under the performed transformation, or, in other words, if it holds that $\mathcal{O}(x, y, z) \equiv \mathcal{O}(x', y', z')$, then it follows from Eq. (6-13b) that

$$\mathscr{T} \mathcal{O} = \mathcal{O} \mathscr{T} \tag{6-14}$$

Thus, if $\mathscr{T}$ is an operator corresponding to the symmetry transformation of operator $\mathcal{O}$, this property will be manifested by commutation of the two operators.

As an example of the determination of symmetry operations of an operator, the properties of a Hamiltonian corresponding to an electron, moving in the electrostatic field of four protons, located at the corners of a rectangle whose orientation with respect to the coordinate system is defined by the figure in the first column of Table 6-1, will be considered. The Hamiltonian for this system can be expressed as

$$\mathscr{H} = -\frac{\hbar^2}{2m} \Delta - \sum_{I=1}^{4} \frac{e^2}{4\pi\varepsilon_0 \left| \boldsymbol{r} - \boldsymbol{R}_I \right|}, \tag{6-15}$$

where $\boldsymbol{r} \equiv (x, y, z)$ gives the position of the electron under study and the summation expresses its electrostatic interaction with the four protons, whose positions are determined by vectors $\boldsymbol{R}_I$. First, the behaviour of the Laplace operator under the rotation operation, for example, under rotation about the z-axis, can be established. According to the elementary rules of differentiation,

$$\mathscr{R}(z, \alpha) \left\{ \frac{\partial^2}{\partial x^2} + \frac{\partial^2}{\partial y^2} + \frac{\partial^2}{\partial z^2} \right\} =$$

$$= \left\{ \left[\left(\frac{\partial x'}{\partial x} \right)^2 + \left(\frac{\partial x'}{\partial y} \right)^2 \right] \frac{\partial^2}{\partial x'^2} + \left[\left(\frac{\partial y'}{\partial x} \right)^2 + \left(\frac{\partial y'}{\partial y} \right)^2 \right] \frac{\partial^2}{\partial y'^2} + \right.$$

$$\left. + 2 \left[\frac{\partial x'}{\partial x} \frac{\partial y'}{\partial x} + \frac{\partial x'}{\partial y} \frac{\partial y'}{\partial y} \right] \frac{\partial^2}{\partial x' \partial y'} + \frac{\partial^2}{\partial z'^2} \right\} \mathscr{R}(z, \alpha), \tag{6-16}$$

where it is assumed that the dependence of the primed coordinates on the unprimed ones is known and has the form of Eq. (6-12); for rotation by angle α about the z-axis, the relationships

$$x' = \cos \alpha \cdot x + \sin \alpha \cdot y \qquad (6\text{-}17a)$$

$$y' = -\sin \alpha \cdot x + \cos \alpha \cdot y \qquad (6\text{-}17b)$$

are obtained and, from these equations, the expressions necessary for derivation can be obtained. It is seen that

$$\Delta(x, y, z) = \Delta(x', y', z') \qquad (6\text{-}18)$$

or, in other words, that the Laplace operator is invariant to rotation about the z-axis; in addition, it can be shown that this operator is invariant to rotation about an arbitrary axis passing through the origin.

If the transformation relations for other symmetry operations presented in Table 6-1 [see Eq. (6-10)] are also taken into consideration, it can be concluded that operator Δ is invariant to all the operations of a rectangle symmetry, and the second term of the Hamiltonian (6-15) remains to be investigated. First, the potential originating from four protons at an arbitrary point in the $[x, y, z]$ space must be expressed. It is obvious that any change in the position of these protons, expressed by the transformations shown in Table 6-1, does not change this potential, and therefore the potential energy for the mutual interaction between the electron and protons can be described in the form

$$\sum_{I=1}^{4} \frac{e^2}{4\pi\varepsilon_0 \, |\mathbf{r} - \mathbf{R}_I|} = \sum_{I=1}^{4} \frac{e^2}{4\pi\varepsilon_0 \, |\mathbf{r}' - \mathbf{R}_I'|}, \qquad (6\text{-}19)$$

where $\mathbf{r}' \equiv (x', y', z')$ is the position vector of the electron and $\mathbf{R}_I'$ the position vector of the nucleus I after the symmetry transformation.

In summary it can be said that the Hamiltonian (6-15) is invariant under any symmetry operation of a rectangle (see Table 6-1), or that

$$\mathscr{H}(x, y, z) = \mathscr{H}(x', y', z') \qquad (6\text{-}20)$$

In accordance with this finding, *a symmetry transformation of the Hamiltonian* will be generally defined as a linear transformation of the coordinates that leaves the Hamiltonian unchanged in the sense of Eq. (6-20).

The above consideration concerned a system of a single electron which moves in an electrostatic field of symmetrically arranged nuclei. It is obvious that a similar approach can be employed in studying the symmetry properties of the Hamiltonian corresponding to the independent electron model [see Eq. (5-37)], since the symmetry of the effective potential $\mathscr{V}$ is consistent with the configuration of the nuclei of atoms which form a molecule. Apart from one-electron contributions, a complete quantum-chemical Hamiltonian also contains the operators of electrostatic interaction between electrons, i.e. the expression

$$\sum_{i<j} \mathscr{g}(i,j) = \sum_{i<j} \frac{e^2}{4\pi\varepsilon_0 |r_i - r_j|} \tag{6-21}$$

It is evident that the distance between two points in space is not changed by the symmetry transformations and consequently that

$$|r_i - r_j| = |r_i' - r_j'| \tag{6-22}$$

Therefore, even sum (6-21) is invariant to the simultaneous orthogonal transformation of the coordinates of all the electrons. Thus even for a precise Hamiltonian, the spatial symmetry is determined by the configuration of the atomic nuclei forming the molecule. In this connection it is necessary to realize that, in all these considerations on the symmetry properties of the Hamiltonian, the validity of the Born-Oppenheimer approximation, according to which atomic nuclei, responsible for the molecular geometry, are considered to be a rigid configuration, is tacitly assumed.

The electron system of an atom that is exposed to the spherically symmetrical potential of an atomic nucleus must be considered as a special case. This, in its own way, the highest kind of spatial symmetry, manifested in invariance of the Hamiltonian to rotation around any axis passing through the atomic nucleus, has already been taken into account by classifying states in atomic systems using eigenvalues of the angular momentum operators. It can be demonstrated that there is a close relationship between angular momentum operators and operators of infinitesimal rotations[1].

It will be recalled that the angular momentum was established as one of the constants of motion. On comparing Eq. (6-14), where the Hamiltonian of the system can be substituted for the general operator $\mathcal{O}$. with Eq. (4-56) it can be concluded that the operators of symmetry transformations of the Hamiltonian also act as constants of motion and can be used to classify various states. It can easily be proved that all symmetry transformations of the Hamiltonian satisfy axioms a) to d) in Section 6.1, and that they therefore form a group. In this connection we speak about the symmetry group of the Hamiltonian.

6.3 The principal symmetry groups and their notation

In addition to groups of finite order there are also groups of infinite order. Thus, for example, there is an infinite number of symmetry operations on the sphere corresponding to *the full three-dimensional rotation group*.

If some of the elements of a group in themselves satisfy the group axioms, the set of these elements is called *a subgroup* of the corresponding group. From this point of view, all symmetry groups related to molecular symmetry are subgroups of the full rotational group, since the symmetry operations of every molecule are also included in the symmetry operations of a sphere.

The symmetry properties of a molecule can be described if all the possible symmetry operations under which the molecule is not physically changed are given. Thus simple symbols (*the Schönflies notation*) were introduced to denote the most important symmetry groups.

Among the simplest point groups are those corresponding to operations of rotation about a single axis. If the molecule is invariant under rotation through $2\pi/n$, about a particular axis ξ, then this is termed an n-fold axis, and the corresponding symmetry group is denoted C_n. If the symbols given in the introduction are used, the rotation operator can be defined as

$$\mathscr{C}_n \equiv \mathscr{R}\left(\xi, \frac{2\pi}{n}\right), \tag{6-23}$$

for which n-fold repetition leads to the identity operation:

$$(\mathscr{C}_n)^n = \mathscr{E} \tag{6-24}$$

Each of the integral multiples k of the elementary rotation $\mathscr{C}_n$ ($k = 1, 2, \ldots, n$) represents an element of the group. Groups whose elements satisfy property (6-24) are termed *cyclic*.

Another cyclic group of the n-th order is the group S_n, which is composed of the multiples of element $\mathscr{S}_n \equiv \sigma_h \mathscr{C}_n$, which refers to rotation through $2\pi/n$ ($\mathscr{C}_n$) with reflection (σ_h) in the plane perpendicular to the symmetry axis; this kind of rotation is called *improper rotation*. Groups of this type are defined for even n only.

Other groups are derived from the above-described cyclic groups by supplying them with further symmetry elements. *Symmetry elements*, such as, for example, different types of symmetry axes, can be distinguished from symmetry operations, e.g., the operations of rotation corresponding to an active rotation through a particular angle about the axis. Dihedral groups also have, in addition to the principal rotational axis (which is the axis of highest order compared to the other symmetry axes), twofold axes, which are perpendicular to the principal axis. The operations of rotation about these axes will be denoted by primes, e.g., $\mathscr{C}'_2$, $\mathscr{C}''_2$, and the respective symmetry elements by C'_2, C''_2. Further symmetry elements can be reflection planes σ with a different orientation with respect to the principal axis:

    a) perpendicular to it: σ_h,

    b) passing through it: σ_v,

    c) passing through it and also bisecting the angles between minor twofold axes: σ_d.

The notation of indices on the symbols for reflection σ can be derived from the following model: if the principal axis is orientated

Table 6-3

Selection of Point Groups

Group	Symmetry elements	Examples of symmetrical molecules for $n = 2$	
		Molecule	Graphical representation
C_n	n-fold axis of rotation	H_2O_2	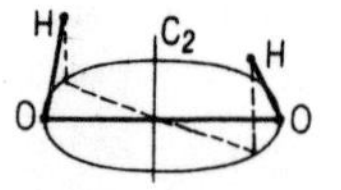
S_n	n-fold axis of improper rotations	*trans* ClBrHC$-$CHBrCl	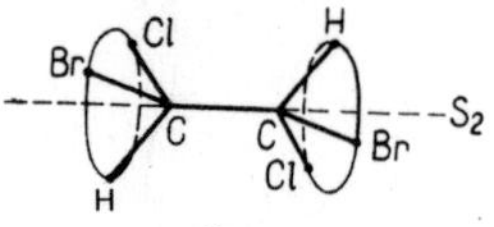
C_{nh}	n-fold axis of rotation, σ_h plane	*trans* ClHC$=$CHCl	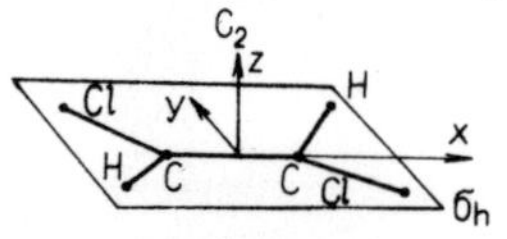
C_{nv}	n-fold axis of rotation, n σ_v planes	H_2O	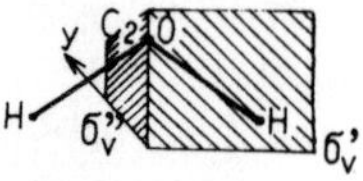
D_n	n-fold axis of rotation, n two fold axes C_2', C_2'', ...	partially deformed $H_2C\cdots CH_2$	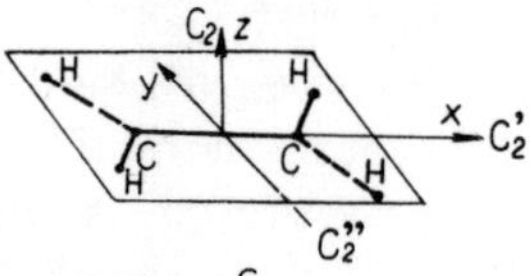
D_{nd}	as D_n, $n\sigma_d$ planes	deformed H_2C-CH_2	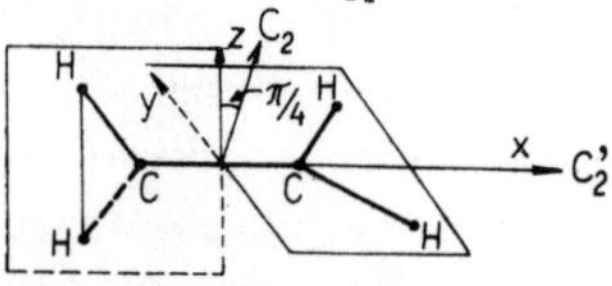
D_{nh}	as D_n, σ_h plane	$H_2C=CH_2$	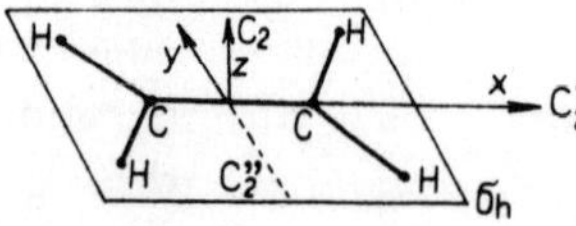

vertically, the symmetry plane σ_h is orientated horizontally, the σ_v plane vertically and the σ_d plane is, in a certain sense, diagonal.

Different types of groups, which can be formed in this way, are given in Table 6-3, together with examples of molecules whose symmetry corresponds to the $n = 2$ term of a certain type of symmetry group, where n denotes the order of the principal rotational axis. In this connection it is necessary to note that there exist symmetrical configurations corresponding only to a limited number of point groups, which can be deduced from the general properties of the space group of which the point groups are subgroups. On the other hand, it is possible to consider some continuous groups, to which, for instance, diatomic molecules correspond in their rotational symmetry, to be a limiting case for $n \to \infty$. Thus to the CO molecule corresponds the group $C_{\infty v}$, and, to the H_2 molecule, the group $D_{\infty h}$, where the symbol ∞ denotes the presence of a symmetry axis of "infinite order".

The graphical representation of molecules in Table 6-3 is supplemented by denotation of the various symmetry elements of the respective groups, which are again and more illustratively given in the diagrams in Fig. 6-2. The diagrams are termed *stereographic projections* of the

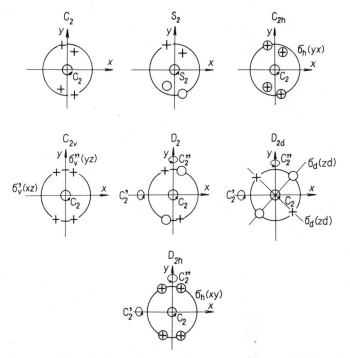

Fig. 6-2. Stereographic projections of some point groups with two-fold principal axis.

point groups and enable a quick and easy estimate of the relevant symmetry properties. In constructing these diagrams, a convention according to which a + sign denotes the points above the projection plane and ◯ denotes points below it is employed. In some instances the projection plane is identical with the σ_h plane; the principal symmetry axis then passes through the centre of a circle perpendicular to the σ_h plane. The meaning of the other symbols in the diagram is self-evident.

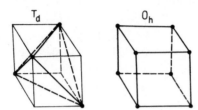

Fig. 6-3. A tetrahedron and a cube.

In conclusion, two important symmetry groups, T_d and O_h, corresponding to a tetrahedron and a cube, respectively (see Fig. 6-3), can be mentioned. The first consists of 24 symmetry elements $(E, 8C_3, 3C_2, 6S_4, 6\sigma_d)$, and methane is an example of a molecule with this symmetry. The O_h group, consisting of 48 elements $(E, 3C_2, 6C_4, 6C_2', 8C_3, i, 3iC_2, 6iC_4, 6iC_2', 8iC_3)$ assumes an important position in the theory of inorganic complex compounds.

6.4 Matrix representation of symmetry groups

The discussion begun in Section 6.2 can now be continued, starting with the general Schrödinger equation

$$\mathscr{H}(r)\,\Psi_1(r) = E\Psi_1(r), \qquad (6\text{-}25)$$

where the symbol r denotes dependence on all the space coordinates of the system. If operator $\mathscr{T}$, which belongs among the operators of symmetry transformations of the Hamiltonian, acts on both sides of Eq. (6-25), the relationship

$$\mathscr{H}(r')\,\Psi_2(r') = E\Psi_2(r') \qquad (6\text{-}26a)$$

is obtained, in which use is made of the invariance of the Hamiltonian under the symmetry operation. However, function Ψ_1 can in general change its form, denoted by a change in the subscript. The denotation of variables is, of course, arbitrary and therefore Eq. (6-26a) can be expressed using unprimed coordinates:

$$\mathcal{H}(r)\,\Psi_2(r) = E\Psi_2(r), \tag{6-26b}$$

whence it can be seen that $\Psi_2(r)$ is also an eigenfunction of the Hamiltonian and possesses the same eigenvalue as $\Psi_1(r)$. Thus the set of all symmetry transformations of the Hamiltonian can be used to determine different eigenfunctions corresponding to one energy level. This fact permits determination of the degree of degeneracy of the energy level, which can be defined as the number of linearly independent functions Ψ_i.

If level E is m-fold degenerate, then functions Ψ_i $(i = 1, 2, ..., m)$ can be considered to form a set of orthonormalized functions. The action of operator $\mathcal{T}$ on one of the functions Ψ_i must necessarily be expressible in the form of a linear combination of functions of the given set

$$\mathcal{T}\Psi_i = \sum_{j=1}^{m} A_{ji}^{(T)}\Psi_j, \tag{6-27}$$

where $\| A_{ji}^{(T)} \| = \mathbf{A}^{(T)}$ is a matrix whose elements are coefficients of linear expansion (6-27). The orthonormality of functions, written using the notation introduced in Section 5.4 as

$$\langle \Psi_i | \Psi_j \rangle = \delta_{ij}; \quad (i, j = 1, 2, ..., m),$$

results in the relationship

$$\mathcal{T}\langle \Psi_i | \Psi_j \rangle = \sum_{k=1}^{m} \sum_{l=1}^{m} (A_{ki}^{(T)})^* A_{lj}^{(T)}\delta_{kl} = \sum_{k=1}^{m} (A_{ki}^{(T)})^* A_{kj}^{(T)} = \delta_{ij}, \tag{6-28}$$

indicating that $\mathbf{A}^{(T)}$ is a unitary matrix [cf. Eq. (4-110)]. When deriving relationship (6-28) use was made of the fact that the expression $\langle \Psi_i | \Psi_j \rangle$ is a number and thus cannot be influenced by transformation $\mathcal{T}$. An operator leaving a scalar product invariant is termed a unitary operator; it is represented by a unitary matrix.

Equation (6-27) can be rewritten using the matrix formalism introduced in Section 4.5:

$$\mathcal{T}\boldsymbol{\Psi} = \boldsymbol{\Psi}\mathbf{A}^{(T)}, \tag{6-29}$$

where $\boldsymbol{\Psi}$ must be taken as the one-row matrix

$$\boldsymbol{\Psi} = \| \Psi_1, \Psi_2, ..., \Psi_m \| \tag{6-30}$$

This notation will enable easy investigation of the successive action of the two operators $\mathcal{T}$ and $\mathcal{V}$, belonging among the symmetry transformations of the Hamiltonian, on the set of functions Ψ_i, which refer to a degenerate level. From Eq. (6-29) it is evident that

$$\mathcal{W}\mathbf{\Psi} = \mathcal{V}\mathcal{T}\mathbf{\Psi} = \mathcal{V}\mathbf{\Psi}\mathbf{A}^{(T)} = \mathbf{\Psi}\mathbf{A}^{(V)}\mathbf{A}^{(T)} = \mathbf{\Psi}\mathbf{A}^{(VT)} = \mathbf{\Psi}\mathbf{A}^{(W)}, \qquad (6\text{-}31a)$$

whence it follows that

$$\mathbf{A}^{(W)} = \mathbf{A}^{(VT)} = \mathbf{A}^{(V)}\mathbf{A}^{(T)}, \qquad (6\text{-}31b)$$

where the operation

$$\mathcal{W} = \mathcal{V}\mathcal{T} \qquad (6\text{-}32)$$

has been introduced, which according to the definition of a group also belongs among the symmetry transformations of the Hamiltonian. It is worth noting that the unit matrix $\mathbf{A}^{(E)} = \mathbf{1}$ corresponds to the identity operation.

Equations (6-31) and (6-32) imply that the original group of symmetry transformations $\mathcal{T}$, $\mathcal{V}$, ..., has been replaced by a set of unitary matrices $\mathbf{A}^{(T)}$, $\mathbf{A}^{(V)}$, ..., which also form a group in accordance with the definition in Section 6.1, provided that the operation of matrix multiplication is introduced as group multiplication. Such a set of matrices satisfying Eqs. (6-31) and (6-32) is said to constitute an m-dimensional *matrix representation* of the original group of symmetry operations, and the set of functions $\mathbf{\Psi}_i$ ($i = 1, 2, ..., m$) is called the *basis of this representation*.

This important conclusion will enable us to denote the energy level and the corresponding eigenfunctions by the representation which is related to it. Specification of the representation affords information on the symmetry properties of eigenfunctions, which must be known, for example, in determining the selection rules for various types of matrix elements, as will be seen below.

Let us assume that the matrix representation $\mathbf{A}^{(T)}$, $\mathbf{A}^{(V)}$, $\mathbf{A}^{(W)}$, ... of a certain group of symmetry operations is known. It is also assumed that the set of functions ψ_i ($i = 1, 2, ..., m$) is known, constituting a basis for this representation, which guarantees that the action of a symmetry operation on any of these functions forms new functions "remaining" in the space of ψ_i functions. For some purposes it can be useful to pass from one set of functions ψ_i to another set of orthonormalized functions φ_i ($i = 1, 2, ..., m$), which can be carried out by the transformation

$$\psi = \varphi\mathbf{U}, \qquad (6\text{-}33)$$

where $\mathbf{U}$ is a unitary matrix [see Eq. (4-124)]. The new basis can be taken as physically fully equivalent to the original basis. Substituting

Eq. (6-33) into Eq. (6-29) gives

$$\mathscr{T}\psi = \mathscr{T}\varphi\mathbf{U} = \varphi\mathbf{U}\mathbf{A}^{(T)} \tag{6-34a}$$

and, after multiplying from the right by matrix $\mathbf{U}^{-1}$, the relationship

$$\mathscr{T}\varphi = \varphi\mathbf{U}\mathbf{A}^{(T)}\mathbf{U}^{-1} \tag{6-34b}$$

is obtained. Equation (6-34b) justifies introduction of the matrices

$$\mathbf{B}^{(T)} = \mathbf{U}\mathbf{A}^{(T)}\mathbf{U}^{-1}, \tag{6-35a}$$

which determine the transformation properties of the new basis φ_i ($i =$ $= 1, 2, ...$) with respect to the symmetry operations under consideration. The inverse transformation to Eq. (6-35a),

$$\mathbf{A}^{(T)} = \mathbf{U}^{-1}\mathbf{B}^{(T)}\mathbf{U}, \tag{6-35b}$$

permits substitution into Eq. (6-31b):

$$\mathbf{U}^{-1}\mathbf{B}^{(V)}\mathbf{U}\mathbf{U}^{-1}\mathbf{B}^{(T)}\mathbf{U} = \mathbf{U}^{-1}\mathbf{B}^W\mathbf{U}, \tag{6-36a}$$

whence it follows that

$$\mathbf{B}^{(V)}\mathbf{B}^{(T)} = \mathbf{B}^{(W)} \tag{6-36b}$$

Thus it has been shown that the matrices $\mathbf{B}^{(T)}$, $\mathbf{B}^{(V)}$, $\mathbf{B}^{(W)}$, ... also form a representation of the group of symmetry operations under consideration. The representations whose mutual relationship is carried out using equations of the type (6-35) (similarity transformation) are termed *equivalent representations*.

Let us introduce now a matrix of the type

$$\mathbf{B} = \left\|\begin{matrix} \mathbf{B}_1, & & & 0 \\ & \mathbf{B}_2, & & \\ & & \cdot & \\ & & & \cdot \\ 0, & & & \mathbf{B}_k \end{matrix}\right\|, \tag{6-37}$$

which will be termed *a block matrix*. Non-zero matrix elements occur here only in submatrices $\mathbf{B}_1$, $\mathbf{B}_2$, ... along the main diagonal, while zeros are everywhere else.

When passing from one representation to another using the equivalence relation (6-35a), it can happen that the new matrix representation will be such that all matrices $\mathbf{B}^{(K)}$ ($K = T, V, W, ...$) are block matrices of the same type, i.e. the homothetic submatrices have the same dimensions. It can easily be demonstrated that, for the matrix product of such block matrices, it is valid that

$$\mathbf{B}^{(V)}\mathbf{B}^{(T)} = \begin{Vmatrix} \mathbf{B}_1^{(V)}, & & 0 \\ & \mathbf{B}_2^{(V)}, & \\ & & \cdot \\ & & & \cdot \\ 0, & & & \mathbf{B}_k^{(V)}, \end{Vmatrix} \begin{Vmatrix} \mathbf{B}_1^{(T)}, & & 0 \\ & \mathbf{B}_2^{(T)}, & \\ & & \cdot \\ & & & \cdot \\ 0, & & & \mathbf{B}_k^{(T)} \end{Vmatrix} =$$

$$= \begin{Vmatrix} \mathbf{B}_1^{(W)}, & & 0 \\ & \mathbf{B}_2^{(W)}, & \\ & & \cdot \\ & & & \cdot \\ 0, & & & \mathbf{B}_k^{(W)}, \end{Vmatrix} = \mathbf{B}^{(W)}, \tag{6-38}$$

where

$$\mathbf{B}_i^{(W)} = \mathbf{B}_i^{(V)}\mathbf{B}_i^{(T)}, \quad i = 1, 2, ..., k \tag{6-39}$$

Thus submatrices $\mathbf{B}_i^{(T)}$, $\mathbf{B}_i^{(V)}$, $\mathbf{B}_i^{(W)}$, ... for a given subscript i also form a matrix representation of the corresponding group, and this representation generally has a lower dimension than the original. If the resulting situation is interpreted using the basis functions, then the new set of functions φ_i can be divided into subsets with the dimensions of the submatrices. The functions belonging to a certain subset are only mutually transformed by the action of the symmetry operations of the group under consideration. If the original m-dimensional representation is denoted by the symbol Γ, then it has been *decomposed* into representations Γ_1, Γ_2, ..., Γ_k, which is normally denoted as *a direct sum*, and is written in the form

$$\Gamma = \Gamma_1 + \Gamma_2 + ... + \Gamma_k, \tag{6-40}$$

where it cannot be excluded that each new representation will be contained more than once, for example,

$$\Gamma = k_1\Gamma_1 + k_2\Gamma_2 + ... \tag{6-41}$$

The sum of the dimensions of the components must, of course, equal the dimension of the original representation Γ.

A matrix representation for which there is a unitary transformation [i.e. a transformation of the type (6-35)] such that it transforms it into block form [see Eq. (6-38)] is denoted as *reducible*. A matrix representation that cannot be reduced is called *irreducible*. As will be seen later, the concept of reducibility of the representation is of fundamental importance for applying group theory in quantum mechanics. As an example, a statement sometimes also introduced as an axiom can be formulated:

Theorem 6-1. The eigenfunctions of the Hamiltonian belonging to the same energy level form the basis for an irreducible representation of the symmetry group under which the Hamiltonian is invariant (if accidental degeneracies are excluded). Accidental degeneracy occurs when the order of the degeneracy cannot be accounted for by symmetry considerations.

Similarly as the concept of equivalance was introduced in the form of the relationship of two representations, the equivalence relationship between elements of the same group can also be introduced. Two *elements* of a given group, X and Y, are *equivalent* (or *conjugate*) when in the group appears an element Z such that the following relation is valid:

$$X = Z^{-1}YZ \qquad (6\text{-}42)$$

All equivalent elements of a given group form *a class of equivalent elements*. As a rule, a group is composed of several classes. If the symmetry operations are group elements, it is possible, from analogy with Eqs. (6-33) and (6-35a), to conclude that, in Eq. (6-42), X denotes the operation which results from operation Y by similarity transformation through symmetry operation Z. Thus, equivalent operations X and Y can, in principle, be taken as identical, possessing, however, a different system of coordinates in which the operation is carried out. An example of the division of group elements into classes is the recording of the symmetry operations of groups T_d and O_h, given at the end of Section 6.3 (p. 114), where it can be seen that group T_d has five and group O_h ten classes of equivalent elements.

The number, N_{ir}, of non-equivalent irreducible representations of a finite group is closely connected with the number, N_c, of classes in the group. It can be proved[2] that

$$N_{ir} = N_c \qquad (6\text{-}43)$$

Further considerations will be based on the relationship between the matrix elements $\{\mathbf{B}_i^{(T)}\}_{\mu\nu}$ of irreducible representations. In expression (6-44), $\{\mathbf{B}_i^{(T)}\}_{\mu\nu}$ denotes the matrix element of the i-th irreducible representation, which lies at the point of intersection of the μ-th row and ν-th column of matrix $\mathbf{B}_i^{(T)}$, which corresponds to the symmetry operation $\mathcal{T}$ of symmetry group G. In group representation theory it has been shown[3] that

$$\sum_{T \in G} \{\mathbf{B}_i^{(T)}\}_{\mu\nu}^* \{\mathbf{B}_j^{(T)}\}_{\varkappa\lambda} = \frac{g}{m_i} \delta_{ij}\delta_{\mu\varkappa}\delta_{\nu\lambda}, \qquad (6\text{-}44)$$

where the symbol $T \in G$ under the summation sign indicates that the summation proceeds over all group elements (symmetry operations); g denotes the order of the group and m_i *the dimension of the i-th matrix representation.* Equation (6-44) expresses *the orthogonality relation between the matrix elements of representations* and indicates that only the sum of the squares of moduli of homothetic matrix elements of a given irreducible representation equals g/m_i and that all other types of products are equal to zero.

It has so far been shown that all necessary data on the symmetry properties of a certain symmetry group are stored in the sets of matrices forming the irreducible representations of that group. It appears, however, that this information can be stored in a still more concise form. The *character* $\chi^{(T)}$ *of element* $\mathcal{T}$ of the group under study, corresponding to matrix representation $\mathbf{A}^{(T)}$, is defined as the trace of this matrix [see Eq. (4-127)]; thus

$$\chi^{(T)} = \operatorname{Tr} \mathbf{A}^{(T)} = \sum_{\mu=1}^{m} A_{\mu\mu}^{(T)} \tag{6-45}$$

The *character of a representation* is understood to be a set of characters $\chi^{(T)}$, $T \in G$, corresponding to all the group elements.

Since the trace of a matrix is invariant under a unitary transformation [see Eq. (4-130)], all equivalent representations [i.e. those which satisfy relationship (6-35)] have the same character. Moreover, it is evident that, for the same reasons [cf. Eq. (6-42)], the elements of the same class of a group have identical character and that, consequently, this character is a property of a class of equivalent elements. It is further evident that the character of a reducible matrix can be expressed as the sum of the characters of its components, as follows directly from Eqs. (6-38) and (6-40).

Character tables of irreducible representations of all the required point groups are listed in numerous text-books on quantum chemistry[4-8] and group theory [9-12]. For illustration, in Tables 6-4 to 6-6 are given the character tables of representations of some groups discussed earlier (denotation of symmetry elements is the same as in Fig. 6-2), of the group D_{6h} (corresponding to the symmetry of the benzene molecule), and of the group O, both of which will be needed in the ensuing discussion. From Tables 6-4 to 6-6 it can be seen that a standard notation is assigned to the characters: the letters A and B correspond to one-dimensional, E to two-dimensional, and T to three-dimensional representations. If inversion is involved in a group, the index g (gerade) or u (ungerade) is attached to the letter according to whether the sign is preserved or changed during the inversion, respectively.

In a one-dimensional representation, the matrix element is directly equal to the character, which enables us to verify the validity of the

Table 6-4

Characters of Irreducible Representations of Some Point Groups

C_2, S_2				
C_2			E	C_2
	S_2		E	i
A	A_g		1	1
B	A_u		1	-1

C_{2h}, C_{2v}, D_2						
C_{2h}			E	C_2	i	σ_h
	C_{2v}		E	C_2	σ_v'	σ_v''
		D_2	E	C_2	C_2''	C_2'
A_g	A_1	A	1	1	1	1
A_u	A_2	B_1	1	1	-1	-1
B_g	B_1	B_2	1	-1	1	-1
B_u	B_2	B_3	1	-1	-1	1

D_{2h}								
D_{2h}	E	$C_2(z)$	$C_2''(y)$	$C_2'(z)$	i	σ_{xy}	σ_{xz}	σ_{yz}
A_g	1	1	1	1	1	1	1	1
A_u	1	1	1	1	-1	-1	-1	-1
B_{1g}	1	1	-1	-1	1	1	-1	-1
B_{1u}	1	1	-1	-1	-1	-1	1	1
B_{2g}	1	-1	1	-1	1	-1	1	-1
B_{2u}	1	-1	1	-1	-1	1	-1	1
B_{3g}	1	-1	-1	1	1	-1	-1	1
B_{3u}	1	-1	-1	1	-1	1	1	-1

general equation, (6-44): at first glance it is obvious that the rows (i.e. the characters of irreducible representations) correspond to orthogonal vectors.

An additional useful *orthogonality relation for the characters* can readily be derived from Eq. (6-44). By substituting $\mu = \nu$ and $\varkappa = \lambda$ into this equation, the relationship

$$\sum_{T \in G} \{\mathbf{B}_i^{(T)}\}^*_{\mu\mu} \{\mathbf{B}_j^{(T)}\}_{\lambda\lambda} = \frac{g}{m_i} \delta_{ij} \delta_{\mu\lambda} \tag{6-46}$$

is obtained. This relation can further be employed to express the product

$$\sum_{T \in G} [\chi_i^{(T)}]^* \chi_j^{(T)} = \sum_{T \in G} \sum_{\mu=1}^{m_i} \sum_{\lambda=1}^{m_i} \{\mathbf{B}_i^{(T)}\}_{\mu\mu}^* \{\mathbf{B}_j^{(T)}\}_{\lambda\lambda} =$$

$$= \sum_{\mu=1}^{m_i} \frac{g}{m_i} \delta_{ij}\delta_{\mu\mu} = g\delta_{ij}, \tag{6-47}$$

where labels i and j indicate that the characters correspond to any two irreducible representations. The validity of orthogonality relation (6-47) for irreducible representations can easily be verified on the particular examples given in Table 6-4.

So far only one way of constructing matrix representations of a group from their representations (irreducible ones, for instance), namely in the form of a direct sum, has been discussed. Now another way,

Table 6-5

Characters of Representations of the D_{6h} Group

D_{6h}	E	$2C_6$	$2C_6^2$ $\equiv 2C_3$	C_6^3 $\equiv C_2''$	$3C_2$	$3C_2'$	σ_h	$3\sigma_v$	$3\sigma_d$	$2S_6$	$2S_3$	S_2 $\equiv i$
A_{1g}	1	1	1	1	1	1	1	1	1	1	1	1
A_{1u}	1	1	1	1	1	1	−1	−1	−1	−1	−1	−1
A_{2g}	1	1	1	1	−1	−1	1	−1	−1	1	1	1
A_{2u}	1	1	1	1	−1	−1	−1	1	1	−1	−1	−1
B_{1g}	1	−1	1	−1	1	−1	−1	−1	1	1	−1	1
B_{1u}	1	−1	1	−1	1	−1	1	1	−1	−1	1	−1
B_{2g}	1	−1	1	−1	−1	1	−1	1	−1	1	−1	1
B_{2u}	1	−1	1	−1	−1	1	1	−1	1	−1	1	−1
E_{1g}	2	1	−1	−2	0	0	−2	0	0	−1	1	2
E_{1u}	2	1	−1	−2	0	0	2	0	0	1	−1	−2
E_{2g}	2	−1	−1	2	0	0	2	0	0	−1	−1	2
E_{2u}	2	−1	−1	2	0	0	−2	0	0	1	1	−2
$(\chi_{E_{1g}}^{(T)})^2$	4	1	1	4	0	0	4	0	0	1	1	4
$\chi_{E_{1g}}^{(TT)}$	2	−1	−1	2	2	2	2	2	2	−1	−1	2
$[\chi_{E_{1g}}^2]_S^{(T)}$	3	0	0	3	1	1	3	1	1	0	0	3
$[\chi_{E_{1g}}^2]_{AS}^{(T)}$	1	1	1	1	−1	−1	1	−1	−1	1	1	1

Table 6-6

Characters of Irreducible Representations of the O Group

O	E	$8C_3$	$3C_2$	$6C_4$	$6C_2'$
A_1	1	1	1	1	1
A_2	1	1	1	−1	−1
E	2	−1	2	0	0
T_1	3	0	−1	1	−1
T_2	3	0	−1	−1	1

based on the so-called *direct product of* two (or more) *representations,* symbolically denoted by $\Gamma = \Gamma_1 \otimes \Gamma_2$, will be introduced.

In Chapter 5 many-electron systems were described using product functions, consisting of the products of one-electron functions – atomic or molecular orbitals. These one-electron functions are, as a rule, the solution of a problem within the independent particle approximation (see Section 5.5) and, according to Theorem 6-1, form the basis for a representation of the symmetry group of the corresponding Hamiltonian.

Let us assume that two sets of functions ψ_i $(i = 1, 2, ..., m)$ and φ_i $(i = 1, 2, ..., m')$ are available, each of which forms a basis for the matrix representation of the same group, which can be expressed using the general group element $\mathcal{T}$ [cf. Eq. (6-29)] as follows:

$$\mathcal{T}\psi = \psi \mathbf{A}^{(T)} \tag{6-48a}$$

$$\mathcal{T}\varphi = \varphi \mathbf{B}^{(T)}, \quad T \in G \tag{6-48b}$$

The product space of functions ψ_i and φ_i will be constructed so that all possible products of the type $\psi_i \varphi_j$, the number of which is $u = m \cdot m'$, are formed. For the sake of consistency with the previous discussion, it should be remembered that such a product space would be suitable for the description of a two-electron system. Let us first verify that a set of product functions forms a basis for the representation of group G. Taking into account Eqs. (6-27) and (6-48), it holds that

$$\mathcal{T}\psi_i \varphi_j = \sum_{k=1}^{m} \sum_{l=1}^{m'} A_{ki}^{(T)} B_{lj}^{(T)} \psi_k \varphi_l \tag{6-49}$$

If an auxiliary index r is introduced to designate a pair of indices i, j and indices k, l are replaced by the new index s, then the expressions occuring in Eq. (6-49) can be rewritten to give

$$\left. \begin{array}{l} \psi_1 \varphi_1, \psi_1 \varphi_2, ..., \psi_1 \varphi_{m'}, \psi_2 \varphi_1, ..., \psi_i \varphi_j, ..., \psi_k \varphi_l, ..., \psi_m \varphi_{m'} \\ \Phi_1, \quad \Phi_2, \quad ..., \qquad\qquad\qquad \Phi_r, \quad ..., \Phi_s, \quad ..., \Phi_u \end{array} \right\} \tag{6-50}$$

$$D_{sr}^{(T)} = A_{ki}^{(T)} B_{lj}^{(T)} \tag{6-51}$$

Thus, Eq. (6-49) can be rewritten in the form

$$\mathcal{T}\Phi_r = \sum_{s=1}^{u} D_{sr}^{(T)} \Phi_s, \tag{6-52}$$

whence, on comparison with Eqs. (6-27) to (6-32), it follows that matrices $\mathbf{D}^{(T)}$, $T \in G$, form a new representation obtained as a direct product of the two original representations. The relationship of the product representation to the original representations is worth noting.

It is expressed by Eq. (6-51) which, for lucidity, can be rewritten explicitly:

$$\begin{Vmatrix} D_{11}^{(T)}, & D_{12}^{(T)}, & ..., & D_{1u}^{(T)} \\ D_{21}^{(T)}, & D_{22}^{(T)}, & ..., & D_{2u}^{(T)} \\ \vdots & & & \\ D_{u1}^{(T)}, & D_{u2}^{(T)}, & ..., & D_{uu}^{(T)} \end{Vmatrix} = \begin{Vmatrix} A_{11}^{(T)}B_{11}^{(T)}, & A_{11}^{(T)}B_{12}^{(T)}, & ..., & A_{1m}^{(T)}B_{1m'}^{(T)} \\ A_{11}^{(T)}B_{21}^{(T)}, & A_{11}^{(T)}B_{22}^{(T)}, & ..., & A_{1m}^{(T)}B_{2m'}^{(T)} \\ \vdots & & & \\ A_{m1}^{(T)}B_{m'2}^{(T)}, & A_{m1}^{(T)}B_{m'2}^{(T)}, & ..., & A_{mm}^{(T)}B_{m'm'}^{(T)} \end{Vmatrix} \qquad (6\text{-}53)$$

Equations (6-53) and (6-51) represent the definition of *the direct product of two matrices* (in contrast to the matrix product introduced in Section 4.5).

The direct product can be obtained for any two representations, i.e. also for irreducible representations. It can be expected that the direct product will generally afford a representation which can be decomposed in the sense of Eqs. (6-40) and (6-41) into irreducible components. The fact that the character of the product representation is equal to the product of the characters of the original representations, as follows from Eqs. (6-45) and (6-53), can be used here:

$$\chi_D^{(T)} = \chi_A^{(T)}\chi_B^{(T)}, \quad T \in G \qquad (6\text{-}54)$$

It has been mentioned that, for irreducible representations, the characters $\chi_A^{(T)}$ and $\chi_B^{(T)}$ are known, so that $\chi_D^{(T)}$, $T \in G$, can be easily calculated.

According to the basic character properties, the following relation for the character $\chi_D^{(T)}$ of a representation reducible in the sense of Eq. (6-41) holds:

$$\chi_D^{(T)} = \sum_i k_i \chi_i^{(T)}, \qquad (6\text{-}55)$$

where, as in Eq. (6-47), index i expresses the relationship to the irreducible representation i. If Eq. (6-55) is multiplied by the complex conjugate character of the j-th irreducible representation $[\chi_j^{(T)}]^*$ and if summation is carried out over all the group elements, then it follows that

$$\sum_{T \in G} [\chi_j^{(T)}]^* \chi_D^{(T)} = \sum_i k_i \sum_{T \in G} [\chi_j^{(T)}]^* \chi_i^{(T)} = g \sum_i k_i \delta_{ij} = g k_j \qquad (6\text{-}56)$$

This equation indicates that the reducible representation described by its character $\chi_D^{(T)}$, $T \in G$, contains the j-th irreducible representation as a component k_j times. Thus Eq. (6-56), based on knowledge of the characters alone, permits *decomposition of the reducible representation into its irreducible components*.

It should be noted, in conclusion, that basic equations have been introduced in this section which will be necessary when utilizing the symmetry properties of the studied systems for their quantum chemical solution. If the reader has not yet noticed the usefulness of some particular concepts, it will become more evident in further sections of the book where their application in practice will be discussed.

6.5 Selection rules for matrix elements

It was mentioned in the previous chapter that one of the steps in quantum chemical calculations is the evaluation of integrals of the type

$$M_{ij} = \int \psi_i^* \mathcal{M} \psi_j' \, d\tau = \langle \psi_i | \mathcal{M} | \psi_j' \rangle, \qquad (6\text{-}57)$$

where the reader should reacquaint himself with the symbols introduced in Section 5.4 for matrix elements of the operator $\mathcal{M}$ between functions ψ_i and ψ_j'. The action of an operator on a function generally leads to another function,

$$\mathcal{M} \psi_j' = \varphi_j, \qquad (6\text{-}58)$$

and, after substituting in Eq. (6-57),

$$M_{ij} = \langle \psi_i \| \varphi_j \rangle \qquad (6\text{-}59)$$

It should be emphasized in this connection that only the symmetry properties of the functions are now relevant and it is not necessary for the functions to be normalized; this fact is manifested by using two strokes in the scalar product symbol. ψ_i is assumed to be one of the functions forming the basis for the irreducible representation Γ_1, corresponding to matrices $\mathbf{A}^{(T)}$, $T \in G$, of group G, while φ_j belongs to the basis of the irreducible representation Γ_3, which corresponds to matrices $\mathbf{B}^{(T)}$, $T \in G$. Should it happen that both irreducible representations are the same, they will be assumed to be identical and not merely equivalent. Expression (6-59) is a scalar product, i.e., a number, and the action of an operator of symmetry transformation $\mathcal{T}$ on M_{ij} can therefore not change its value and, using Eq. (6-49), the expression

$$M_{ij} = \mathcal{T} M_{ij} = \sum_{k=1}^{m} \sum_{l=1}^{m'} (A_{ki}^{(T)})^* \, B_{lj}^{(T)} \langle \psi_k \| \varphi_l \rangle \qquad (6\text{-}60)$$

can be written. Operation $\mathcal{T}$ in Eq. (6-60) is, of course, arbitrary; the symmetry operations $\mathcal{V}$, $\mathcal{W}$, ... belonging to the given group might equally well have been chosen. Summing all such cases, then the relationship

$$\sum_{T \in G} \mathcal{T} M_{ij} = g M_{ij} = \sum_{k=1}^{m} \sum_{l=1}^{m'} \sum_{T \in G} (A_{ki}^{(T)})^* B_{lj}^{(T)} \langle \psi_k \| \varphi_l \rangle \qquad (6\text{-}61)$$

results. Eq. (6-61) can be simplified using Eq. (6-44), so that the final result can be written in the form

$$M_{ij} = \frac{1}{m} \delta_{AB} \delta_{ij} \sum_{k=1}^{m} \langle \psi_k \| \varphi_k \rangle \qquad (6\text{-}62)$$

According to Eq. (6-62), integral (6-57) is nonvanishing only when:
1. $A \equiv B$, i.e. irreducible representations Γ_1 and Γ_3 are identical.
2. If condition 1 is fulfilled, it must also hold that $i = j$ (i.e. $\psi_i \equiv \varphi_j$).
If conditions 1 and 2 hold, then Eqs. (6-59) and (6-62) yield the relationship

$$M_{ii} = \langle \psi_i \| \psi_i \rangle = \frac{1}{m} \sum_{k=1}^{m} \langle \psi_k \| \psi_k \rangle \qquad (6\text{-}63)$$

Since the right-hand side of Eq. (6-63) is independent of index i, Eq. (6-63) holds if all the values $\langle \psi_k \| \psi_k \rangle$, $k = 1, 2, \ldots, m$, are the same.

The derived relations will now be applied to particular operators $\mathcal{M}$. Two cases will be discussed separately:
  a) the operators $\mathcal{T}$ of the given group, G, correspond to the symmetry transformations of operator $\mathcal{M}$,
  b) condition *a*) is not fulfilled.

The first category includes, of course, the trivial case when $\mathcal{M} \equiv 1$ (more generally $\mathcal{M} \equiv$ a constant). Relation (6-62) then ensures the orthogonality of some functions purely on the basis of their symmetry properties. The case when $\mathcal{M}$ is the Hamiltonian (many-electron, Hartree-Fock or some other one-electron Hamiltonian) is, however, typical for this category. The Hamiltonian $\mathcal{H}$ is invariant under all symmetry operations of the given group and, therefore corresponds to the irreducible representation A_{1g}, for which it is characteristic that all matrix elements of the one-dimensional representation are equal to one (cf. Table 6-4). Since the symmetry properties of φ_j are, according to Eq. (6-58), given in the form of a direct product of irreducible representations A_{1g} (corresponding to $\mathcal{M}$) and Γ_2 [corresponding to the basis ψ_i', ($i = 1, 2, \ldots$)], it must necessarily hold that even the functions $\varphi_1, \varphi_2, \ldots, \varphi_j$ form the basis of Γ_2; in short, from the relationship

$$A_{1g} \otimes \Gamma_2 = \Gamma_3 \qquad (6\text{-}64)$$

it follows that

$$\Gamma_2 = \Gamma_3 \qquad (6\text{-}65)$$

Thus, the application of Eq. (6-62) to this case can be summarized as follows:

$$\langle \psi_i | \mathscr{H} | \psi_j' \rangle \neq 0, \tag{6-66}$$

provided the following conditions are fulfilled:

1. irreducible representations Γ_1 and Γ_2, to which bases ψ_i ($i = 1, 2, \ldots$) and ψ_i' ($i = 1, 2, \ldots$) correspond, are equal.

2. $i = j$.

Equation (6-63) then acquires the following form

$$\langle \psi_i | \mathscr{H} | \psi_i \rangle = \frac{1}{m} \sum_{k=1}^{m} \langle \psi_k | \mathscr{H} | \psi_k \rangle, \tag{6-67}$$

from which it follows that the level whose energy is given by the value $\langle \psi_i | \mathscr{H} | \psi_i \rangle$ is m-fold degenerate; this is an alternate formulation of Theorem 6-1 (see Section 6.4).

Now the second category of operators, i.e. those which do not satisfy the above condition a) will be treated. In general, it can be assumed that operator $\mathscr{M}$ has symmetry properties such that it is transformed according to representation Γ_M, which need not be irreducible. The functions φ_j, $j = 1, 2, \ldots$ then correspond to the product representation

$$\Gamma = \Gamma_M \otimes \Gamma_2 = \sum_i k_i \Gamma_i, \tag{6-68}$$

where its decomposition has already been given in Eq. (6-41). Regarding the validity of Eq. (6-62), it is immediately evident that matrix element M_{ij} is nonvanishing only when $k_1 \neq 0$, where k_1 is the coefficient of the irreducible representation Γ_1, according to which function ψ_i is transformed.

One of the most important ways of applying the given result is the determination of *selection rules* for spectroscopic transitions between two states. Here the selection rules imply prediction, on the basis of symmetry considerations, of whether the intensity of the transition is strictly zero or differs from zero. We shall see in the chapter on molecular spectroscopy that the basic quantity for calculation of the intensity of a transition is the *transition moment*, which is expressed by integrals of the type

$$\langle \psi_i | x | \psi_j' \rangle, \quad \langle \psi_i | y | \psi_j' \rangle, \quad \langle \psi_i | z | \psi_j' \rangle \tag{6-69}$$

where x, y, z are the Cartesian coordinates of the position vector of the electron and the matrix elements represent the components of the one-electron dipole transition moment vector. At least one vector component must be nonvanishing if the transition is to be allowed for symmetry reasons. If we consider the transition between the ground state of the molecule, which is usually symmetrical with respect to all operations of the symmetry group of the Hamiltonian (and thus corresponds to the irreducible representation A_{1g}) and the excited state, and take Eq. (6-68) into account, then the relationship

$$A_{1g} = \Gamma_x \otimes \Gamma_2 \qquad (6\text{-}70)$$

is valid for expressing the condition that component x of the transition moment does not vanish. Γ_x in Eq. (6-70) denotes the irreducible representation of the symmetry group of the Hamiltonian according to which coordinate x is transformed under the given symmetry conditions. Equation (6-70) can be satisfied only if $\Gamma_2 \equiv \Gamma_x$, because only then it is possible to represent the symmetry of A_{1g} by a direct product. If the fact that coordinate x changes sign when passing through the origin is taken into account, it readily follows from Fig. 6-2 and Table 6-4 that, for example, for groups C_2, S_2, C_{2h}, C_{2v}, D_2 and D_{2h}, the x coordinate gradually corresponds to the irreducible representations B, A_u, B_u, B_1, B_3 and B_{3u} and that the excited states for which the x component of the transition moment differs from zero correspond to the same symmetry type.

The discussion so far has been general in as much as it comprises the interpretation of electronic and vibrational spectra, i.e. spectra in the UV region, visible spectra and IR spectra. For UV and visible spectra, the wave functions ψ_i and ψ'_j from Eq. (6-69) must be understood as electronic functions, for vibrational spectra functions ψ_i and ψ'_j denote the vibrational wave functions.

Determination of the selection rules for Raman spectra resembles the previous cases except that, instead of integrals of the type (6-69), it is necessary to investigate integrals such as

$$\langle \psi_i | \alpha | \psi'_j \rangle, \qquad (6\text{-}71)$$

where α is one of the components of the polarizability tensor which represents a symmetrical matrix of order 3. It is not as easy to determine the symmetry properties of the polarizability tensor as it was in the previous case with Cartesian coordinates, but they are usually given as supplementary information in character tables of irreducible representations.

6.6 Symmetry and hybrid orbitals

All considerations in the previous section were based on the assumption that functions occuring in the matrix elements belong to the bases spanning irreducible representations. The simplification achieved in a number of problems is connected with the fact that the functions of these properties can be considered to be eigenfunctions of the symmetry operators of the Hamiltonian, i.e. of operators which commute with the Hamiltonian. It is evident from this viewpoint that the partial result, presented in the form of Eq. (6-65) and of the additional conditions given in Section 6.5, corresponds to Theorem 5 in Section 4.3.

It often happens in quantum-chemical studies that, in the beginning of the calculation, functions forming the basis of a reducible representation are available. The MO-LCAO method discussed in Section 5.6 is a typical example. It is assumed in this method that the basis set of atomic orbitals, used for the construction of the molecular orbitals, is the basis for a reducible representation of the symmetry group of the Hamiltonian. Otherwise the basis would have no physical meaning, as can be demonstrated by the example of an electron moving in the electrostatic field of four protons and described by Hamiltonian (6-15). The most obvious course is to look for a wave function [cf. Eq. (5-63)] in the form of a linear combination of atomic orbitals $(1s)_i$, $i = 1, 2, 3, 4$, located on all nuclei, $i = 1, 2, 3, 4$ represented by the figure in the first column of Table 6-1, i.e.

$$\varphi = \sum_{i=1}^{4} c_i(1s)_i \tag{6-72}$$

On the other hand, there is no point in assigning 1s orbitals to certain nuclei and, for example, p orbitals to others. It is also clear that the requirement of a reducible basis is essentially identical with the requirement that physically equivalent atoms (here four protons) supply the same atomic orbitals to the total set of atomic functions. The atomic orbitals $(1s)_i$, $i = 1, 2, 3, 4$, arranged in the correct sequence in a one-row matrix of the type (6-30), are transformed by the symmetry operations of a rectangle as follows:

$$\mathbf{A}^{(E)} = \begin{Vmatrix} 1 & 0 & 0 & 0 \\ 0 & 1 & 0 & 0 \\ 0 & 0 & 1 & 0 \\ 0 & 0 & 0 & 1 \end{Vmatrix} \tag{6-73a}$$

$$\mathbf{A}^{(\sigma_{xz})} = \mathbf{A}^{(C_2')} = \begin{Vmatrix} 0 & 1 & 0 & 0 \\ 1 & 0 & 0 & 0 \\ 0 & 0 & 0 & 1 \\ 0 & 0 & 1 & 0 \end{Vmatrix} \tag{6-73b}$$

$$\mathbf{A}^{(\sigma_{yz})} = \mathbf{A}^{(C_2'')} = \begin{Vmatrix} 0 & 0 & 0 & 1 \\ 0 & 0 & 1 & 0 \\ 0 & 1 & 0 & 0 \\ 1 & 0 & 0 & 0 \end{Vmatrix} \tag{6-73c}$$

$$\mathbf{A}^{R(z,\pi)} = \mathbf{A}^{(i)} = \begin{Vmatrix} 0 & 0 & 1 & 0 \\ 0 & 0 & 0 & 1 \\ 1 & 0 & 0 & 0 \\ 0 & 1 & 0 & 0 \end{Vmatrix} = \mathbf{A}^{(C_2)} \qquad (6\text{-}73\text{d})$$

It can easily be verified that binary products of matrices (6-73) conform with the multiplication table (Table 6-2); thus the matrices form a representation which can obviously be reduced (as follows from Table 6-4 according to the character table for the group D_{2h}, corresponding to the symmetry of a rectangle) and the atomic orbitals form the respective basis.

Thus functions are available which belong to the basis of the reducible representation, and functions which form the bases of the irreducible representations of the symmetry group G of the Hamiltonian must be constructed. Let $\{\mathbf{B}_i^{(T)}\}_{\mu\nu}$ be the matrix element of irreducible representation i, satisfying Eq. (6-44) and Φ — a function which belongs to the basis of the reducible representation. At arbitrary but then fixed value of ν, function $\varphi_\mu^{(i)}$ can be defined by the equation

$$\varphi_\mu^{(i)} = \sum_{T \in G} \{\mathbf{B}_i^{(T)}\}_{\mu\nu}^* \mathscr{T}\Phi, \qquad (6\text{-}74)$$

where the summation includes all the elements of G. In order to succeed in the investigation of the symmetry properties of $\varphi_\mu^{(i)}$, a further symmetry operation $\mathscr{V}$ will be introduced, so that Eq. (6-32) is valid and consequently

$$\mathscr{T} = \mathscr{V}^{-1}\mathscr{W} \qquad (6\text{-}75)$$

It then holds that

$$\mathscr{V}\varphi_\mu^{(i)} = \sum_{T \in G} \{\mathbf{B}_i^{(T)}\}_{\mu\nu}^* \mathscr{V}\mathscr{T}\Phi = \sum_{W \in G} \{\mathbf{B}_i^{(V^{-1}W)}\}_{\mu\nu}^* \mathscr{W}\Phi, \qquad (6\text{-}76)$$

as in the summation the product $\mathscr{V}\mathscr{T}$ again runs through all the elements of the group. Because the matrices of the representation [see Eqs. (6-28) and (4-110)] are unitary it holds that

$$\mathbf{B}_i^{(V^{-1})} = [\mathbf{B}_i^{(V)}]^{-1} = [\mathbf{B}_i^{(V)}]^H, \qquad (6\text{-}77)$$

which allows modification of the matrix element in Eq. (6-76) to give

$$\{\mathbf{B}_i^{(V^{-1}W)}\}_{\mu\nu}^* = \sum_\varkappa \{\mathbf{B}_i^{(V^{-1})}\}_{\mu\varkappa}^* \{\mathbf{B}_i^{(W)}\}_{\varkappa\nu}^* = \sum_\varkappa \{\mathbf{B}_i^{(V)}\}_{\varkappa\mu} \{\mathbf{B}_i^{(W)}\}_{\varkappa\nu}^* \qquad (6\text{-}78)$$

By substituting Eq. (6-78) into Eq. (6-76), the final form is obtained,

$$\mathscr{V}\varphi_\mu^{(i)} = \sum_{W \in G} \sum_\varkappa \{\mathbf{B}_i^{(V)}\}_{\varkappa\mu} \{\mathbf{B}_i^{(W)}\}_{\varkappa\nu}^* \mathscr{W}\Phi =$$

$$= \sum_\varkappa \{\mathbf{B}_i^{(V)}\}_{\varkappa\mu} \sum_{W \in G} \{\mathbf{B}_i^{(W)}\}_{\varkappa\nu}^* \mathscr{W}\Phi = \sum_\varkappa \{\mathbf{B}_i^{(V)}\}_{\varkappa\mu} \varphi_\varkappa^{(i)}, \qquad (6\text{-}79)$$

where relation (6-74) was again employed. If this equation is compared with Eq. (6-27), it will be seen that all the conditions are satisfied for functions $\varphi_\mu^{(i)}$, $\mu = 1, 2, ...,$ defined by Eq. (6-74) to span the i-th irreducible representation of G. The given problem is, thus solved. Unfortunately, the whole matrices of the representation are not generally known, but only their characters. Because, however, index v in Eq. (6-74) has an arbitrary value, it can be set equal to μ. If, in addition, the summation is carried out over μ, the relationship

$$\varphi^{(i)} = \sum_{T \in G} (\chi_i^{(T)})^* \mathscr{T} \Phi \tag{6-80}$$

is obtained from Eq. (6-74), where the symbol of the character is substituted for the sum of the diagonal elements [cf. Eq. (6-45)]. It is clear that function $\varphi^{(i)}$ in Eq. (6-80) belongs to the space of functions spanning the i-th irreducible representation, and this equation can be considered as a recipe for construction of the functions forming the bases of irreducible representations. It should be stressed that the thus-obtained functions will not in general be normalized.

Returning to the model of the electron in the electrostatic field of four protons, it is necessary first of all to be aware that the symmetry operations of group D_2 are sufficient for describing the symmetry properties of a rectangle, as follows from Table 6-4; it is superfluous to attribute full D_{2h} symmetry to the rectangle, because, due to the planarity of the figure, some operations of group D_{2h} become identical. In order to decompose the matrix representation given by Eqs. (6-73a) to (6-73d), the irreducible representations contained as its components must be discovered. This problem is easily solved using Eq. (6-56), as it is sufficient to sum the products

$$k_i = \tfrac{1}{4} \sum_{T \in D_2} (\chi_i^{(T)})^* \chi_D^{(T)} \tag{6-81}$$

(where $g = 4$ is the order of group D_2) to determine the "participation number" k_i of the individual irreducible representations. The traces of matrices (6-73a) to (6-73d) yield the characters

$$\chi_D^{(E)} = 4; \qquad \chi_D^{(C_2)} = 0; \qquad \chi_D^{(C_2')} = 0; \qquad \chi_D^{(C_2')} = 0 \tag{6-82}$$

These, together with the characters of the irreducible representations of group D_2 (Table 6-4), enable decomposition of the reducible representation into its irreducible components:

$$\Gamma = A + B_1 + B_2 + B_3 \tag{6-83}$$

As a consequence, four new functions can be formed using 1s orbitals located on four centres, which establish the bases of the given irreducible representations. Equation (6-80) can then be rewritten to give

$$\varphi^{(i)} = \chi_i^{(E)}\mathscr{E}(1s)_1 + \chi_i^{(C_2)}\mathscr{C}_2(1s)_1 + \chi_i^{(C_2'')} \mathscr{C}_2''(1s)_1 + \chi_i^{(C_2')} \mathscr{C}_2'(1s)_1, \qquad (6\text{-}84)$$

where it is borne in mind that the characters are real numbers and $i = A$, B_1, B_2 and B_3. Using Table 6-1, the effect of the operators on the $(1s)_1$ function [e.g., $\mathscr{C}_2'(1s)_1 = (1s)_2$] is determined and, after substituting the characters in Eq. (6-84), four functions with the required properties are obtained as the final result:

$$\varphi^{(A)} = (1s)_1 + (1s)_3 + (1s)_4 + (1s)_2$$

$$\varphi^{(B_1)} = (1s)_1 + (1s)_3 - (1s)_4 - (1s)_2$$

$$\varphi^{(B_2)} = (1s)_1 - (1s)_3 + (1s)_4 - (1s)_2 \qquad (6\text{-}85)$$

$$\varphi^{(B_3)} = (1s)_1 - (1s)_3 - (1s)_4 + (1s)_2$$

These functions are not normalized but determination of the corresponding normalization constants is a simple matter:

$$K = \frac{1}{\sqrt{(\langle\varphi^{(i)} | \varphi^{(i)}\rangle)}} \qquad (6\text{-}86)$$

The described symmetrization of the functions will prove very useful in simplifying the solution of the secular determinant, encountered in calculations using one-electron LCAO methods [see Eq. (5-65)]. This simplification follows from the validity of the selection rules for matrix elements. For instance, in the illustrative example the calculation of the wave function and the energy considering the original formulation [see Eqs. (5-64) and (5-65)] would lead to a fourth-order secular determinant. If the basis set of atomic orbitals is replaced by the basis set of symmetry orbitals, expressed, for example, by Eq. (6-85), molecular orbitals can be sought in the form

$$\varphi^{(i)} = \sum_k c_k \varphi_k^{(i)}, \qquad (6\text{-}87)$$

where all the functions $\varphi_k^{(i)}$ belong to the i-th irreducible representation. Thus the property that the matrix elements of the Hamiltonian vanish when the functions surrounding the operator belong to different irreducible representations is employed. Instead of the original secular determinant, it is necessary to solve a number of secular determinants of lower order, each corresponding to one irreducible representation of the corresponding symmetry group: this is termed *factorization of the secular determinant* by using the spatial symmetry of the system.

In the case under study, the inclusion of symmetry considerations leads to a complete factorization of the secular problem so that the form of the molecular orbitals can be entirely determined from the symmetry. The energies corresponding to the molecular orbitals are determined as the diagonal elements of the normalized functions (6-85).

A complete determination of the wave functions through symmetry considerations is, of course, only possible in special cases. Generally the molecular orbitals are sought in the form of (6-87); the practical procedure will be demonstrated using the ethylene molecule (presented in Table 6-3), which has D_{2h} symmetry in the ground state.

First, the special case of the LCAO approximation will be discussed, including in the basis set only those atomic orbitals which are occupied by electrons in the free (i.e. not bound) atoms in the ground state. This *basis* set of atomic orbitals is termed *the minimum basis* set. For the C_2H_4 molecule the minimum basis set is formed by the following orbitals:

for each carbon: $(1s)^C, (As), (2p_x), (2p_y), (2p_z)$
for each hydrogen: $(1s)$

Thus the basis set of atomic orbitals has a dimension of 14 and the secular determinant without considering symmetry would also be of order 14. In the first stage the inclusion of symmetry considerations requires separation of the atomic orbitals into subsets of equivalent atomic orbitals which are interchangeable during symmetry operations of the molecule.

The following subsets are evidently of this type,

$$\{(1s)^C_1, (1s)^C_2\}, \ \{(2s)_1, (2s)_2\}, \ \{(2p_x)_1, (2p_x)_2\}, \ \{(2p_y)_1, (2p_y)_2\},$$
$$\{(2p_z)_1, (2p_z)_2\}, \ \{(1s)_1, (1s)_2, (1s)_3, (1s)_4\},$$

which span reducible representations of group D_{2h}. By decomposing these representations the following symmetry functions are obtained:

A_g: $(1s)^C_1 + (1s)^C_2$

B_{3u}: $(1s)^C_1 - (1s)^C_2$

A_g: $(2s)_1 + (2s)_2$

B_{3u}: $(2s)_1 - (2s)_2$

B_{3u}: $(2p_x)_1 + (2p_x)_2$

A_g: $(2p_x)_1 - (2p_x)_2$

B_{2u}: $(2p_y)_1 + (2p_y)_2$

B_{1g}: $(2p_y)_1 - (2p_y)_2$

B_{1u}: $(2p_z)_1 + (2p_z)_2$

B_{2g}: $(2p_z)_1 - (2p_z)_2$

$$A_g: \quad (1s)_1 + (1s)_2 + (1s)_3 + (1s)_4$$
$$B_{3u}: \quad (1s)_1 - (1s)_2 - (1s)_3 + (1s)_4$$
$$B_{2u}: \quad (1s)_1 + (1s)_2 - (1s)_3 - (1s)_4$$
$$B_{1g}: \quad (1s)_1 - (1s)_2 + (1s)_3 - (1s)_4$$

Thus the following distribution of symmetry orbitals has been obtained: $4 \times A_g$, $4 \times B_{3u}$, $2 \times B_{1g}$, $2 \times B_{2u}$, $1 \times B_{1u}$, $1 \times B_{2g}$, where the number of cases corresponding to the same type of irreducible representation simultaneously indicates the order of the secular determinant, the solution of which yields molecular orbitals with the pertinent symmetry. For example, the molecular orbitals of A_g symmetry assume the form

$$\varphi^{(A_g)} = c_1[(1s)_1^C + (1s)_2^C] + c_2[(2s)_1 + (2s)_2] +$$
$$+ c_3[(2p_x)_1 - (2p_x)_2] + c_4[(1s)_1 + (1s)_2 + (1s)_3 + (1s)_4] \quad (6\text{-}88)$$

The symmetry of a problem can also be used to determine orbitals with specific spatial properties, called *hybrid orbitals*. Pauling[13] demonstrated on the basis of orbital hybridization that it is possible to construct linear combinations of atomic wave functions such that equivalent orbitals are formed which, however, are oriented in different directions. When, for example, describing the chemical bonding in the methane molecule using orbitals located in the $C-H$ bonds, it is necessary to begin with four equivalent orbitals directed from the carbon atom to the corners of a regular tetrahedron where the hydrogen atoms are located. Kimball[14] formulated a general procedure for describing hybrid orbitals on the basis of group considerations. This procedure will be clarified for the case when one atom forms six equivalent orbitals within the molecule. Such a situation is encountered when interpreting the properties of transition metal complexes.

The central atom, which has six identical neighbours (e.g. atoms), is located at the origin of the rectangular coordinate system; the neighbours, called ligands, lie on the x, y, z axes at equal distances from the origin (Fig. 6-4). The six equivalent hybrid orbitals on the central atom directed to the ligands are denoted by σ_i, $i = 1, 2, ..., 6$. These orbitals span the reducible representation Γ, which can be decomposed into irreducible representations of group O (see Table 6-6):

$$\Gamma = A_1 + E + T_1 \quad (6\text{-}89)$$

In solving the problem it suffices here (cf. the case of the rectangle) to assign the system to the subgroup O of symmetry group O_h. The symmetry orbitals corresponding to the given irreducible representations will be constructed using relation (6-80). Since determination of the functions spanning a multidimensional representation has not yet been discussed, it will be carried out here for the representation E. Using Eq. (6-80) and the character table for group O in Table 6-6, the relationship (the characters are real numbers)

$$\sum_{T \in O} \chi_E^{(T)} \mathcal{T} \sigma_1 = 4\sigma_1 - 2\sigma_2 - 2\sigma_3 - 2\sigma_4 - 2\sigma_5 + 4\sigma_6 \qquad (6\text{-}90)$$

is obtained. Since the other linearly independent function (representation E is two-dimensional) is to be determined, the operator is applied to still another function, σ_i, for example

$$\sum_{T \in O} \chi_E^{(T)} \mathcal{T} \sigma_2 = 4\sigma_2 - 2\sigma_1 - 2\sigma_3 - 2\sigma_5 - 2\sigma_6 + 4\sigma_4 \qquad (6\text{-}91)$$

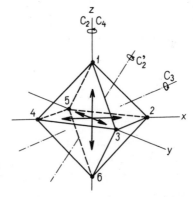

Fig. 6-4. Schematic representation of octahedral complex symmetry.

If the result is a function linearly dependent on the original function (6-90), it is necessary to try application of the operator on still another function σ_i. Here, however, this is not necessary. For reasons which will become obvious later, that function is chosen as the second required function which is obtained by multiplying Eq. (6-91) by two and adding to Eq. (6-90). Thus, the following function orthogonal to Eq. (6-90) is obtained:

$$2(\sigma_2 + \sigma_4 - \sigma_3 - \sigma_5)$$

The functions spanning representation T_1 would also be constructed in a similar manner. Since the hybrid orbitals are assumed to be ortho-normalized functions, the resulting functions spanning the corresponding irreducible representations can be written in the form

$$A_1: \quad \sigma_1' = \frac{1}{\sqrt{6}}(\sigma_1 + \sigma_2 + \sigma_3 + \sigma_4 + \sigma_5 + \sigma_6)$$

$$E: \quad \sigma_2' = \frac{1}{\sqrt{12}}(2\sigma_1 - \sigma_2 - \sigma_3 - \sigma_4 - \sigma_5 + 2\sigma_6)$$

$$\sigma'_3 = \frac{1}{2}(\sigma_2 - \sigma_3 + \sigma_4 - \sigma_5)$$

$$T_1: \quad \sigma'_4 = \frac{1}{\sqrt{2}}(\sigma_2 - \sigma_4)$$

$$\sigma'_5 = \frac{1}{\sqrt{2}}(\sigma_3 - \sigma_5)$$

$$\sigma'_6 = \frac{1}{\sqrt{2}}(\sigma_1 - \sigma_6) \tag{6-92}$$

If the central atom is assumed to have s, p and d orbitals, nine functions s, p_x, p_y, p_z, d_{z^2}, $d_{x^2-y^2}$, d_{xz}, d_{yz}, d_{xy} are available for the calculation. Their symmetry properties with respect to the symmetry operations of group O are such that they span the following irreducible representations:

A_1: s

E: d_{z^2}, $d_{x^2-y^2}$

T_1: p_x, p_y, p_z

T_2: d_{xz}, d_{yz}, d_{xy}

Usually information on the transformation properties of atomic orbitals can be found as supplementary data in character tables, e.g., in reference 4. If the respective atomic orbitals are substituted for functions σ'_i (it would be necessary, of course, to first ensure that the functions σ'_i and the atomic orbitals possess the same transformation properties, in other words that their corresponding matrix representations are identical and not merely equivalent), then Eq. (6-92) can be rewritten in matrix form:

$$
\begin{array}{cc}
A & s \\
E & d_{z^2} \\
& d_{x^2-y^2} \\
T_1 & p_x \\
& p_y \\
& p_z
\end{array}
=
\begin{Vmatrix}
1/\sqrt{6} & 1/\sqrt{6} & 1/\sqrt{6} & 1/\sqrt{6} & 1/\sqrt{6} & 1/\sqrt{6} \\
1/\sqrt{3} & -1/\sqrt{12} & -1/\sqrt{12} & -1/\sqrt{12} & -1/\sqrt{12} & 1/\sqrt{3} \\
0 & 1/2 & -1/2 & 1/2 & -1/2 & 0 \\
0 & 1/\sqrt{2} & 0 & -1/\sqrt{2} & 0 & 0 \\
0 & 0 & 1/\sqrt{2} & 0 & -1/\sqrt{2} & 0 \\
1/\sqrt{2} & 0 & 0 & 0 & 0 & -1/\sqrt{2}
\end{Vmatrix}
\begin{Vmatrix}
\sigma_1 \\ \sigma_2 \\ \sigma_3 \\ \sigma_4 \\ \sigma_5 \\ \sigma_6
\end{Vmatrix}
$$

$$\tag{6-93}$$

The inverse of transformation (6-93) allows expression of the hybrid orbitals in terms of atomic orbitals. Since the transformation matrix is a unitary matrix, replacement of the rows by the columns leads to the inverse of the original matrix [cf. Eq. (4-110)]. Thus, for example, the hybrid σ_1 oriented in the direction of the positive part of the z-axis is given by the expression

$$\sigma_1 = \frac{1}{\sqrt{6}} s + \frac{1}{\sqrt{3}} d_{z^2} + \frac{1}{\sqrt{2}} p_z \qquad (6\text{-}94)$$

Thus it has been shown that the d^2sp^3 electron configuration is capable of forming equivalent orbitals in the direction of three right-angled axes. Kimball[14] has published a table of possible configurations for hybrid orbitals (cf. Table 7-2) corresponding to coordination numbers 2 to 8. Thus, for example, with compounds of the methane type, where four equivalent orbitals must be constructed along the axes of a regular tetrahedron, an s orbital must be combined with three p orbitals. Using the notation for electron configurations this can be expressed as sp^3 hybridization.

6.7 Spin and spatial symmetry of many-electron systems

In Section 5.4, a very general way of expressing the wave function of a many-electron system in the form of a linear combination of Slater determinants was described. Since the molecular Hamiltonian does not usually contain operators depending upon spin variables, the operators of the total spin $\mathscr{S}_z$ and $\mathscr{S}^2$ are constants of motion and it is, therefore, expedient to expand the total wave function by means of linear combinations of the Slater determinants chosen so that they are eigenfunctions not only of $\mathscr{S}_z$, but also of $\mathscr{S}^2$.

In order to investigate the action of the spin operators and symmetry transformations, the general Slater determinant is studied:

$$\Delta_K(1, 2, ..., n) = \left| \varphi_1(1)\, \eta_1(1),\ \varphi_2(2)\, \eta_2(2),\ ...,\ \varphi_n(n)\, \eta_n(n) \right|, \qquad (6\text{-}95)$$

where the notation introduced in Eqs. (5-29) and (5-32) is employed. The function η is a general spin function which can be α or β, according to the circumstances. The definition of $\mathscr{S}_z$ from Eqs. (4-84) and (4-76) permits instant application of the operator to function (6-95). Thus, it is seen that Δ_K is the eigenfunction of $\mathscr{S}_z$:

$$\mathscr{S}_z \Delta_K = \tfrac{1}{2}\hbar(n_\alpha - n_\beta)\, \Delta_K, \qquad (6\text{-}96)$$

where n_α is the number of α spins and n_β is the number of β spins in the Slater determinant Δ_K, for which it holds, of course, that $n = n_\alpha + n_\beta$.

It must simultaneously be stated that the form of the total angular momentum (spin or orbital), as introduced in Eqs. (4-66), (4-76) and (4-85),

138

requires a certain modification in order to investigate its effect upon product functions.

This modification, which we shall demonstrate on relations of the angular momentum, can also be directly applied to the spin momentum. It consists in rearranging the commutation relations of type (4-65) to the form

$$[\mathscr{L}_z, \mathscr{L}_+] = \hbar \mathscr{L}_+ \qquad (6\text{-}97\text{a})$$

$$[\mathscr{L}_z, \mathscr{L}_-] = -\hbar \mathscr{L}_- \qquad (6\text{-}97\text{b})$$

$$[\mathscr{L}_+, \mathscr{L}_-] = 2\hbar \mathscr{L}_z, \qquad (6\text{-}97\text{c})$$

where $\mathscr{L}_+$ and $\mathscr{L}_-$ are called *the shift operators*, defined by equations

$$\mathscr{L}_+ = \mathscr{L}_x + i\mathscr{L}_y \qquad (6\text{-}98\text{a})$$

$$\mathscr{L}_- = \mathscr{L}_x - i\mathscr{L}_y \qquad (6\text{-}98\text{b})$$

Operators $\mathscr{L}_+$ and $\mathscr{L}_-$ are called shift operators because of a further given property. The relation of the functions

$$u_{lm} = \mathscr{L}_+ Y_{lm} \qquad (6\text{-}99)$$

to the operator $\mathscr{L}_z$, where Y_{lm} are the spherical harmonics (see Table 3-1) is of particular interest. Employing commutation relations (6-97) and Eqs. (4-72) and (4-74) it then follows that

$$\mathscr{L}_z(\mathscr{L}_+ Y_{lm}) = \frac{1}{\hbar} \mathscr{L}_z(\mathscr{L}_z \mathscr{L}_+ - \mathscr{L}_+ \mathscr{L}_z) Y_{lm} =$$

$$= \frac{1}{\hbar} \mathscr{L}_z^2 \mathscr{L}_+ Y_{lm} - \mathscr{L}_z \mathscr{L}_+ m Y_{lm} \qquad (6\text{-}100)$$

On multiplying Eq. (6-100) from the left by $\mathscr{L}_z^{-1}$, the expression

$$\mathscr{L}_z(\mathscr{L}_+ Y_{lm}) = \hbar(m + 1) \mathscr{L}_+ Y_{lm} \qquad (6\text{-}101\text{a})$$

is obtained, and, in an analogous way, also obtained is the relationship

$$\mathscr{L}_z(\mathscr{L}_- Y_{lm}) = \hbar(m - 1) \mathscr{L}_- Y_{lm} \qquad (6\text{-}101\text{b})$$

Equations (6-101) indicate that the functions $\mathscr{L}_+ Y_{lm}$ and $\mathscr{L}_- Y_{lm}$ are also eigenfunctions of $\mathscr{L}_z$ with eigenvalues $\hbar(m + 1)$ and $\hbar(m - 1)$, respectively. It follows from Eqs. (4-72) and (4-74) that

$$\mathscr{L}_+ Y_{lm} = A_+ Y_{l,m+1} \qquad (6\text{-}102\text{a})$$

$$\mathscr{L}_- Y_{lm} = A_- Y_{l,m-1}, \qquad (6\text{-}102\text{b})$$

where A_+ and A_- are proportionality constants. These constants can be determined from the conditions that all functions Y_{lm} must be normalized and that the relationship

$$-l \leqq m \leqq l$$

is valid for m:

$$A_+ = \hbar[l(l + 1) - m(m + 1)]^{1/2} = \hbar[(l - m)(l + m + 1)]^{1/2} \qquad (6\text{-}103a)$$

$$A_- = \hbar[l(l + 1) - m(m - 1)]^{1/2} = \hbar[(l + m)(l - m + 1)]^{1/2} \qquad (6\text{-}103b)$$

It can easily be verified using commutation relations (4-65) that the operator of the square of the angular momentum can be alternately expressed as follows:

$$\mathscr{L}^2 = \mathscr{L}_- \mathscr{L}_+ + \mathscr{L}_z^2 + \hbar \mathscr{L}_z \qquad (6\text{-}104a)$$

$$\mathscr{L}^2 = \mathscr{L}_+ \mathscr{L}_- + \mathscr{L}_z^2 - \hbar \mathscr{L}_z \qquad (6\text{-}104b)$$

In comparison with the previous expression [cf. Eq. (4-66)], this relationship has the advantage that the effect of all the operators on the right-hand side of the equations on the one-particle functions is known. This advantage is not yet apparent in the simplest case when

$$\mathscr{L}^2 Y_{lm} = (\mathscr{L}_- \mathscr{L}_+ + \mathscr{L}_z^2 + \hbar \mathscr{L}_z) Y_{lm} =$$
$$= \hbar[l(l + 1) - m(m + 1) + m^2 + m] Y_{lm} = \hbar^2 l(l + 1) Y_{lm}, \qquad (6\text{-}105)$$

where the self-evident result is obtained [cf. Eq. (4-71)]. If $\mathscr{L}^2$ refers to the total angular momentum of n particles, then

$$\mathscr{L}^2 = \mathscr{L}_x^2 + \mathscr{L}_y^2 + \mathscr{L}_z^2 = (\sum_{i=1}^{n} \mathscr{L}_{xi})^2 + (\sum_{i=1}^{n} \mathscr{L}_{yi})^2 + (\sum_{i=1}^{n} \mathscr{L}_{zi})^2 =$$

$$= \sum_{i=1}^{n} \mathscr{L}_i^2 + 2 \sum_{i<j}^{n} \mathscr{L}_{zi} \mathscr{L}_{zj} + \sum_{i<j}^{n} (\mathscr{L}_{+i} \mathscr{L}_{-j} + \mathscr{L}_{-i} \mathscr{L}_{+j}), \qquad (6\text{-}106)$$

where subscripts i and j denote the angular momentum operators of the individual particles. The operator in this form is directly applicable to determinant functions and is currently used in atomic quantum theory.

Analogous relationships apply for the spin momentum operators; the possible values of the spin quantum numbers must, of course, be borne in mind [see Eq. (4-84)]. Thus Eqs. (6-102) and (6-103) for the one-electron spin operators acquire the form

$$\mathscr{S}_+ \alpha = 0 \qquad (6\text{-}107a)$$

$$\mathscr{S}_+ \beta = \hbar \alpha \qquad (6\text{-}107b)$$

$$\mathscr{S}_- \alpha = \hbar \beta \qquad (6\text{-}107c)$$

$$\mathscr{S}_- \beta = 0, \qquad (6\text{-}107d)$$

where $\mathscr{S}_+ = \mathscr{S}_x + i\mathscr{S}_y$ and $\mathscr{S}_- = \mathscr{S}_x - i\mathscr{S}_y$.

For illustration the form of $\mathscr{S}^2$ for a two-electron system will be given:

$$\mathscr{S}^2 = \mathscr{S}_1^2 + \mathscr{S}_2^2 + 2\mathscr{S}_{z1}\mathscr{S}_{z2} + \mathscr{S}_{+1}\mathscr{S}_{-2} + \mathscr{S}_{-1}\mathscr{S}_{+2}, \quad (6\text{-}108)$$

where the numbers denote the dependence of the operators on the particle coordinates. To determine the eigenfunctions of $\mathscr{S}^2$ as linear combination of Slater determinants, the Slater determinant is introduced:

$$|\varphi_1\eta_1, \varphi_2\eta_2| = \frac{1}{\sqrt{2}} \begin{vmatrix} \varphi_1(1)\,\eta_1(1), & \varphi_2(1)\,\eta_2(1) \\ \varphi_1(2)\,\eta_1(2), & \varphi_2(2)\,\eta_2(2) \end{vmatrix} \quad (6\text{-}109)$$

Two cases will be distinguished:

a) $\qquad \eta_1 = \alpha, \; \eta_2 = \alpha \qquad$ or $\qquad \eta_1 = \beta, \; \eta_2 = \beta$

In view of the validity of Eq. (6-107), only three terms yield non-zero contributions when operator (6-108) acts, for example, on the determinant $|\varphi_1\alpha, \varphi_2\alpha|$:

$$\mathscr{S}^2 |\varphi_1\alpha, \varphi_2\alpha| = \hbar^2 \left[\frac{3}{4} + \frac{3}{4} + \frac{2}{4} \right] |\varphi_1\alpha, \varphi_2\alpha| =$$
$$= \hbar^2 \cdot 1(1 + 1) |\varphi_1\alpha, \varphi_2\alpha|, \quad (6\text{-}110)$$

in other words the determinant is an eigenfunction of $\mathscr{S}^2$ with $S = 1$ [cf. Eq. (4-85)] and therefore corresponds to a triplet state*. The following expanded form of the determinant is also worth noting:

$$|\varphi_1\alpha, \varphi_2\alpha| = \frac{1}{\sqrt{2}} [\varphi_1(1)\,\varphi_2(2) - \varphi_1(2)\,\varphi_2(1)]\,\alpha(1)\,\alpha(2) \quad (6\text{-}111)$$

The conclusion would be analogous if both spin functions were β functions. These two determinant functions differ by an eigenvalue with respect to $\mathscr{S}_z$, which, according to Eq. (6-96), has the value $\hbar$ for the first case and $-\hbar$ for the second case.

b) $\qquad \eta_1 = \alpha, \; \eta_2 = \beta \qquad$ or $\qquad \eta_1 = \beta, \; \eta_2 = \alpha$

When applying operator (6-108) to the two determinants, it follows that neither of them is an eigenfunction of $\mathscr{S}^2$. However, on formation of linear combinations, it follows that

$$\mathscr{S}^2 \left\{ \frac{1}{\sqrt{2}} (|\varphi_1\alpha, \varphi_2\beta| \pm |\varphi_1\beta, \varphi_2\alpha|) \right\} =$$
$$= \frac{\hbar^2}{\sqrt{2}} \left\{ \left(\left(\frac{3}{4} + \frac{3}{4} - \frac{2}{4} \pm 1 \right) |\varphi_1\alpha, \varphi_2\beta| + \right.\right.$$
$$\left.\left. + \left(1 \pm \left[\frac{3}{4} + \frac{3}{4} - \frac{2}{4} \right] \right) |\varphi_1\beta, \varphi_2\alpha| \right\}, \quad (6\text{-}112)$$

---

* The quantity $(2S + 1)$ is termed the multiplicity of the state; e.g., $S = 1$ corresponds to a triplet and $S = 0$ to a single state.

where the combination with the positive sign corresponds to the triplet state ($S = 1$) and the combination with the negative sign corresponds to the singlet state ($S = 0$). The two functions have the common property that they are eigenfunctions of $\mathscr{S}_z$ with eigenvalues equal to zero. The expanded form of functions (6-112),

$$\frac{1}{\sqrt{2}} (|\varphi_1\alpha, \varphi_2\beta| \pm |\varphi_1\beta, \varphi_2\alpha|) =$$

$$= \tfrac{1}{2}[\alpha(1)\,\beta(2) \pm \alpha(2)\,\beta(1)]\,[\varphi_1(1)\,\varphi_2(2) \mp \varphi_1(2)\,\varphi_2(1)], \qquad (6\text{-}113)$$

enables comparison with Eq. (6-111), leading to the conclusion that the triplet state consists of the product of a symmetrical spin function and an antisymmetrical spatial function; for the singlet state the symmetries are the reverse.

The given example of a two-electron system demonstrates the method used in studying the symmetry properties of a product function of the Slater type of determinant. The spatial (molecular) orbitals φ_1, φ_2, ..., φ_i, ... in the Slater determinant are assumed to correspond to the irreducible representations Γ_1, Γ_2, ..., Γ_i, ... The symmetry properties of the determinant then correspond to the product representation which is generally reducible and which can be decomposed according to Eq. (6-40):

$$\Gamma_1 \otimes \Gamma_2 \otimes \dots \otimes \Gamma_i \otimes \dots = \Gamma = \sum_j \Gamma_j \qquad (6\text{-}114)$$

The spin function product, which also generates a generally reducible representation,

$$\Gamma_{(1/2)} \otimes \Gamma_{(1/2)} \otimes \dots \otimes \Gamma_{(1/2)} = \sum_S \Gamma_{(S)}, \qquad (6\text{-}115)$$

decomposable into components which correspond to the pure spin states of a many-electron system, can be interpreted similarly. This is, of course, only another way of expressing the vector addition of spin momenta [see Eqs. (4-85) and (4-79)].

The resulting states corresponding to the products of the spatial and spin functions can then be satisfied using the dual notation Γ_j, $\Gamma_{(S)}$, where the first symbol indicates the behaviour of the function under transformation of the space coordinates and the second indicates the properties of the function towards operations affecting the spin coordinates. It is, of course, necessary that the total function satisfy the Pauli principle, i.e. that it be antisymmetrical towards permutation of the electron coordinates.

In order to be able to express the left-hand sides of Eqs. (6-114) and (6-115), the occupation of the one-electron energy levels, i.e. the electron configurations, must be known. The solution will be demonstrated on a particular case, using the benzene molecule in the so-called π-electron approximation, in which only the most freely bonded electrons of the double bonds (see below) are considered. The electronic states are assumed to be given in the form of a linear combination of p orbitals, oriented perpendicularly to the benzene ring plane, i.e. in the orientation given in Fig. 6-5 by the six p_z orbitals located on the nuclei of the carbon atoms. The benzene molecule is considered to be a suitable example since, because of its high symmetry, degenerate one-electron states must be considered, which somewhat complicates the analysis of the symmetry properties of the states.

Six p_z atomic orbitals (denoted p_i where the subscript i indicates the relationship to the atom) form the basis of the reducible representation Γ of the group D_{6h}. Decomposition of this representation using the character table (Table 6-5) leads to following result:

$$\Gamma = A_{2u} + B_{2g} + E_{1g} + E_{2u} \tag{6-116}$$

It must be borne in mind, however, that p_z orbitals are antisymmetric with respect to the $xy(\sigma_h)$ plane. Similarly as in the previous example of an electron in the field of four protons, it is possible to use symmetry considerations to determine the complete form of the molecular orbitals (in non-normalized form):

$$\varphi_1(A_{2u}) = p_1 + p_2 + p_3 + p_4 + p_5 + p_6$$

$$\varphi_2(E_{1g}) = p_1 + p_2 - p_4 - p_5$$

$$\varphi_3(E_{1g}) = p_1 - p_2 - 2p_3 - p_4 + p_5 + 2p_6$$

$$\varphi_4(E_{2u}) = p_1 + p_2 - 2p_3 + p_4 + p_5 - 2p_6 \tag{6-117}$$

$$\varphi_5(E_{2u}) = p_1 - p_2 + p_4 - p_5$$

$$\varphi_6(B_{2g}) = p_1 - p_2 + p_3 - p_4 + p_5 - p_6$$

Any calculation within the framework of the one-electron approximation (for example, using the Hückel method) would afford the sequence of energy levels (from the lowest energy value) $A_{2u}, E_{1g}, E_{2u}, B_{2g}$ (cf. Fig. 6-6). By distributing the six electrons which form the "π-electron" system into one-particle levels, the electron configurations are obtained. States of different spatial symmetry and spin multiplicity can arise from a given electron configuration.

In Fig. 6-6 is given, as the first case, the ground state configuration [cf. Eq. (5-45)] corresponding to the configuration $(A_{2u})^2 (E_{1g})^4$; the

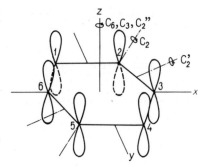

Fig. 6-5. Symmetry operations of the
benzene molecule.

occupation is denoted by the upper indices. In this configuration all the
bonding levels are fully occupied by electrons and the only way of
realizing occupation of the molecular orbitals is complete spin pairing.
For this reason only a single state corresponds to this configuration,
denoted $^1A_{1g}$, where the multiplicity symbol (upper index) and the total
spatial symmetry symbol are combined. Without going into greater
detail, analysis of the doubly excited configuration $(A_{2u})^2 (E_{1g})^2 (E_{2u})^2$,
also given in Fig. 6-6, will be carried out; of all the possible doubly
excited configurations, this one gives rise to the largest number of states
and affords a general description[15] of their determination.

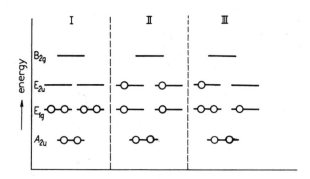

Fig. 6-6. Schematic representation of some electron configurations of the π-electron system
of the benzene molecule.

First it is expedient to divide the total electron system into individual
subgroups which occupy degenerate or non-degenerate energy levels.
These subgroups of electrons correspond to certain electronic states
which can be used for determination of the total electronic states by
forming the direct product of the "partial" electronic states.

The electron pair occupying the A_{2u} level is the first subgroup in the studied doubly excited configuration. Because it is a non-degenerate level according to the Pauli principle, these two electrons have opposite spins and their total spin is zero, i.e. the multiplicity is equal to one. The spatial symmetry of the "partial" state can be expressed as the direct product, for which (cf. Table 6-5)

$$A_{2u} \otimes A_{2u} = A_{1g} \tag{6-118}$$

Thus a single state belongs to the first electronic subgroup, which is expressed symbolically as

$$SG[(A_{2u})^2] = {}^1A_{1g} \tag{6-119}$$

It is easily verified that each level with closed shell character has the same symmetry properties. The second subgroup consists of electrons which partially occupy the degenerate E_{1g} level. The Pauli principle permits two possible states for the total spin, $S = 0$ and $S = 1$, for the pair of electrons in this level, according to whether the electrons have the same or the opposite spin. The situation is similar to the already-described two-electron system starting with Eq. (6-109), where it was found that the triplet state is connected with the antisymmetric space function, whereas the singlet state corresponds to a symmetric space function. It appears that the product representation with basis (6-50) formed from two sets of functions, φ_i, $i = 1, ..., m$, and ψ_i, $i = 1, ..., m$ ($m \geq 2$), which both span the same irreducible representation $\mathbf{A}^{(T)}$, $T \in G$ (so that $\varphi_i \equiv \psi_i$), can be expressed as the direct sum of two representations (not necessarily irreducible), of which one can be combined with the singlet function and the second with the triplet functions. The symmetrical product $(\varphi_i\psi_j + \varphi_j\psi_i)$ spans the first representation, which has the dimension $\frac{1}{2}(m + 1)$, the antisymmetrical product $(\varphi_i\psi_j - \varphi_j\psi_i)$ spans the second representation of dimension $\frac{1}{2}(m - 1)$.

It can now be shown that the characters

$$[\chi_A^2]_S^{(T)} \quad \text{and} \quad [\chi_A^2]_{AS}^{(T)}, \qquad T \in G,$$

connected with the representations defined on the basis of symmetrical (S-) and antisymmetrical (AS-) products, can be expressed in a simple manner. The action of the operator $\mathscr{T}$ of the given symmetry group G, for example, on the general function corresponding to the basis of the S-products can be written according to Eq. (6-49):

$$\mathscr{T}(\varphi_i\psi_j + \varphi_j\psi_i) = \sum_{k=1}^{m} \sum_{l=1}^{m} A_{ki}^{(T)} A_{lj}^{(T)}(\varphi_k\psi_l + \varphi_l\psi_k) =$$

$$= \sum_{k=1}^{m} \sum_{l=1}^{m} A_{kj}^{(T)} A_{li}^{(T)}(\varphi_k\psi_l + \varphi_l\psi_k) =$$

$$= \frac{1}{2} \sum_{k=1}^{m} \sum_{l=1}^{m} (A_{ki}^{(T)} A_{lj}^{(T)} + A_{kj}^{(T)} A_{li}^{(T)})(\varphi_k\psi_l + \varphi_l\psi_k) \qquad (6\text{-}120)$$

It appears from Eqs. (6-50) and (6-51) that in the diagonal element of the product representation $i = k$, $j = l$ and that the character of the representation defined by Eq. (6-120) can thus be expressed as follows:

$$[\chi_A^2]_S^{(T)} = \frac{1}{2} \sum_{i=1}^{m} \sum_{j=1}^{m} (A_{ii}^{(T)} A_{jj}^{(T)} + A_{ij}^{(T)} A_{ji}^{(T)}) \qquad (6\text{-}121)$$

According to Eq. (6-54), the first term in the sum in Eq. (6-121) expresses the square of the character of the representation $\mathbf{A}^{(T)}$, $T \in G$; the second term represents the trace of the matrix product of two matrices $\mathbf{A}^{(T)}$ and therefore corresponds to the character of the matrix $\mathbf{A}^{(V)}$, where $\mathscr{V}$ is the operation corresponding to repeated action of operator $\mathscr{T}$,

$$\mathscr{V} = \mathscr{T}^2 \qquad (6\text{-}122)$$

Therefore, Eq. (6-121) can be written in the final form

$$[\chi_A^2]_S^{(T)} = \tfrac{1}{2}[(\chi_A^{(T)})^2 + \chi_A^{(TT)}] \qquad (6\text{-}123)$$

For the representation corresponding to the AS products, it can similarly be derived that

$$[\chi_A^2]_{AS}^{(T)} = \tfrac{1}{2}[(\chi_A^{(T)})^2 - \chi_A^{(TT)}] \qquad (6\text{-}124)$$

Equations (6-123) and (6-124) enable the symmetric and asymmetric parts of the direct product $E_{1g} \otimes E_{1g}$ to be found. The squares of the characters $(\chi_{E_{1g}}^{(T)})^2$ of the irreducible representation E_{1g} of the group D_{6h} can easily be calculated. They are also given in the lower part of Table 6-5. In determination of the $\chi_{E_{1g}}^{(TT)}$ values, the characters of group D_{6h} are again used, where it is always necessary to determine operation $\mathscr{V}$ according to Eq. (6-122) and to find its respective character. Thus, for example, repetition of operation $\mathscr{C}_6$ leads to operation $\mathscr{C}_3$, repetition of $\mathscr{C}_2$ leads to $\mathscr{E}$, etc. Table 6-5 also gives the resulting values of characters $\chi_{E_{1g}}^{(TT)}$, together with values

$$[\chi_{E_{1g}}^2]_S^{(T)} \qquad \text{and} \qquad [\chi_{E_{1g}}^2]_{AS}^{(T)}$$

determined according to Eqs. (6-123) and (6-124). If the values of the characters in the last line of this table are compared with the values of the characters of representation A_{2g}, they are found to be equal. Since the antisymmetric space product function is combined with the triplet spin state, the conditions for the existence of state $^3A_{2g}$ are fulfilled. The character of the symmetric product representation obviously corresponds to a reducible representation which can be decomposed, using Eq. (6-56), into a direct sum of irreducible representations $(A_{1g} + E_{2g})$. If we bear in mind that these spatial functions are to be combined with the singlet spin functions, the possible states of the second subgroup of electrons can be written in the form

$$SG[(E_{1g})^2] = {}^1A_{1g} + {}^1E_{2g} + {}^3A_{2g} \qquad (6\text{-}125)$$

It could be found, using a similar procedure, that for the third subgroup of electrons the following decomposition is valid

$$SG[(E_{2u})^2] = {}^1A_{1g} + {}^1E_{2g} + {}^3A_{2g} \qquad (6\text{-}126)$$

All the states corresponding to the configuration $(A_{2u})^2 (E_{1g})^2 (E_{2u})^2$ of the six-electron system can be determined by expressing the direct product

$$SG[(A_{2u})^2] \otimes SG[(E_{1g})^2] \otimes SG[(E_{2u})^2] =$$
$$= ({}^1A_{1g} + {}^1E_{2g} + {}^3A_{2g}) \otimes ({}^1A_{1g} + {}^1E_{2g} + {}^3A_{2g}), \qquad (6\text{-}127)$$

where it has already been taken into account that the direct product is not affected by multiplication by the totally symmetric representation $^1A_{1g}$. Multiplication and further decomposition yields

$$^1A_{1g} + {}^1E_{2g} + {}^3A_{2g} + {}^1E_{2g} + {}^1E_{2g} \otimes {}^1E_{2g} + {}^1E_{2g} \otimes {}^3A_{2g} +$$
$$+ {}^3A_{2g} + {}^3A_{2g} \otimes {}^1E_{2g} + {}^3A_{2g} \otimes {}^3A_{2g} = \qquad (6\text{-}128)$$
$$= {}^5A_{1g} + {}^3A_{1g} + 3\,{}^1A_{1g} + 2\,{}^3A_{2g} + {}^1A_{2g} + 2\,{}^3E_{2g} + 3\,{}^1E_{2g},$$

where the rules for vector addition of spin momenta [cf. Eqs. (4-85) and (4-79)] according to which it is possible to decompose two "partial" triplet states into "total" quintet, triplet and singlet states, were also employed.

Analysis of the given configuration, considering all possible states, is so much more complicated than the other cases that search for all possible states of the remaining configurations is, by contrast, an easy matter. Thus, for example, the states which can be derived from the direct product

$$^2E_{1g} \otimes {}^2E_{2u} = {}^3B_{1u} + {}^3B_{2u} + {}^3E_{1u} + {}^1B_{1u} + {}^1B_{2u} + {}^1E_{1u} \qquad (6\text{-}129)$$

correspond to the singly excited configuration $(A_{2u})^2 (E_{1g})^3 (E_{2u})^1$, also schematically represented in Fig. 6-6. The attentive reader has certainly noticed a certain difference in direct multiplication amongst cases where the symmetry "inside" the degenerate level is being studied and cases where the direct product of "partial" states of electron subgroups is expressed [cf., for example, Eqs. (6-126) and (6-129)]. This difference follows from the validity of the Pauli principle and is manifested in the fact that, in the first case, all possible spatial symmetries with all possible spin states cannot be realized, as they are in the second case.

The considerations given so far include almost all cases which we might encounter when studying the symmetry properties of configurations, except for the electron configuration $(T)^3$, where T denotes one of the triply degenerate irreducible representations occuring in some groups of high symmetry. It is sufficient to state here that they can be solved similarly as in the determination of the states of atoms, which makes use of the shift operators which were already introduced in Eqs. (6-98) and (6-107).

6.8 Perturbation treatment for symmetrical systems

In this section a few remarks will be given, supplementing the discussion of the perturbation method in Section 4.6, from the viewpoint of symmetry relations between solution of the original and the perturbed system.

The basic equation of the perturbation treatment [Eq. (4-145)], where the Hamiltonian of the investigated system $\mathscr{H}$ can be separated into the Hamiltonian of the unperturbed system $\mathscr{H}^\circ$ and the perturbation term $\mathscr{V}$, can be written

$$\mathscr{H} = \mathscr{H}^\circ + \lambda\mathscr{V}, \tag{6-130}$$

where λ is only a parameter. Roughly speaking, $\mathscr{H}^\circ$ corresponds to the approximate solution of the problem, ignoring a number of finer effects which are included in contribution $\mathscr{V}$. It can thus be expected that operator $\mathscr{H}^\circ$ has a higher degree of symmetry than Hamiltonian $\mathscr{H}$ and that the group of symmetry transformations G of $\mathscr{H}$ will, therefore, be a subgroup of the group of symmetry transformations $G_\circ$ of the unperturbed Hamiltonian, $\mathscr{H}^\circ$.

Let us consider an eigensolution of the unperturbed Hamiltonian $\mathscr{H}^\circ$ corresponding to a g-fold degenerate energy level E_i°. According to Theorem 6-1 in Section 6.4, the corresponding eigenfunctions $\psi_1^\circ, \psi_2^\circ, ..., \psi_g^\circ$ (cf. last part of Section 4.6) span the irreducible representation Γ° of group $G_\circ$. The decrease in symmetry under the influence of perturbation $\mathscr{V}$ can result in Γ° becoming reducible with respect to the new group G:

$$\Gamma^\circ = \Gamma_1 + \Gamma_2 + ... + \Gamma_m \tag{6-131}$$

148

If it is decomposed into m irreducible representations of group G, then this phenomenon can be interpreted in physical terms by stating that the original energy level, E_i°, is split by the perturbation into m new levels which can be classified according to the irreducible representations of group G. The same result would, of course, be obtained by analysis of the matrix elements of the secular determinant (4-160) from the viewpoint of selection rules [Eq. (6-65)] if the matrix elements of operator $\mathscr{V}$ were expressed in terms of the symmetry functions corresponding to group G.

A practical application of these general considerations can be demonstrated on one of the quantum chemical methods for calculation of the properties of inorganic complexes, the crystal field theory which is based on the model given below. The Hamiltonian of a free atom, considering only electrostatic interactions, is invariant under the simultaneous rotation of the coordinates of all the electrons. This type of symmetry of the Hamiltonian leads to degeneracy of the atomic terms, so that, for example, the energy level of a single electron in the d state is five-fold degenerate — since there are five different d functions. If the atom is exposed to the effect of ligands (i.e. of chemically bonded adjacent atoms) and if the resulting configuration of nuclei corresponds to symmetry group G, the original spherical symmetry of the atom is disturbed and the original degeneracy will be resolved or "lifted". Quantum numbers L and M_L cease to be valid and are replaced by quantum numbers Γ and m_Γ, where Γ denotes the irreducible representation of G and m_Γ denotes a component of the multidimensional irreducible representation Γ. It was found, for example, in Section 6.6 in the discussion of the construction of hybrid orbitals that, if the atom is placed in a ligand field of octahedral symmetry (see Fig. 6-4), splitting of the atomic degenerate d states into two new states, corresponding to irreducible representations E and T of group O, occurs. Thus the originally five-fold degenerate level is split into two new energy levels, one three-fold degenerate and the other doubly degenerate.

REFERENCES

1. Heine V.: *Group Theory in Quantum Mechanics*. Pergamon Press, London 1960.
2. Smirnov V. I.: *Kurs vysshej matematiki*, vol. III, part 1, Gos. izd. fiz.-mat. lit., Moscow 1958, p. 288.
3. Smirnov V. I.: *Kurs vysshej matematiki*, vol. III, part 1, Gos. izd. fiz.-mat. lit., Moscow 1958, p. 278.
4. Eyring H., Walter J., Kimball G. E.: *Quantum Chemistry*. Wiley, New York 1944.
5. Davtyan O. K.: *Kvantovaya chimiya*. Gos. izd. "Visshaya shkola", Moscow 1962.

6. Ballhausen C. J.: *Introduction to Ligand Field Theory*. McGraw-Hill, New York 1962.
7. Griffith J. S.: *The Theory of Transition-Metal Ions*. Cambridge Univ. Press, 1961.
8. Vlček A. A.: *Struktura a vlastnosti koordinačnich sloučenin*. Academia, Praha 1966.
9. McWeeny R.: *Symmetry*. Pergamon Press, Oxford, The Macmillan Company, New York 1963.
10. Altmann S. L.: *Group Theory in Quantum Theory*, vol. II, Aggregates of Particles (ed. Bates D. R.). Academic Press, New York 1962, p. 87.
11. Jaffé H. H., Orchin M.: *Symmetry in Chemistry*. Wiley, New York 1965; Orchin M., Jaffé H. H.: *Symmetry, Orbitals and Spectra*. Wiley-Interscience, New York 1971.
12. Cotton F. A.: *Chemical Applications of Group Theory*. Wiley, New York 1963.
13. Pauling L.: *The Nature of the Chemical Bond*. Cornell University Press, Ithaca 1939.
14. Kimball G. E.: *J. Chem. Phys.* **8**, 188 (1940).
15. Ellis R. L., Jaffé H. H.: *J. Chem. Educ.* **48**, 92 (1971).

7. ATOMIC ORBITALS (AO) AND MOLECULAR ORBITALS (MO)

7.1 The significance of hydrogen-type orbitals; atomic orbitals

The solution of the Schrödinger equation for the hydrogen atom can rarely be used directly in more complex chemical problems. This solution nevertheless forms a basis for the study of more complicated atoms and even for molecules. The possible modes of graphical representation of the radial and angular parts of hydrogen-type functions – the atomic orbitals – have already been described. Thus only a few remarks will be given here in this connection:

(i) The application of computers permits information on the graphical representation of complete AO's to be obtained[1].

(ii) Frequently, the graphical representation of the angular part is particularly useful. It should be noted that the contours in Fig. 3-10b indicate the regions in space in which the electron can be found (with the given probability). This figure is not to be understood as describing a "smearing out" of the electron charge in space.

(iii) The sign of the wave function (+ or −) in the individual AO parts must be specified. This is important when analyzing the symmetry of the studied formations and in the calculation of some integrals. The sign of the wave function is, however, of no physical importance (in the sense of comparison with a physical quantity).

In qualitative considerations it is often expedient to form molecular orbitals – one-electron functions distributed over the entire molecule – on the basis of the principle of effective overlap of the atomic orbitals. In this connection not only the atomic orbitals themselves (hydrogen-type wave functions) but also linear combinations thereof, called hybrid orbitals [linear combinations of orbitals corresponding to a single atom (cf. Section 6.6)] are employed.

151

7.2 Hybridization

The concept of AO hybridization is of particular importance in explaining the spatial arrangement of inorganic and organic molecules. The concepts of hybridization and hybrid orbitals have already been introduced in Section 6.6, where hybrid orbitals were constructed on the basis of symmetry considerations using group theory. In this section the physical aspect of the problem will be discussed; hybridization in the carbon atom will be mentioned in greater detail, as this atom is the basic building unit of an important and extensive group of organic compounds and because it is an especially instructive case.

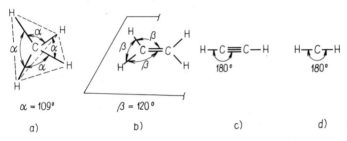

$\alpha = 109°$ $\beta = 120°$

a) b) c) d)

Fig. 7-1. Various types of carbon compounds: (a) methane, (b) ethylene, (c) acetylene, (d) carbene.

First a survey of bonding types in organic compounds will be given (Fig. 7-1). The carbon atom in the ground state (^3P) corresponds to a σ-double-bonded atom, these bonds being perpendicular. These conditions apparently do not exist in any of the mentioned molecules. An explanation for this specific situation can be achieved by assuming the carbon atom to be in an excited state (^5S; Fig. 7-2) and by forming four new orbitals as linear combinations of the four original singly occupied orbitals ($2s$, $2p_x$, $2p_y$, $2p_z$). These new orbitals, called hybrid orbitals, are equivalent and possess quite different directional properties than the

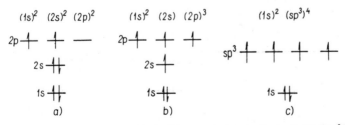

a) b) c)

Fig. 7-2. States of the carbon atom: (a) the ^3P ground state (σ divalent), (b) the ^5S excited state (σ tetravalent non-hybridized), (c) the sp³ hybridized state (σ tetravalent).

initial orbitals. Their spatial arrangement can be depicted by placing the nucleus of the atom in the centre of a regular tetrahedron so that the individual hybrid orbitals are directed toward its corners. These orbitals are called sp^3 or tetragonal orbitals. The four hybrid orbitals are sometimes denoted te_1, te_2, te_3, te_4. They have the following form (Fig. 7-3):

$$te_1 = \tfrac{1}{2}(s + p_x + p_y + p_z)$$
$$te_2 = \tfrac{1}{2}(s + p_x - p_y - p_z)$$
$$te_3 = \tfrac{1}{2}(s - p_x + p_y - p_z)$$
$$te_4 = \tfrac{1}{2}(s - p_x - p_y + p_z)$$

$$(7\text{-}1)$$

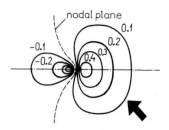

Fig. 7-3. "Contours" of electron densities for the sp^3 hybrid orbital: nodal surface (----), symmetry axis (-·-·-). The thick arrow indicates the part which is usually depicted (considerably deformed) in graphical representations (see Fig. 7-4).

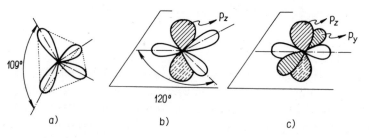

Fig. 7-4. Hybrid orbitals of the carbon atom (not cross-hatched): (a) sp^3, (b) sp^2, (c) sp. Non-hybridized orbitals are cross-hatched (below the indicated plane by dashed lines).

The four hybrid sp^3 orbitals are depicted in Fig. 7-4; for lucidity only the larger parts of the orbitals* in which the presence of the electron is more probable are given. These hybrid orbitals are used for the theoretical description of saturated organic compounds. For the description of conjugated compounds, trigonal sp^2 hybridization and linear (digonal) sp hybridization (Fig. 7-4b, c) are of particular importance. As

---

* On the axis of this larger part but in the opposite direction lies a smaller part of the orbital (separated by a nodal plane), which is, however, not shown for the sake of simplicity (cf. Fig. 7-3).

will be shown later, hybrid orbitals (sp^2, sp) are used for the description of C−C and C−H σ-bonds and the remaining non-hybridized orbitals (p$_z$; p$_z$ and p$_y$) are used for the description of π bonds, which are responsible for the characteristic behaviour of conjugated compounds (conjugated double or triple bonds) (Table 7-1).

Table 7-1

Survey of Hybridization of the Carbon Atom Orbitals[a]

s	p$_x$	p$_y$	p$_z$	Hybrid orbitals (number)	Non-hybrid orbitals	Θ[b]
1	1	1	1	sp^3 (4)	none	109°
1	1	1	1	sp^2 (3)	p$_z$	120°
1	1	1	1	sp (2)	p$_y$, p$_z$	180°
1	1	1	1	−	all	−

[a] Framed values are AO's used for construction of the hybrid orbitals.
[b] The angle formed by the hybrid orbitals.

Table 7-2

Survey of Hybridization in the Central Atom in Complexes

Geometrical arrangement	Coordination number (number of hybrid orbitals)	Hybrid orbital
linear	2	sp
		dp
nonlinear	2	ds
trigonal plane (120° angles)	3	sp^2
		dp^2
trigonal pyramid	3	p^3
		d^{2}p
tetrahedron	4	sp^3
		d^{3}s
square	4	dsp^2
		d^{2}p^2
trigonal bipyramid	5	dsp^3
		d^{3}sp
pentagonal plane	5	d^{3}p^2
octahedron	6	d^{2}sp^3
dodekahedron	8	d^{4}sp^3

154

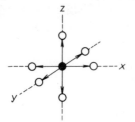

Fig. 7-5. Model of an octahedral
complex: central atom ●, ligands ○.
The d^2sp^3 hybrid orbitals are
represented by arrows.

In connection with transition metal compounds, the coordination
number and the geometric properties of the central atom are of interest
(Table 7-2), cf. Sections 6.6 and 10.6. Most important is octahedral d^2sp^3
hybridization, in which the individual hybrid orbitals are directed (towards
the ligands) in the direction of the axes of rectangular space coordinates
(Fig. 7-5), cf. Sections 6.6 and 10.6. For example Eq. (6-94) is valid for the
hybrid orbital oriented in the direction of the positive part of the z-axis.

7.3 Molecular orbitals

In Sections 5.5 and 5.6 the concept of the one-electron wave function
(a function depending upon the coordinates of a single electron), describing
the electronic state in molecules within the framework of the one-electron
approximation, was introduced. These functions are called molecular
orbitals and were denoted by symbol φ. An effective procedure for
construction of these MO's from atomic orbitals in the form of a linear
combination (LC) of atomic orbitals χ (hence the abbreviated name of the
method MO-LCAO) was also shown:

$$\varphi = \sum_{\mu=1}^{m} c_\mu \chi_\mu \qquad (7\text{-}2)$$

In this equation c_μ denotes parameters to be determined and the χ_μ's are
the individual AO's that are considered when forming the MO. The
assumed shape of the molecular orbitals has been discussed in detail in
Section 5.6.

The atoms combine to form molecules, provided this process is
connected with a decrease in the total energy. It appeared that the
process of covalent bond formation can be described by two apparently
very different methods – the MO method and the valence bond (VB)
method. It can be shown, however, that these methods are equivalent in
many ways (cf. Section 10.5). For this reason and also because the
application of the VB method is much more complicated with more
complex molecules, only the MO method will be discussed here.

Covalent bonds have been shown to be formed when an effective overlap of the atomic orbitals occurs (the conditions for the formation of ionic bonds will be discussed later). Therefore this property can be used for a qualitative estimation of the bonding conditions in molecules without performing calculations based on the variation or perturbation treatments.

A quantitative measure of the overlap of the atomic orbitals can be obtained using the overlap integral. Since we are chiefly interested in the overlap of atomic orbitals, the overlap integral $S_{\mu\nu}$ can be defined as

$$S_{\mu\nu} = \int \chi_\mu^* \chi_\nu \, d\tau, \tag{7-3}$$

where χ_μ and χ_ν are atomic orbitals localized on the same atom or on two different atoms. The overlap integral can be defined, of course, as well as for χ_μ and χ_ν also for other types of functions. When the functions in the integrand are identical, Eq. (7-3) becomes the normalization condition

$$\int |\chi_\mu|^2 \, d\tau = 1 \tag{7-4}$$

The value of integral (7-3) lies within the limits, -1, 1. It is important for the formation of the chemical bond that $S_{\mu\nu}$ be positive and if the bond is to be sufficiently strong, $S_{\mu\nu}$ must be relatively large, as it is assumed, according to the principle of maximum overlap, that the strength of the bond is directly proportional to the value of the overlap integral.

In contrast to the calculation of the other integrals which were discussed earlier (matrix elements), calculation of the overlap integrals should, in principle, present no difficulties (cf. Section 9.3). The calculation need not be carried out, however, for atoms of the first series of the periodic table of elements, since tabulated values[3] of overlap integrals for Slater-type orbitals (cf. Section 8.1) are available.

If the overlap between two atomic orbitals is relatively large, the formation of molecular orbitals

$$\varphi = c_1 \chi_1 + c_2 \chi_2 \tag{7-5}$$

describing a strong bond can be assumed. By means of linear combinations exactly the same number of independent new functions can be formed from m original functions. Two molecular orbitals can be formed from two atomic orbitals. In order to retain a general approach and yet to be able to work with simple expressions, the expansion coefficients will be denoted by two subscripts, generally $c_{\mu i}$, where the first subscript (μ) denotes the atomic orbital and the second subscript (i) denotes the molecular orbital. Thus the two molecular orbitals can be written as

$$\varphi_1 = c_{11}\chi_1 + c_{21}\chi_2 \qquad (7\text{-}6)$$

$$\varphi_2 = c_{12}\chi_1 + c_{22}\chi_2 \qquad (7\text{-}7)$$

Now the conditions for the formation of effective overlap can be specified. The approach of the orbitals to a small distance (mostly 0.1 to 0.2 nm) is a necessary condition (Fig. 7-6). To obtain a bonding overlap, the AO's must be of the same symmetry with respect to rotation or reflection (Fig. 7-7). If one of the AO's is symmetric and the other antisymmetric, the resulting overlap is zero (Fig. 7-7)[2,4]. If the value of the overlap integral is to be sufficiently large, it is necessary, of course, that both atomic orbitals have sufficiently large values in the same region of space (i.e. their product as a function of the space coordinates must not have a value close to zero). This condition is connected with the spatial distribution of the two participating atomic orbitals and it can be expected, considering the radial distribution of atomic orbitals, that

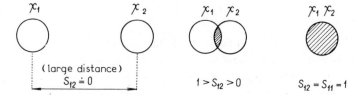

Fig. 7-6. Dependence of the overlap integral S_{12} between two 1s (χ_1, χ_2) atomic orbitals on the interorbital distance.

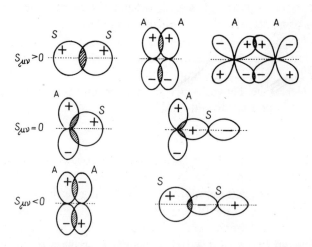

Fig. 7-7. Examples of effective ($S_{\mu\nu} > 0$) and ineffective ($S_{\mu\nu} = 0$, $S_{\mu\nu} < 0$) overlap of s, p and d atomic orbitals, from the viewpoint of covalent bond formation. Letters S and A denote orbitals symmetrical and antisymmetrical with respect to rotation or reflection.

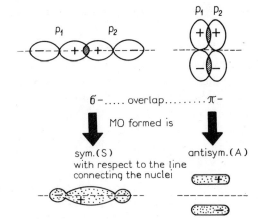

Fig. 7-8. The overlap of two p orbitals in π and σ orientation.

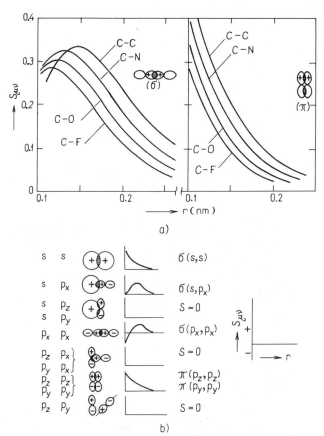

a)

b)

Fig. 7-9. (*a*) Dependence of the $S_{\mu\nu}$ overlap integrals on interatomic distance r for four bonds: σ and π overlap. (*b*) The shape of dependencies of $S_{\mu\nu}$ on r for various types of overlaps between the s and p orbitals (cf. Fig. 7-7).

the situation will be favourable if the energies of the two atomic states are similar.

It appears that p-type orbitals (and higher orbitals) can overlap in two ways, as shown in Fig. 7-8 where the forms of the resulting MO's and their symmetries are also indicated. According to the symmetry, two types of MO can be distinguished: σ-MO's (symmetric with respect to reflection in the plane in which the two atomic nuclei lie) and π-MO's (antisymmetric). A further important difference is that the orbitals of the first group can generally be transformed into two-centre orbitals (they

Table 7-3

Survey of Effective and Ineffective Overlap between χ_μ and χ_ν Orbitals (the x-axis is the molecular axis; from Ref. 2)

χ_μ	χ_ν	
	Overlap	
	effective	ineffective
s	$s, p_x, d_{x^2-y^2}, d_{z^2}$	$p_y, p_z, d_{xy}, d_{xz}, d_{yz}$
p_x	$s, p_x, d_{x^2-y^2}, d_{z^2}$	$p_y, p_z, d_{xy}, d_{xz}, d_{yz}$
p_y	p_y, d_{xy}, d_{xz}	$s, p_x, p_z, d_{yz}, d_{x^2-y^2}, d_{z^2}$
d_{xy}	p_y, d_{xy}	$s, p_x, p_z, d_{yz}, d_{xz}, d_{x^2-y^2}, d_{z^2}$
d_{yz}	d_{yz}	$s, p_x, p_y, p_z, d_{xy}, d_{xz}, d_{x^2-y^2}, d_{z^2}$
$d_{x^2-y^2}$	$s, p_x, d_{x^2-y^2}, d_{z^2}$	$p_y, p_z, d_{xy}, d_{xz}, d_{yz}$

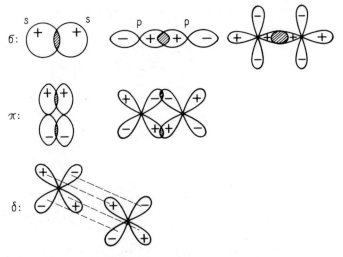

Fig. 7-10. Types of overlap of s, p and d atomic orbitals leading to σ, π and δ type molecular orbitals.

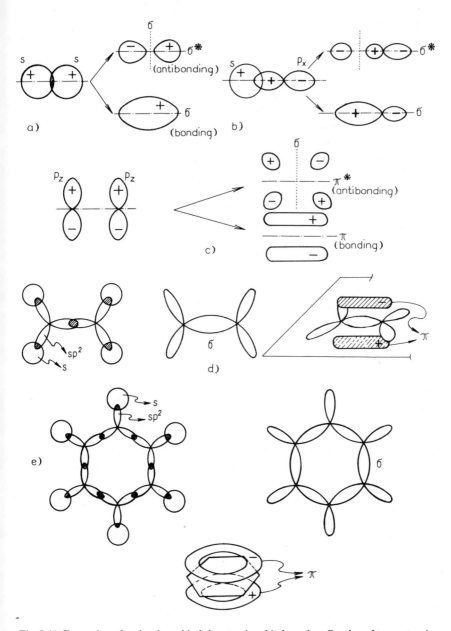

Fig. 7-11. Formation of molecular orbitals by atomic orbital overlap. Overlap of two s atomic orbitals (*a*). Overlap of s and p atomic orbitals (*b*). Overlap of the p_z atomic orbitals of ethylene (*c*). Formation of the σ skeleton of ethylene by overlap of the s and sp^2 atomic orbitals; the bonding π-molecular orbital of ethylene (*d*). Formation of the σ skeleton of benzene by overlap of the s and sp^2 atomic orbitals. In the last figure the lowest energy π-molecular orbital of benzene is indicated.

are located mainly in the region between the two atomic nuclei) whereas it is typical for π-MO's that they are polycentric and often smeared out over a large number of atoms. They are therefore called delocalized MO's (in contrast to localized MO's). In Fig. 7-9 the dependence of the overlap integrals (for overlap of the σ-and π-type orbitals) on the distance between the atoms is depicted. Finally the effective $(S > 0)$ and the ineffective $(S = 0)$ AO combinations (ineffectiveness caused by symmetry) are summarized in Table 7-3. If the overlap integral equals zero, then the respective functions are said to be orthogonal. Finally, examples of overlap leading to σ, π, and δ molecular orbitals are given in Fig. 7-10 (cf. Section 9.4).

Fig. 7-11 indicates the form of the molecular orbitals (bonding and antibonding, cf. Sections 9.1 and 9.4) for several combinations of atomic orbitals. For several simple hydrocarbons, the formation of bonding MO's (the antibonding MO's are not given) of the σ and π types, by overlap of various AO's, is depicted schematically.

REFERENCES

1. Cromer D. T.: *J. Chem. Educ.* **45**, 626 (1968).
2. Coulson C. A.: *Valence*. Clarendon Press, Oxford 1952.
3. Mulliken R. S., Rieke C. A., Orloff D., Orloff H.: *J. Chem. Phys.* **17**, 1248 (1949).
4. Heilbronner E., Bock H.: *The HMO-Model and Its Application*. Wiley, New York and Verlag Chemie, Weinheim 1976. (Three volumes.)

8. MANY-ELECTRON ATOMS

8.1 The one-electron approximation and the periodic system of the elements[1-3]

A system consisting of a nucleus of charge $+Ze$ and n electrons (1, 2, ..., i, j, ..., n) will be chosen as a model. Two types of interactions occur: mutual repulsion of the electrons and attraction between the nucleus and the electrons. The expression for the potential energy of the i-th electron assumes the form

$$V_i(\mathbf{r}_1, \mathbf{r}_2, ..., \mathbf{r}_n) = -\frac{Ze^2}{4\pi\varepsilon_0 |\mathbf{r}_i|} + \sum_{j(\neq i)}^{n} \frac{e^2}{4\pi\varepsilon_0 |\mathbf{r}_i - \mathbf{r}_j|}, \qquad (8\text{-}1)$$

where $\mathbf{r}_i$ is the position vector of the i-th electron. It is assumed that the immobile atomic nucleus is located at the origin of the coordinate system. The two-electron part of the operator can be simplified (cf. Section 5.5) if the i-th electron is assumed to be exposed to the effect of an averaged potential of all the remaining electrons: V_i then depends only on $\mathbf{r}_i$:

$$V_i(\mathbf{r}_i) \approx -\frac{Ze^2}{4\pi\varepsilon_0 |\mathbf{r}_i|} + \sum_{j(\neq i)}^{n} \frac{e^2}{4\pi\varepsilon_0 |\mathbf{r}_i - \mathbf{r}_j|} \qquad (8\text{-}2)$$

A further assumption of averaging in all directions in space can be introduced and a spherically symmetric potential $V_i(r_i)$ (r_i is no longer a vector) is then obtained. These two assumptions permit functions for higher atoms to be obtained in the same form as for the hydrogen atom. The Schrödinger equation then has the form

$$\left\{ \sum_i \Delta_i + \frac{2m}{\hbar^2} [E - \sum_i \mathscr{V}_i(r_i)] \right\} \Psi = 0 \qquad (8\text{-}3)$$

Hartree and Fock[2] introduced the reduction of the many-electron Schrödinger equation to this form and suggested a technique for its solution — the self-consistent field method (SCF method), the general theory of which was explained in Section 5.5. It should be noted that the one-electron functions sought are atomic functions, the atomic orbitals.

A great advantage of this procedure using the one-electron approximation lies in the fact that the solution (orbital) can be expressed in the form

$$\Psi = R(r)\, Y(\Theta, \Phi) \tag{8-4}$$

The atomic SCF orbitals are often algebraically expressed in the form

$$\Psi = \left(\sum_{n,\xi} c_{n\xi} r^{n-1} e^{-\xi r} \right) Y_{lm}(\Theta, \Phi) \tag{8-5}$$

The symbols n, l and m denote the quantum numbers and, similarly as in hydrogen AO's, the principal quantum number n is connected with the number of nodal planes, for $r = 0$ to ∞, the total number being $(n - l - 1)$; here, however, the electron energy is not a simple function of the principal quantum number. Manipulation of these functions is considerably simplified if the summation in Eq. (8-5) is confined to a single term

$$\Psi = c_{n\xi} r^{n-1} e^{-\xi r} Y_{lm}(\Theta, \Phi), \tag{8-6}$$

where $c_{n\xi}$ is the normalization constant, n is the principal quantum number and

$$\xi = \frac{Z - S}{n} \tag{8-7}$$

In this equation Z denotes the atomic number and S is the screening constant.

The electrons are divided into three groups according to their principal quantum number (considering only $n = 1, 2, 3$):

$$1s$$
$$2s, 2p$$
$$3s, 3p, 3d$$

Inside the individual groups, the radial part of the function is the same for all orbitals. Quantity S is calculated for an electron with a certain quantum number n as follows:

1. Electrons in the higher shells are ignored.

2. A contribution of 0.35 is taken for each electron in the same shell; for the 1s electron this value is taken to be 0.30.

3. Contributions from electrons in the lower shells are included in such a way that for the s and p electrons a value of 0.85 is attributed to the nearest lower shell and 1.00 for each electron in the farther shell. For d electrons the value 1.00 is taken for each internal electron.

According to these rules, for example, this quantity for the 1s orbital of the hydrogen atom has the value

$$_H\xi_{1s} = \frac{1-0}{1} = 1$$

and for occupied orbitals of the carbon atom

$$_C\xi_{1s} = \frac{6-0.30}{1} = 5.70$$

$$_C\xi_{2s} = {_C\xi_{2p}} = \frac{6-2\times0.85-3\times0.35}{2} = 1.625$$

These rules are termed *the Slater rules* and are widely applied in quantum chemistry (but must not be confused with the rules introduced by the same author for calculation of the matrix elements of the Hamiltonian). The functions constructed according to these rules are called Slater orbitals.

The periodicity of the ordering of the elements (*Mendeleev's periodic system*) can easily be understood by considering the arrangement of the energy levels in the atom and by recalling that, according to the Pauli principle, no two electrons in the same atom may have all four quantum numbers (n, l, m, s) identical.

For an atom with one electron, the orbital energies are given merely by the principal quantum number so that, for example, the same energy level corresponds to the 2s and 2p atomic orbitals. The transition from the potential of a point charge to the general spherically symmetric potential in the Hamiltonian is manifested by removal of the degeneracy in states which have the same principal quantum number (Fig. 3-15).

In the ground state of atoms, the electrons occupy the atomic orbitals of the lowest energy (the Aufbau principle) according to the Pauli principle. The np orbitals correspond to higher energy than do the ns orbitals, but this is still substantially lower than that corresponding to the $(n + 1)$s orbitals. The electrons in the nd orbitals have roughly the same energy as electrons in the $(n + 1)$s orbitals.

After occupation of the 3s and 3p atomic orbitals, the 4s level is occupied rather than the 3d level; after filling the 4s orbital (K, Ca; in K the 3d level lies about 2.7 eV higher than the 4s level), electrons fill the 3d level. Elements with incompletely filled 3d levels are called transition elements. The small energy difference between the 3d and 4s states is demonstrated by the fact that, for example, chromium does

Table 8-1

Mendeleev's Periodic Table of the Elements [Ref. 1]

Main groups

	1	2a	3b	4b	5b	6b	7b	0
1s	H							He
2s	Li	Be						
2p			B	C	N	O	F	Ne
3s	Na	Mg						
3p			Al	Si	P	S	Cl	Ar
4s	K	Ca						
4p			Ga	Ge	As	Se	Br	Kr
5s	Rb	Sr						
5p			In	Sn	Sb	Te	I	Xe
6s	Cs	Ba						
6p			Tl	Pb	Bi	Po	At	Rn
7s	Fr	Ra						

Transition metals

	3a	4a	5a	6a	7a	8			1b	2b
3d	Sc	Ti	V	Cr	Mn	Fe	Co	Ni	Cu	Zn
4d	Y	Zr	Nb	Mo	Tc	Ru	Rh	Pd	Ag	Cd
5d	Lu	Hf	Ta	W	Re	Os	Ir	Pt	Au	Hg

Rare earths

4f	La	Ce	Pr	Nd	Pm	Sm	Eu	Gd	Tb	Dy	Ho	Er	Tm	Yb

Actinides

5f	Ac	Th	Pa	U	Np	Pu	Am	Cm	Bk	Cf	Es	Fm	Md	No

Table 8-2

Approximate Differences in Energy between the 2s and 2p One-Electron States in eV [Ref. 1]

	Li	Be	B	C	N	O	F
$E(2p) - E(2s)$	1.9	2.7	3.6	4.2	10.9	15.6	20.8

not have the expected configuration $4s^2 3d^4$, but $4s3d^5$. After occupation of the 3d orbitals (Sc ... Zn) the 4p levels (Ga ... Kr) are gradually filled and the process is repeated in the 5s, 4d and 5p levels. It is characteristic for the second series of transition elements (Y ... Cd) that the 4d AO's are incompletely occupied. After occupation of the 6s orbitals (Cs, Ba), the 4f level is gradually filled; these are the rare earth elements (La ... Yb). A similar situation at the 5f level exists for the actinoids (Ac ... No). These relationships are summarized in Table 8-1. Finally, in order to obtain a more quantitative notion, the differences in the energies of one-electron 2p and 2s states are given in Table 8-2 for the elements of the first series of the periodic system.

8.2 The total angular momentum[4,5]

The introduction of the total electron angular momentum in atomic systems has already been discussed in several chapters (cf. Sections 4.4 and 6.7). Therefore, only the important conclusions from the above text and several applications of the rule of vector addition of the individual angular momenta to give the total angular momentum will be given here.

The magnitude of the orbital angular momentum is given by $\hbar \sqrt{[L(L + 1)]}$; the resulting orbital quantum number L can be obtained from the values of the quantum numbers of the individual electrons by vector addition. Only electrons in incomplete shells contribute (for example, in the ground state of the sodium atom only one of eleven electrons need be considered; there is no contribution from the $1s^2$, $2s^2$ and $2p^6$ electrons). It should be added that the magnitude of the orbital angular momentum of the individual electrons is $\hbar \sqrt{[l(l + 1)]}$.

Quantum number L defines the energy of atomic states, which are called terms; we denote the individual terms by capital letters according to the values of L as follows:

L: 0 1 2 3 4 ...
notation of the state: S P D F G ...

The analogy with the symbols used for the individual electrons (s, p, d, ...) is obvious. The component of the angular momentum in the direction of the z-axis equals $\hbar M_L$; the only possible values of quantum number M_L are $L, L-1, ..., 0, ..., -L$.

Figure 8-1 depicts the vector addition[3] (i.e. the calculation of L) for p^2, p^3 and d^2 configurations. The addition for p^2 (Fig. 8-1a) is clear: three states arise, D, P and S. It is evident from Fig. 8-1b that p^3 is considered as $p^2 + p$. Many more terms appear here: F, D (twice), P (three times) and S.

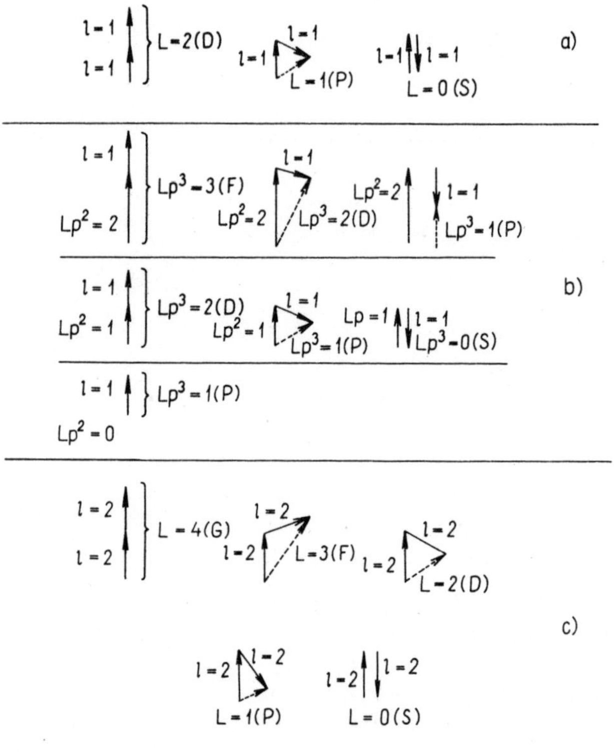

a)

$l=1$ $l=1$ $\}$ $L=2$(D) $l=1$ $l=1$ $L=1$(P) $l=1$ $l=1$ $L=0$(S)

b)

$l=1$ $Lp^2=2$ $\}$ $Lp^3=3$(F) $l=1$ $Lp^2=2$ $Lp^3=2$(D) $Lp^2=2$ $l=1$ $Lp^3=1$(P)

$l=1$ $Lp^2=1$ $\}$ $Lp^3=2$(D) $l=1$ $Lp^2=1$ $Lp^3=1$(P) $Lp=1$ $l=1$ $Lp^3=0$(S)

$l=1$ $Lp^2=0$ $\}$ $Lp^3=1$(P)

c)

$l=2$ $l=2$ $\}$ $L=4$(G) $l=2$ $l=2$ $L=3$(F) $l=2$ $l=2$ $L=2$(D)

$l=2$ $l=2$ $L=1$(P) $l=2$ $l=2$ $L=0$(S)

Fig. 8-1. Determination of quantum number L for (a) two p electrons, (b) three p electrons, (c) two d electrons.

In the graphical representation the procedure is simplified in that vectors of length l are employed and the result amounts to L.

The same rules apply for the addition of the electron spins to give the total spin momentum. For a system with two electrons, for example,

$S = 1$, if the spins are parallel, and

$S = 0$, if the spins are antiparallel.

For a system of three electrons,

$S = \frac{3}{2}$, if the spins are parallel, and

$S = \frac{1}{2}$, if one spin is antiparallel to the other two.

Thus, for two p electrons, six terms come into consideration:

$$^3D, \, ^1D, \, ^3P, \, ^1P, \, ^3S, \, ^1S$$

These states are allowed only for two non-equivalent electrons (e.g., 3p and 2p); if the electrons are equivalent (e.g. $2p^2$) then in consequence of the Pauli exclusion principle (cf. Sections 4.4 and 6.7) the only allowed states are

$$^1D, \, ^3P, \, ^1S$$

If the spin-orbit coupling term is included in the Hamiltonian [cf. Eq. (4-88) and Section 5.2] L ceases to be a "good" quantum number and it is necessary to introduce (cf. Section 4.4) the total angular momentum, the magnitude of which is

$$\hbar\sqrt{[J(J + 1)]},$$

where J is the quantum number of the total angular momentum. J acquires the values

$$L + S, L + S - 1, ..., |L - S|$$

It must be stressed that it is assumed that the spin-orbit coupling can be considered to be a small perturbation in comparison with the electrostatic interaction among the electrons (called the Russell-Saunders coupling) and that the term wave functions are considered to be eigenfunctions of the zeroth order. This assumption is not fulfilled for some atoms of higher atomic number where the so called $j-j$ coupling approximation is more suitable; this takes into account the fact that the spin-orbit interaction energy exceeds the electrostatic interaction energy. It can be said on the whole that the actual energy levels of all atoms lie between these extreme cases.

In Fig. 8-2 an example of the calculation of J for two non-equivalent p electrons is given. Strictly speaking, the initial vectors of the orbital angular momentum and of the spin have the magnitude

$$\hbar\sqrt{[L(L + 1)]} \quad \text{and} \quad \hbar\sqrt{[S(S + 1)]}$$

In the graphical representation again vectors **L** and **S** are used and thus quantum number J is obtained directly. The number of possible J values has a simple relation to S, being equal to $(2S + 1)$, *the multiplicity of the state*. The value J is given by this expression only if $L \geqq S$ (cf. Fig. 8-2). If $L < S$, the number of values of J is given by the expression $(2L + 1)$.

For qualitative information the energy levels of a many-electron atom are given in the so-called term diagram. The differences between the energy of the ground state and the energy of the individual excited states gives the positions of the absorption and emission lines in the spectrum of the free atom.

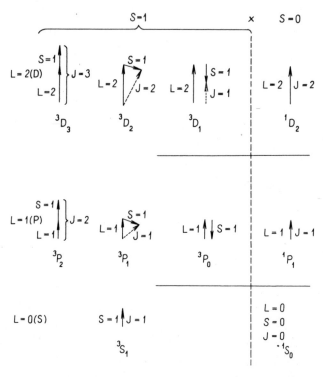

Fig. 8-2. Determination of quantum number J. S as a term symbol $(L = 0)$ should not be confused with S denoting the total spin.

In order to calculate the energy of a many-electron atom, a many-electron wave function must be found that satisfies the conditions imposed on wave functions. It is expedient to construct this n-electron wave function from one-electron functions, i.e. from the atomic orbitals. The construction of such a function can be carried out using Slater determinants or a linear combination thereof. The construction of these determinants for molecular systems has been described in Sections 5.4 and 5.5, where the discussion given also applies to atoms.

REFERENCES

1. Murrell J. N., Kettle S. F. A., Tedder J. M.: *Valence Theory*. Wiley, London 1965.
2. Hartree D. R.: *The Calculation of Atomic Structures*. Wiley, New York 1957.
3. Peacock T. E.: *Foundation of Quantum Chemistry*. Wiley, London 1968.
4. Sutton D.: *Electronic Spectra of Transition Metal Complexes*. McGraw-Hill, London 1968.
5. Condon E. U., Shortley G. H.: *The Theory of Atomic Spectra* Cambridge at the University Press, 1963.

9. DIATOMIC MOLECULES

9.1 Introductory comments; the hydrogen molecular ion, H_2^+

Diatomic molecules can be divided into homonuclear and heteronuclear molecules; examples are molecular hydrogen and carbon monoxide, respectively. The simplest known diatomic molecule, the molecular hydrogen ion, also belongs in the first group. This system consists of two protons and one electron. The simplicity of this system makes it a suitable link between the theories of atoms and molecules (Fig. 3-4).

Therefore an attempt will be made to calculate the energy of the H_2^+ ion. A molecular orbital constructed as a linear combination of atomic orbitals can be used as a wave function for the description of the electron in the ion, so that[1]

$$\varphi = c_1\chi_1 + c_2\chi_2 \qquad (9\text{-}1)$$

[cf. Eq. (5-63)], where orbital χ_1 corresponds to atom A and orbital χ_2 corresponds to atom B. The electronic energy is given by the relationship

$$E = \int \varphi^* \mathscr{H}\varphi \, d\tau = \langle \varphi | \mathscr{H} | \varphi \rangle, \qquad (9\text{-}2)$$

where it is assumed that orbital φ is a normalized function. The explicit form of the Hamiltonian defined in Section 3.2.4 is not used here because in the following qualitative considerations the matrix elements will be handled as compact wholes. The application of the variation method for the determination of the optimum values of the expansion coefficients c_1 and c_2 (i.e. of coefficients which afford minimum energy) leads to a system of linear equations for the coefficients [cf. Eqs. (5-64) and (4-141)]:

$$c_1(H_{11} - E) + c_2(H_{12} - ES_{12}) = 0 \qquad (9\text{-}3a)$$

$$c_1(H_{12} - ES_{12}) + c_2(H_{22} - E) = 0 \qquad (9\text{-}3b)$$

In the equations

$$H_{\mu\nu} = \langle \chi_\mu | \mathcal{H} | \chi_\nu \rangle,$$

$$S_{\mu\nu} = \langle \chi_\mu | \chi_\nu |,$$

the hermicity of the Hamiltonian and the normality of the atomic orbitals is employed. It has already been mentioned [cf. Eqs. (5-65) and (4-142)] that non-trivial values of c_1 and c_2 (the trivial values $c_1 = c_2 = 0$ are not important here) can be obtained only for quite definite energy values. These values can be obtained by solving a quadratic equation in E, which is obtained from the condition that the determinant of system (9-3) must vanish:

$$\begin{vmatrix} H_{11} - E, & H_{12} - ES_{12} \\ H_{12} - ES_{12}, & H_{22} - E \end{vmatrix} = 0 \qquad (9\text{-}4)$$

By rearrangement of the determinant an explicit expression for the quadratic equation can be obtained:

$$E^2(1 - S_{12}^2) - E(H_{11} + H_{22} - 2H_{12}S_{12}) + H_{11}H_{22} - H_{12}^2 = 0 \quad (9\text{-}5)$$

Solution then gives

$$E_{1,2} = \frac{H_{11} + H_{22} - 2H_{12}S_{12}}{2(1 - S_{12}^2)} \pm$$

$$\pm \frac{\sqrt{[(H_{11} + H_{22} - 2H_{12}S_{12})^2 - 4(1 - S_{12}^2)(H_{11}H_{22} - H_{12}^2)]}}{2(1 - S_{12}^2)} \qquad (9\text{-}6)$$

It appears that qualitatively correct results can be obtained even with such drastic simplification as neglecting the overlap integral:

$$S_{12} = 0$$

(We shall return to this unexpected approximation later.)
   Thus it follows that

$$E_{1,2} = \frac{H_{11} + H_{22} \pm \sqrt{[(H_{11} + H_{22})^2 - 4(H_{11}H_{22} - H_{12}^2)]}}{2} \qquad (9\text{-}7)$$

A new denotation can be introduced here:

$$H_{11} = H_{22} = \alpha \qquad (9\text{-}8)$$

(centres A and B are identical, so that $H_{11} = H_{22}$)

$$H_{12} = \beta \qquad (9\text{-}9)$$

$$E_{1,2} = \frac{2\alpha \pm \sqrt{[(2\alpha)^2 - 4(\alpha^2 - \beta^2)]}}{2} = \alpha \pm \beta \qquad (9\text{-}10)$$

This result can now be interpreted. Quantity α [Eq. (9-8)] is termed the Coulomb integral and quantity β [Eq. (9-9)] the resonance integral. Because of the definition

$$\alpha_\mu = \langle \chi_\mu | \mathcal{H} | \chi_\mu \rangle \qquad (9\text{-}11)$$

the Coulomb integral obviously represents the energy of the electron in the field of both nuclei, situated in the atomic orbital χ_μ (function χ_μ is normalized). The interpretation of the resonance integral is not as simple:

$$\beta_{\mu\nu} = \langle \chi_\mu | \mathcal{H} | \chi_\nu \rangle \qquad (9\text{-}12)$$

It suffices, however, to state that β also has the dimension of energy and is connected with the strength of the bond which it describes (the greater its absolute value, the stronger the respective bond) and that, similar to α, it is a negative quantity:

$$\alpha < 0; \quad \beta < 0 \qquad (9\text{-}13)$$

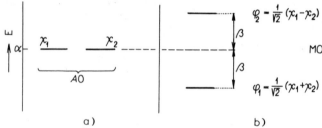

Fig. 9-1. Interaction of two atomic orbitals (with the same energy) leading to bonding and antibonding molecular orbitals: (a) state prior to interaction, (b) state following interaction.

Apparently the value of one calculated energy level in the molecular ion is lower and the other is higher than the energy of an electron situated in the atomic orbital [cf. Eq. (9-10) and Fig. 9-1]. As has already been mentioned, the energy corresponding to the atomic orbital equals α.

The expansion coefficients of the MO's can be calculated by substituting the calculated energy values into system of equations (9-3) ($S = 0$) to give (first for $E = \alpha + \beta$)

$$-c_1\beta + c_2\beta = 0$$

$$c_1 = c_2$$

and (for $E = \alpha - \beta$)

$$c_1\beta + c_2\beta = 0$$

$$c_1 = -c_2$$

Thus for the unnormalized MO's it holds that

$$\varphi_1 = \chi_1 + \chi_2 \qquad (9\text{-}14)$$

$$\varphi_2 = \chi_1 - \chi_2 \qquad (9\text{-}15)$$

Neglecting the overlap between the atomic orbitals, the normalization condition for the molecular LCAO orbital has the form

$$\langle \varphi_i | \varphi_i \rangle = \int (\sum_{\mu=1}^{2} c_{\mu i} \chi_\mu)^2 \, d\tau = \sum_{\mu=1}^{2} c_{\mu i}^2 = 1 \qquad (9\text{-}16)$$

for $i = 1$ and 2. Here is necessary that

$$c_1^2 + c_2^2 = 1,$$

so that the normalization factor is then

$$N = \frac{1}{\sqrt{2}}$$

The normalized MO's will therefore have the following form:

$$\varphi_1 = \frac{1}{\sqrt{2}} (\chi_1 + \chi_2) \qquad (9\text{-}17)$$

$$\varphi_2 = \frac{1}{\sqrt{2}} (\chi_1 - \chi_2) \qquad (9\text{-}18)$$

For the original, physically correct assumption that $S_{12} \neq 0$, the following expressions for the MO's and their energies can be obtained:

$$\varphi_1 = \frac{1}{\sqrt{[2(1 + S_{12})]}} (\chi_1 + \chi_2); \qquad E_1 = \frac{\alpha + \beta}{1 + S_{12}} \qquad (9\text{-}19)$$

$$\varphi_1 = \frac{1}{\sqrt{[2(1 - S_{12})]}} (\chi_1 - \chi_2); \qquad E_2 = \frac{\alpha - \beta}{1 - S_{12}} \qquad (9\text{-}20)$$

It should be noted that, on the basis of the symmetry considerations in Section 6.6, coefficients c_1 and c_2 could be determined directly without calculation.

If the total energy of the system is to be determined, then the dependence of the electron energy of both states (the bonding state and the antibonding state), and the dependence of the energy of the proton repulsion on the internuclear distance are equally important. By the superposition of the two quantities, the dependence of the total energy on the internuclear distance is obtained. This dependence for the bonding state has a minimum (indicating the equilibrium internuclear distance and the magnitude of the dissociation energy), whereas the dependence for the antibonding state is monotonic (Fig. 9-2).

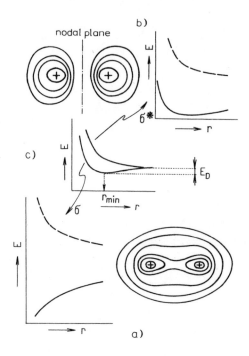

Fig. 9-2. The molecular hydrogen ion: dependence of the electron energy (————) and the proton repulsion energy (— — — —) on the distance r between the protons, for the bonding σ (*a*) and antibonding σ* (*b*) states. (*c*) Superposition of the two dependences. r_{min} is the equilibrium distance, E_D is the dissociation energy.

9.2 The H_2 molecule

The H_2^+ molecular ion represents a model system for illustrating the effects leading to the bonding properties of molecules. However, the transition to real and chemically interesting systems, to many-electron systems, makes it necessary to include contributions of mutual electron repulsion which do not occur in the Hamiltonian of H_2^+. New specific problems arise when the interelectron interaction is included and can best be demonstrated on a related system — the H_2 molecule.

The hydrogen molecule[2] is composed of two protons which will be denoted by μ and ν and two electrons. Provided the relativistic effects are not taken into consideration, the electronic Hamiltonian of the H_2 molecule can be written within the limits of the Born-Oppenheimer approximation in the form [cf. Eq. (5-18)]

$$\mathcal{H} = \sum_{i=1}^{2} \hbar(i) + g(1, 2), \qquad (9-21)$$

where the one-electron operator $\hbar(i)$ (depending only on the coordinates of a single electron) equals the sum of the kinetic energy operator of the electron and of the potential energy of Coulomb interaction between the electron and the nuclei:

$$\hbar(i) = -\frac{\hbar^2}{2m}\Delta_i - \frac{e^2}{4\pi\varepsilon_0}\left(\frac{1}{|\mathbf{r}_i - \mathbf{R}_\mu|} + \frac{1}{|\mathbf{r}_i - \mathbf{R}_\nu|}\right) \qquad (9\text{-}22)$$

Operator $g(1,2)$ depends on the distance between the two electrons [cf. Eq. (5-19b)] and corresponds to their Coulomb interaction:

$$g(1,2) = \frac{e^2}{4\pi\varepsilon_0|\mathbf{r}_1 - \mathbf{r}_2|} \qquad (9\text{-}23)$$

The problem can first be considered from the aspect of the one-electron approximation. The behaviour of each electron i can then be described by a one-electron wave function $\varphi(i)$ (cf. Section 5.6), by an orbital which can best be expressed in the form (9-1). Since the H_2 molecule is symmetric with respect to a plane perpendicular to the bond and lying midway between the nuclei, the values of coefficients c_1 and c_2 in Eq. (9-1) can be determined directly (cf. Section 6.6). Two molecular orbitals, φ_1 and φ_2, are obtained [cf. Eq. (9-19) and (9-20)], of which one is symmetric (φ_1) and the other antisymmetric (φ_2). The electron wave function of the H_2 molecule in the ground state, which has the character of a closed shell system, can be written [cf. Eq. (5-43)] in the form of a Slater determinant,

$$\Delta_0 = \frac{1}{\sqrt{2}}\begin{vmatrix} \varphi_1(1)\,\alpha(1), & \varphi_1(1)\,\beta(1) \\ \varphi_1(2)\,\alpha(2), & \varphi_1(2)\,\beta(2) \end{vmatrix} = |\,\varphi_1\alpha, \; \varphi_1\beta\,| =$$

$$= \frac{1}{\sqrt{2}}\varphi_1(1)\,\varphi_1(2)\,[\alpha(1)\,\beta(2) - \alpha(2)\,\beta(1)], \qquad (9\text{-}24)$$

which can be intepreted as occupation of φ_1 by two electrons, one in spin state α and the second in spin state β.

Since the wave function of the ground state is known, an attempt can be made to calculate the electron energy for a molecule in the ground state:

$$E_0 = \langle \Delta_0 | \mathscr{H} | \Delta_0 \rangle \qquad (9\text{-}25)$$

Substituting Eqs. (9-21) and (9-24) into Eq. (9-25) gives

$$E_0 = \frac{1}{2}\langle \varphi_1(1)\,\varphi_1(2) | \sum_{i=1}^{2} \hbar(i) + g(1,2) | \varphi_1(1)\,\varphi_1(2)\rangle_r \times$$

$$\times \langle \alpha(1)\,\beta(2) - \alpha(2)\,\beta(1) | \alpha(1)\,\beta(2) - \alpha(2)\,\beta(1)\rangle_s, \qquad (9\text{-}26)$$

where use was made of the fact that Hamiltonian $\mathcal{H}$ does not depend on the spin coordinates, allowing separation of the space and spin variables. The variables are denoted by the index at the Dirac bracket. Elsewhere the symbols introduced in Chapter 5 are used, which permit simpler expression of multiple integrals. As spin functions $\alpha(i)$ and $\beta(i)$ are orthonormal functions [cf. Eqs. (4-82) and (4-83)], it follows that

$$\langle \alpha(1)\,\beta(2) - \alpha(2)\,\beta(1)\,|\,\alpha(1)\,\beta(2) - \alpha(2)\,\beta(1)\rangle_s = 2,$$

so that Eq. (9-26) can be rewritten in the form

$$E_0 = \langle \varphi_1(1)\,|\,\hbar(1)\,|\,\varphi_1(1)\rangle + \langle \varphi_1(2)\,|\,\hbar(2)\,|\,\varphi_1(2)\rangle +$$
$$+ \langle \varphi_1(1)\,\varphi_1(2)\,|\,g(1,2)\,|\,\varphi_1(1)\,\varphi_1(2)\rangle, \qquad (9\text{-}27)$$

where limitation of the integrations to the space coordinates of the electrons alone is not denoted, as this can no longer lead to error. It is worth noting that the first two integrals are identical; they differ merely in the denotation of the integration variable: expression (9-27) represents the result which could be determined directly by employing the Slater rules [see Table 5-2 and Eqs. (5-33) and (5-34)]. If the expression for the molecular orbital φ_1 [cf. Eq. (9-19)] is also considered,

$$\varphi_1 = \frac{1}{\sqrt{[2(1 + S_{\mu\nu})]}}\,(\chi_\mu + \chi_\nu) \qquad (9\text{-}28)$$

then the energy can be expressed in terms of integrals containing only atomic orbitals. The numerical calculation of the integrals required assumption of the analytical form of atomic orbitals χ_μ and χ_ν. If they are chosen in the form of Slater 1s functions (cf. Section 8.1), the relationship

$$\chi_\mu(1) = \frac{1}{\sqrt{\pi}}\,e^{(-r_{\mu_1})}, \qquad (9\text{-}29)$$

is obtained, where r_{μ_1} is the radial coordinate of the first electron related to the nucleus μ as the origin. One-electron integrals lead to the following types of integrals (without including the multiplication factors):

$$\langle \chi_\varkappa(1)\,|\,\Delta_1\,|\,\chi_\lambda(1)\rangle \qquad (9\text{-}30a)$$

$$\left\langle \chi_\varkappa(1)\,\left|\,\frac{1}{r_{1\varkappa}}\,\right|\,\chi_\lambda(1)\right\rangle \qquad (9\text{-}30b)$$

$$\left\langle \chi_\lambda(1)\,\left|\,\frac{1}{r_{1\varkappa}}\,\right|\,\chi_\lambda(1)\right\rangle \qquad (9\text{-}30c)$$

Here either $\varkappa = \lambda$ or $\varkappa \neq \lambda$. Similarly, the last integral (called the two-electron Coulomb integral) on the right-hand side of Eq. (9-27) can be expressed in terms of various integrals of the type

$$\langle \chi_\varkappa(1)\, \chi_\lambda(2) \,|\, \mathscr{g}(1,2) \,|\, \chi_\gamma(1)\, \chi_\delta(2) \rangle \qquad (9\text{-}31)$$

In our particular case, when the energy of the diatomic molecule is to be calculated in terms of atomic orbitals which are spherically symmetric towards the centre on which they are located [cf. Eq. (9-28)], the given types of integrals can be expressed in analytical form (cf. Section 9.3). This is not generally true and in the calculation of some types of integrals it is necessary to resort to numerical integration.

So far only the expression and calculation of the electronic energy of the ground state of the H_2 molecule has been discussed, which, from the viewpoint of the one-electron approximation, corresponds to occupation of φ_1 by two electrons. To extend these considerations to electronically excited states, the distribution of the two electrons of the H_2 molecule between orbitals φ_1 and φ_2 must be known. The situation can be schematically represented as follows:

$$(9\text{-}32)$$

The direction of the arrow represents the spin state of the electron; an arrow directed upwards denotes spin α, downwards spin β. Each of the six electron configurations can be represented by the Slater determinant indicated below scheme (9-32) [cf. Eqs. (5-29) and (9-24)]. If the corresponding Slater determinants are pure spin states (i.e. if they are eigenfunctions of $\mathscr{S}^2$), their multiplicity is expressed by the index at the top left in the symbol of the Slater determinant of the given configuration. Of the functions given, Slater determinants Δ_3 and Δ_4 do not correspond to pure spin states, but suitable linear combinations thereof do [cf. Eq. (6-112)], one leading to the singlet state ($^1\Delta_3'$) and the other to the triplet state ($^3\Delta_4'$).

In the framework of the one-particle approximation, Slater determinants (or suitable linear combinations thereof, for example, $^1\Delta_3'$ and $^3\Delta_4'$) of the given electron configurations can be considered to be wave functions describing electronically excited states of the H_2 molecule. The respective energy levels can be calculated in a manner analogous to that employed for the energy E_0 of the ground state [cf. Eqs. (9-25), (9-30) and (9-31)]. The total energy of the molecule, E_{tot}, can be determined when the nuclear interaction energy is added to the total electron energy, this term corresponding to the Coulomb repulsion of two point charges of magnitude $+e$, independent of the electronic state within the Born-Oppenheimer approximation. The dependence of the total energies for various states on the distance between the atoms is given in Fig. 9-3; minima appear on the curves for the ground state (at the equilibrium distance of the two atoms corresponding to formation of the chemical bond) and for the singly excited singlet state. It is also worth noting that the curve describing the ground state does not converge, on separation of the atoms, to the energy of the two individual atoms, but to a larger value. This important fact will be discussed in Section 10.5, where the integrals responsible for the origin of bonding states in molecules will be given.

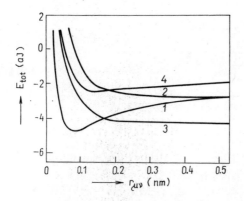

Fig. 9-3. Dependence of E_{tot} of the H_2 molecule as a function of the interatomic distance $(r_{\mu\nu})$ for various electron configurations $(^1\Delta_0 \equiv 1, ^1\Delta_5 \equiv 2, ^3\Delta \equiv 3, ^1\Delta'_3 \equiv 4)$.

It remains to be added that the numerical results can be improved by the method of configuration interaction (cf. Section 5.4). On the basis of the arguments in Sections 5.4 and 5.5 the wave function of the ground state Ψ can be sought in the form

$$\Psi = c_0\,^1\Delta_0 + c_1\,^1\Delta_5, \qquad (9\text{-}33)$$

where the variation principle is used for determination of the expansion coefficients. Other configurations of scheme (9-32) do not interact with the ground state configuration $(^1\Delta_0)$ because of inconvenient space and spin symmetry properties (cf. Section 6.7). The matrix elements of the corresponding secular problem, leading to determination of the allowed energy values, can again be expressed by means of integrals of the type (9-30) and (9-31) [cf. Table 5-2 and Eqs. (5-33) and (5-34)].

9.3 Calculation of the molecular integrals

In all the problems solved so far, the results were expressed in terms of integrals whose integrand was composed of an operator surrounded by atomic orbitals. This was also true in the previous section for the discussion of the H_2 molecule. The calculation of these integrals is actually only applied mathematics. Furthermore, a number of these integrals can be found in tables (almost solely for Slater-type atomic orbitals), and the computer programmes for their calculation are available within the international exchange programme. Nonetheless, we consider it expedient to describe the calculation of at least the simplest integrals in order to give the reader an idea of the approach taken in the calculation of the molecular integrals. The integrals used in the calculation for the H_2 molecule are particularly useful for this purpose.

Similarly as in the previous section, it will be assumed that the molecular orbitals are a linear combination of Slater functions (1s), i.e.

$$(1s) = \sqrt{\left(\frac{\alpha^3}{\pi}\right)} e^{-\alpha r}, \qquad (9\text{-}34)$$

where, for the time being, the centre to which the atomic orbital (1s) is related is not designated because one-centre integrals will be discussed first. In contrast to Eq. (9-29), the value of the exponential factor is not yet specified in Eq. (9-34). It is first necessary to show that orbital (9-34) is a normalized function. In spherical polar coordinates it holds,

$$\langle (1s)\,|\,(1s)\rangle = \frac{\alpha^3}{\pi} \int e^{-2\alpha r} r^2 \sin\Theta \; d\Phi \, d\Theta \; dr, \qquad (9\text{-}35)$$

for the following integration limits:

$$\Phi \in \langle 0, 2\pi\rangle$$
$$\Theta \in \langle 0, \pi\rangle \qquad (9\text{-}36)$$
$$r \in \langle 0, \infty\rangle$$

After integration over angular variables, the expression

$$\langle (1s)\,|\,(1s)\rangle = 4\alpha^3 \int e^{-2\alpha r} r^2 \, dr \qquad (9\text{-}37)$$

is obtained. Integration by parts yields the relationship

$$\int_0^\infty e^{-ax} x^n \, dx = \frac{n!}{a^{n+1}}, \qquad (9\text{-}38)$$

where a is a positive and n a non-negative integral constant. It then follows from Eq. (9-38) that

$$\langle (1s) | (1s) \rangle = 4\alpha^3 \frac{2}{(2\alpha)^3} = 1 \qquad (9\text{-}39)$$

The expression for the one-centre nuclear attraction integral can also be derived:

$$\left\langle (1s) \left| \frac{1}{r} \right| (1s) \right\rangle = 4\alpha^3 \int e^{-2\alpha r} r \, dr = \alpha \qquad (9\text{-}40)$$

The one-centre kinetic energy integral can conveniently be expressed using the Laplace operator in spherical polar coordinates:

$$\Delta = \frac{1}{r^2} \frac{\partial}{\partial r} \left[r^2 \frac{1}{\partial r} \right] + \frac{1}{r^2} \left(\frac{1}{\sin \Theta} \frac{\partial}{\partial \Theta} \left[\sin \Theta \frac{\partial}{\partial \Theta} \right] + \frac{1}{\sin^2 \Theta} \frac{\partial^2}{\partial \Phi^2} \right) \qquad (9\text{-}41)$$

Because the (1s) function does not depend on the angular coordinates it follows that

$$\Delta(1s) = \sqrt{\left(\frac{\alpha^3}{\pi} \right)} \frac{1}{r^2} \frac{\partial}{\partial r} \left(r^2 \frac{\partial}{\partial r} [e^{-\alpha r}] \right) = \sqrt{\left(\frac{\alpha^3}{\pi} \right)} \alpha \left[\alpha e^{-\alpha r} - \frac{2e^{-\alpha r}}{r} \right] \qquad (9\text{-}42)$$

and after substituting into expression (9-30a),

$$\langle (1s) | \Delta | (1s) \rangle = 4\alpha^3 [\alpha^2 \int e^{-2\alpha r} r^2 \, dr - 2\alpha \int e^{-2\alpha r} r \, dr] = -\alpha^2, \qquad (9\text{-}43)$$

where Eq. (9-38) was employed.

Now the one-centre two-electron integral expressing the Coulomb interaction between electrons occuring on the same atom can be calculated:

$$\langle (1s)_1 (1s)_2 | g(1,2) | (1s)_1 (1s)_2 \rangle =$$

$$= \frac{\alpha^6}{\pi^2} \int\int \frac{e^{-2\alpha r_1} e^{-2\alpha r_2}}{|r_1 - r_2|} \sin \Theta_1 \sin \Theta_2 r_1^2 r_2^2 \, d\Phi_1 \, d\Phi_2 \, d\Theta_1 \, d\Theta_2 \, dr_1 \, dr_2, \qquad (9\text{-}44)$$

where the dependence of the coordinates on the position of the first or second electron is denoted by the indices. As the interaction occurs between two spherically symmetric charge distributions, integration over angular variables can easily be carried out. If the integration is carried out stepwise, for example, first over the coordinates of the first electron, then the fact that the classical electrostatic potential outside the charge distribution of a spherical shell is equal to the potential caused by the same total charge located at the origin and that the potential inside a spherical shell is equal to its value on the surface can be used.

It therefore follows that

$$\langle (1s)_1 (1s)_2 \mid \mathcal{g}(1,2) \mid (1s)_1 (1s)_2 \rangle =$$

$$= 16\alpha^6 \int_0^\infty e^{-2\alpha r_2} \left[\int_{r_2}^\infty e^{-2\alpha r_1} r_1 \, dr_1 + \frac{1}{r_2} \int_0^{r_2} e^{-2\alpha r_1} r_1^2 \, dr_1 \right] r_2^2 \, dr_2 \quad (9\text{-}45)$$

Integration by parts of the expressions inside the bracket yields

$$\frac{1}{r_2} \int_0^{r_2} e^{-2\alpha r_1} r_1^2 \, dr_1 = -\frac{e^{-2\alpha r_2}}{2} \left[\frac{r_2}{\alpha} + \frac{1}{\alpha^2} + \frac{1}{2r_2 \alpha^3} \right] + \frac{1}{4r_2 \alpha^3} \quad (9\text{-}46)$$

$$\int_{r_2}^\infty e^{-2\alpha r_1} r_1 \, dr_1 = \frac{e^{-2\alpha r_2}}{2\alpha} \left[r_2 + \frac{1}{2\alpha} \right] \quad (9\text{-}47)$$

and substitution into Eq. (9-45) yields the relationship

$$16\alpha^6 \left[-\frac{1}{4\alpha^2} \int_0^\infty r_2^2 \, e^{-4\alpha r_2} \, dr_2 - \frac{1}{4\alpha^3} \int_0^\infty r_2 \, e^{-4\alpha r_2} \, dr_2 + \right.$$

$$\left. + \frac{1}{4\alpha^3} \int_0^\infty r_2 \, e^{-2\alpha r_2} \, dr_2 \right] = \frac{5}{8} \alpha, \quad (9\text{-}48)$$

where Eq. (9-38) has again been used.

It therefore follows that the calculation of one-centre integrals can easily be performed if spherical polar coordinates are used. The calculation of two-centre integrals is more difficult. One approach to their solution lies in the introduction of ellipsoidal coordinates, where the centres on which the atomic orbitals are located act as foci for the ellipsoidal coordinate system.

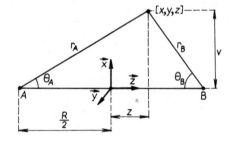

Fig. 9-4. Graphical representation of relationship of elliptical, rectangular and spherical polar coordinates for the two-centre problem.

Let A and B be two centres and r_A, r_B and R be determined according to Fig. 9-4. The ellipsoidal coordinates ξ, η and Φ of the point given by the Cartesian coordinates x, y, z are defined as follows:

$$\xi = \frac{r_A + r_B}{R}$$

$$\eta = \frac{r_A - r_B}{R}$$

Φ is the angle between the plane defined by the triangle A, B, $[x, y, z]$ and the x-axis.

It is obvious from the definition that the values of the ellipsoidal coordinates lie in the following intervals:

$$\xi \in \langle 1, \infty \rangle \tag{9-49a}$$

$$\eta \in \langle -1, 1 \rangle \tag{9-49b}$$

$$\Phi \in \langle 0, 2\pi \rangle \tag{9-49c}$$

The relationship of the ellipsoidal coordinate system to the Cartesian system, with its origin in the centre of the line connecting points A and B, can be derived on the basis of simple geometrical considerations. It follows from Fig. 9-4 that

$$r_A^2 = v^2 + \left(\frac{R}{2} + z\right)^2, \tag{9-50a}$$

$$r_B^2 = v^2 + \left(\frac{R}{2} - z\right)^2, \tag{9-50b}$$

where v is the height of the (rotating) triangle A, B, $[x, y, z]$, and, after substracting these equations,

$$z = \frac{1}{2R}(r_A + r_B)(r_A - r_B) = \frac{R}{2}\xi\eta \tag{9-51a}$$

Because the remaining Cartesian coordinates depend simply on v,

$$x = v \cos \Phi$$

$$y = v \sin \Phi,$$

it follows from Eqs. (9-50a) and (9-50b) that

$$x = \frac{R}{2}\sqrt{[(\xi^2 - 1)(1 - \eta^2)]} \cos \Phi \tag{9-51b}$$

$$y = \frac{R}{2}\sqrt{[(\xi^2 - 1)(1 - \eta^2)]} \sin \Phi \tag{9-51c}$$

All the relationships for the two-centre spherical polar and ellipsoidal coordinates can be obtained just as easily; in summarized form,

$$r_A = \frac{\xi + \eta}{2} R \tag{9-52a}$$

$$r_B = \frac{\xi - \eta}{2} R \qquad (9\text{-}52b)$$

$$\cos \Theta_A = \frac{1 + \xi\eta}{\xi + \eta} \qquad (9\text{-}52c)$$

$$\cos \Theta_B = \frac{1 - \xi\eta}{\xi - \eta} \qquad (9\text{-}52d)$$

The volume element for integration in three-dimensional space can then be written as

$$d\tau \equiv dx\, dy\, dz \equiv \left(\frac{R}{2}\right)^3 (\xi^2 - \eta^2)\, d\xi\, d\eta\, d\Phi \qquad (9\text{-}53)$$

The usefulness of the coordinate transformations can be demonstrated on the calculation of overlap integrals between (1s)-type Slater orbitals, one of which is located on centre A and the second on centre B, assuming that, in general, the atomic orbitals differ in their exponent; one will be denoted α and the other β. Therefore the integral

$$\langle (1s)_A \,|\, (1s)_B \rangle = \frac{\sqrt{(\alpha^3\beta^3)}}{\pi} \int e^{-\alpha r_A}\, e^{-\beta r_B}\, d\tau \qquad (9\text{-}54)$$

must be calculated. By introducing relations (9-52a), (9-52b) and (9-53) into Eq. (9-54) and integrating over Φ, the relationship

$$2\sqrt{(\alpha^3\beta^3)} \left(\frac{R}{2}\right)^3 \times$$

$$\times \int \left[\xi^2\, e^{-\zeta(R/2)(\alpha+\beta)}\, e^{-\eta(R/2)(\alpha-\beta)} - e^{-\zeta(R/2)(\alpha+\beta)}\eta^2\, e^{-\eta(R/2)(\alpha-\beta)} \right] d\xi\, d\eta =$$

$$= 2\sqrt{(\alpha^3\beta^3)} \left(\frac{R}{2}\right)^3 [A_2^{(a)} B_0^{(b)} - A_0^{(a)} B_2^{(b)}] \qquad (9\text{-}55)$$

is obtained, where new constants

$$a = \frac{R}{2}(\alpha + \beta) \qquad (9\text{-}56a)$$

and

$$b = \frac{R}{2}(\alpha - \beta) \qquad (9\text{-}56b)$$

were introduced, as were the auxiliary integrals

$$A_n^{(a)} = \int_1^{\infty} \xi^n\, e^{-a\xi}\, d\xi \qquad (9.57a)$$

and

$$B_n^{(b)} = \int_{-1}^{1} \eta^n\, e^{-b\eta}\, d\eta \qquad (9\text{-}57b)$$

184

The introduction of various types of auxiliary integrals is common in calculations of molecular integrals, as it enables rational calculation or tabulation of the results (cf., for example, the tables by Miller, Gerhauser and Matsen[3]). Calculation of integrals of type $A_n^{(a)}$ and $B_n^{(b)}$ is simple and is carried out according to the value of parameters a or b either using recurrence formulas derived by integration by parts or numerically by expanding the exponential function into a power series. It is worth noting that even two-centre Coulomb integrals with spherically symmetrical charge distributions can be expressed by integrals of type $A_n^{(a)}$ and $B_n^{(b)}$.

In conclusion it should be mentioned that the calculation of two-centre molecular integrals with charge distributions that are not characterized by spherical symmetry and, in particular, the calculation of many-centre (three- and four-centre) integrals are much more complicated than calculation of the integrals discussed in this section. An introduction to these problems and the respective references can be found in review articles (for example Ref. 4).

9.4 General diatomic molecules and correlation diagrams

First, the homonuclear diatomic molecules composed of elements of the first series of the periodic table of the elements will be discussed. On the two atoms, which will be denoted by A and B, 1s, 2s, $2p_x$, $2p_y$ and $2p_z$ atomic orbitals are available. AO's of the same energy will be combined, taking their symmetry properties into consideration. The interaction between AO's corresponding to different energy levels will not be considered. By combination of two AO's, wto MO's are always obtained, of which one is bonding in relation to the initial AO's and the other is antibonding. It is necessary to be aware of the relativity of these concepts; for example, the antibonding MO formed from two 1s AO's has a more favourable energy than the bonding MO derived from the 2s AO's. The expressions for the MO in Fig. 9-5 have the form given by Eqs. (9-19) and (9-20). For illustration, the MO's formed by combination of the 1s AO's can be given:

$$\sigma(1s) = \frac{1}{\sqrt{[2(1 + \langle (1s)_A | (1s)_B \rangle)]}} [(1s)_A + (1s)_B] \qquad (9\text{-}58a)$$

$$\sigma^*(1s) = \frac{1}{\sqrt{[2(1 - \langle (1s)_A | (1s)_B \rangle)]}} [(1s)_A - (1s)_B] \qquad (9\text{-}58b)$$

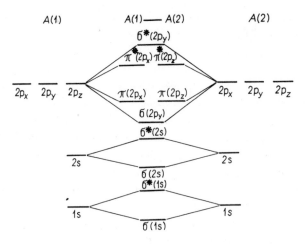

Fig. 9-5. Formation of σ- and π-MO's by interactions between s and p atomic orbitals localized on atoms A(1) and A(2).

The difference in the energies of the σ- and π-MO's formed by the overlap of p-type AO's is not, in practice, large and their energies can therefore lie in the opposite sequence, depending on the specific conditions in the individual molecule. The example given in Fig. 9-5 corresponds, for example, to the O_2 molecule, but it does not describe the conditions in the B_2 molecule.* The oxygen molecule is correctly described inasmuch as the reason why the oxygen has a triplet ground state is evident (the presence of two electrons in degenerate orbitals with parallel spins). If the 16 electrons of the corresponding O_2 molecule are placed in the MO's according to increasing energy (Aufbau principle), 14 of them will completely occupy seven MO's and the 15th and 16th electron, in agreement with Hund's rule (cf. Section 10.6.2), will be placed, with parallel spins, in the degenerate $\pi^*(2p_z)$- and $\pi^*(2p_x)$-MO's.

Studies of *correlation diagrams*[5-8] provide useful information on the bonding conditions in diatomic molecules. A correlation diagram characterizes the gradual transition of one-electron states from a system of two infinitely distant atoms to a system where the atoms coincide (the united atom). The states of diatomic molecules obviously lie between these two extremes.

Before discussing correlation diagrams, some preliminary considerations on the symmetry of a system composed of two atoms on the *y*-axis will be mentioned. Three cases can be distinguished.

---

* The B_2 molecule has a triplet ground state, whereas it should have a singlet ground state according to Fig. 9-5.

a) If the atoms are infinitely far apart, each is spherically symmetrical and n, l and m are valid quantum numbers.

b) If the nuclei approach each other, the spherical symmetry of the partial systems, the atoms, is lost. The system is invariant under arbitrary rotation about the y-axis, i.e. the formation is characterized by axial symmetry. Classical mechanics shows that, for such a system, the projection of the total angular momentum onto the y-axis is retained (termed the constant of motion). In quantum mechanics, this type of symmetry is manifested by quantization of the projection of the total angular momentum onto the y-axis (cf. Section 4.4). Because we are interested in classification of the one-electron levels, quantum number λ is introduced to denote the value of the projection of the angular momentum of the electron (in multiples of $\hbar$) in the given state. If l is the quantum number of the electron in the atom, it is evident that λ can assume the values

$$\lambda = l, l - 1, \ldots, 2, 1, 0, \tag{9-59}$$

where, e.g., it is sufficient to consider only the value $\lambda = l$, as the state with $\lambda = -l$ (momentum with the opposite orientation) has the same energy and thus corresponds to a doubly degenerate level. Levels with different values of quantum number λ are usually denoted by small Greek letters; for states $\lambda = 0, 1, 2, 3, \ldots$ the notation σ, π, δ, φ, ... is used.

If the molecule is homonuclear, the symmetry of the system is further increased by inversion in the point which divides the line connecting the two atoms. This type of symmetry permits classification of the states of homonuclear molecules according to whether the inversion does or does not produce a change in the sign of the wave function (cf. Section 6.2); the symmetric states are denoted by g and the antisymmetric states by u.

c) If the given atoms combine to form a single, united atom (it must be stressed that this is purely a mental process) a spherically symmetric system is again obtained for which the same classification of states applies as for the atom.

If certain rules are taken into account, it is possible to ensure that the states in cases *a*) to *c*) are connected and are a continuous function of the distance between the atoms. Information on the energy distribution corresponding to the molecular orbitals in a diatomic molecule is obtained by connecting the levels at the two extremes by lines (infinitely remote and united atoms). The following rules must be obeyed:

A. *Symmetry conservation* is manifested in the fact that quantum number λ and the u- or g-property of the orbital do not change.

B. *The non-crossing rule*[7], according to which it is inadmissible for energy lines correspoding to orbitals of the same symmetry to cross, because interaction occurs between these states; no limitation applies on the other hand if levels of different symmetry intersect.

The occupation of these one-particle states by electrons must obey the Pauli principle and, if the degenerate levels are incompletely occupied, Hund's rule of maximum multiplicity must be applied.

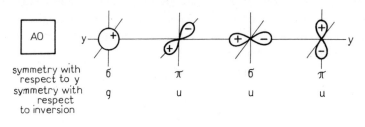

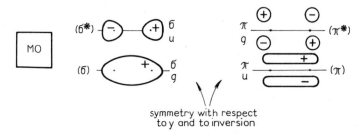

symmetry with respect
to y and to inversion

Fig. 9-6. Classification of atomic orbitals and molecular orbitals according to their symmetry elements.

The classification of some atomic and molecular orbitals based on their symmetry elements, according to point b) is depicted in Fig. 9-6. A number of useful conclusions can be obtained on the basis of graphical representation of orbitals. For example, the bonding and antibonding combinations of 1s orbitals can be represented as follows:

$$(1s)_A + (1s)_B \quad \rightarrow \quad \oplus \ \oplus \tag{9-60a}$$

$$(1s)_A - (1s)_B \quad \rightarrow \quad \oplus \ \ominus \tag{9-60b}$$

Because the quantum number for separated atoms is $l = 0$, then $\lambda = 0$, so that only σ states can arise in both cases. The symmetrical combination (9-60a) remains spherically symmetric on united atom formation so that the transition, separated atoms→molecule→united atom, can be written as follows:

$$\sigma_g: \quad 1s \quad \rightarrow \quad \sigma(1s)_g \quad \rightarrow \quad (1s)_g \tag{9-61a}$$

The antisymmetrical combination (9-60b) in the limiting transition to the united atom has the symmetry properties of a p orbital orientated in the direction of the y-axis. It therefore holds that

$$\sigma_u: \quad 1s \quad \rightarrow \quad \sigma^*(1s)_u \quad \rightarrow \quad (2p)_u \quad [\text{i.e. } (2p)_y] \tag{9-61b}$$

All s-type orbitals behave analogously. It is, of course, more difficult to study the symmetry properties of orbitals with higher values of quantum number l. Evidently, however, for example, antibonding combinations of p orbitals orientated parallel to each other (for example, in the direction of the x-axis) have the symmetry properties of d orbitals (concretely of the d_{xy} orbital) on transition to the united atom.

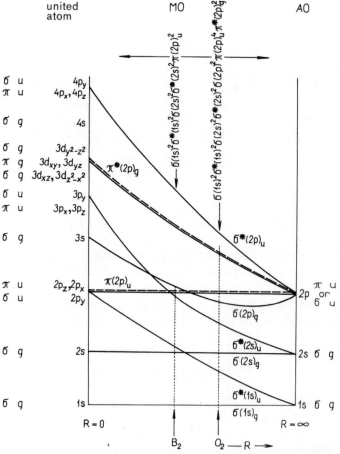

Fig. 9-7. Qualitative correlation diagram for homonuclear diatomic molecules. Right ordinate: separated atoms, left ordinate: united atoms; the region between these extremes characterizes conditions in molecules. Examples: B_2, O_2 (according to Heilbronner and Bock[8]).

189

In Fig. 9-7 a qualitative correlation diagram for diatomic homonuclear molecules is given. In contrast to Eqs. (9-61a) and (9-61b), the right-hand side of the diagram, corresponding to the separated atoms, represents the symmetry properties of the atomic orbitals of the partial systems. In the diagram the positions which correspond to the O_2 molecule and the B_2 molecule are indicated. This procedure is most useful, as information on molecules is obtained from knowledge of the symmetry properties of the participating orbitals alone. In recent years correlation diagrams have been widely applied, for example, with complex compounds, in the analysis of the mechanism of some stereospecific cyclizations and in the interpretation of collision processes of atoms. It should be mentioned that correlation diagrams can also be constructed for heteronuclear molecules (cf., for example, Ref. 9).

REFERENCES

1. Coulson C. A.: *Valence*. Clarendon Press, Oxford 1952.
2. Dewar M. J. S., Kelemen J.: *J. Chem. Educ.* **48**, 494 (1971).
3. Kotani M., Amemiya A., Ishiguro E., Kimura T.: *Tables of Molecular Integrals*. Maruzen Co., Ltd., Tokyo 1955.
   Miller J., Gerhauser J. M., Matsen F. A.: *Quantum Chemistry Integrals and Tables*. University of Texas Press, Austin 1959.
   Preuss H.: *Integraltafeln zur Quantenchemie*. Springer, Berlin 1956.
4. Harris F. E., Michels H. H.: *Advan. Chem. Phys.* **13**, 205 (1967).
5. Hund F.: *Z. Physik* **40**, 742 (1927).
6. Mulliken R. S.: *Phys. Rev.* **32**, 186 (1928).
7. Landau L. D., Lifshits E. M.: *Kvantovaya mekhanika*, Nerelativistitskaya teoria. Gos. izd. fiz.-mat. lit. Moscow 1963, p. 329. *Quantum Mechanics, Non-relativistic Theory*. Addison-Wesley, New York 1965.
8. Heilbronner E., Bock H.: *The HMO-Model and Its Application*. Wiley, New York and Verlag Chemie, Weinheim 1976. (Three volumes.)
9. Barat M., Lichten W.: *Phys. Rev.* **A6,** 211 (1972).

10. CALCULATION METHODS IN THE THEORY OF THE CHEMICAL BOND

10.1 Introductory remarks

It was seen in Chapter 5 that the one-electron approximation represents a basic approach to the study of the electronic structure of atoms and molecules, either as a self-contained model or as a starting point for more accurate calculations.

In analyzing the solvability of the Hartree-Fock equations, relations (5-59a) to (5-59d), which apply when the ground state is described by a Slater determinant of type (5-43) corresponding to a closed shell system, will be used as a starting point. Such systems are of particular interest here. From the viewpoint of the variation principle, the optimum one-electron functions (orbitals), depending on the space coordinates of the given electron, can be of two kinds (depending on whether atoms or molecules are considered). These are either a) atomic orbitals χ, which are located on the selected atom establishing the origin of the local coordinate system in which the electronic coordinates are defined, or b) molecular orbitals φ, which are distributed on a greater number of centres in the many-nuclear system — the molecule. The construction of molecular orbitals by expansion in terms of atomic orbitals located on the atoms forming the molecule is most expedient [cf. Eq. (5-63)]; in other words, these atomic orbitals form a basis set for the expansion of the molecular orbitals. If the number of AO's is such that they describe (in a minimum number) only electrons of atoms in the ground state, we speak of *a minimum basis set* (cf. Section 6.6). An example of *an extended basis set* is the Slater "double zeta" (DZ) basis set, where two Slater functions (see below) with different exponents correspond to one atomic orbital (the exponents denoted here by ξ are sometimes denoted by ζ).

It was shown in Section 5.6 that, within the framework of the Hartree-Fock scheme, the determination of the optimum linear combination of atomic orbitals (i.e. those fulfilling the variation principle)

$$\varphi_i = \sum_{\mu=1}^{m} c_{\mu i} \chi_\mu \qquad (10\text{-}1)$$

is connected with the solution of the Roothaan equations

$$\sum_\nu c_{\nu i} \left\{ \int \chi_\mu^* \mathscr{F} \chi_\nu \, d\tau_1 - \varepsilon_i \int \chi_\mu^* \chi_\nu \, d\tau_1 \right\} = 0, \quad \mu = 1, 2, \ldots, m, \qquad (10\text{-}2)$$

where $\mathscr{F}$ is the Hartree-Fock operator. Because a non-trivial solution for coefficients $c_{\mu i}$ must be found, the calculation of the secular determinant [cf. Eq. (4-142)],

$$\det \left\| \langle \chi_\mu | \mathscr{F} | \chi_\nu \rangle - \varepsilon \langle \chi_\mu | \chi_\nu \rangle \right\| = 0, \qquad (10\text{-}3)$$

becomes necessary, where the notation used in Section 5.6 is again used. Considering Eqs. (5-59b) and (5-59c) and the expansion of the molecular orbitals in the form (10-1), the matrix element of operator $\mathscr{F}$ from Eq. (10-3) can be written as

$$F_{\mu\nu} = \langle \chi_\mu | \mathscr{F} | \chi_\nu \rangle = \langle \chi_\mu | h | \chi_\nu \rangle +$$
$$+ \sum_{\varrho=1}^{m} \sum_{\sigma=1}^{m} P_{\varrho\sigma} \left\{ (\mu\nu | \varrho\sigma) - \frac{1}{2} (\mu\varrho | \nu\sigma) \right\}, \qquad (10\text{-}4)$$

where the charge- and bond-order matrix was introduced, whose general element is defined by the expression

$$P_{\varrho\sigma} = 2 \sum_{j=1}^{n/2} c_{\varrho j}^* c_{\sigma j} \qquad (10\text{-}5)$$

(summation is carried out over the occupied molecular orbitals) and the shortened form for the two-electron integrals is given by

$$(\mu\nu | \varrho\sigma) = \iint \chi_\mu^*(1) \chi_\nu(1) \, \mathscr{g}(1,2) \, \chi_\varrho^*(2) \chi_\sigma(2) \, d\tau_1^* \, d\tau_2 =$$
$$= \langle \chi_\mu \chi_\varrho | \mathscr{g} | \chi_\nu \chi_\sigma \rangle \qquad (10\text{-}6)$$

[cf. Eq. (5-31b)].

Eqs. (10-2) to (10-4) represent the exact formulation of the Hartree-Fock n-electron problem for a closed shell system in the MO-LCAO approximation. In principle the basis set of the atomic orbitals can be chosen extensive enough that the calculated value of the total energy is the lowest within the framework of the Hartree-Fock model, this value being referred to as *the Hartree-Fock energy limit*. When starting such a calculation, it is necessary to choose the AO basis set. As a rule, atomic orbitals are used which have identical angular parts (spherical harmonics) and different radial parts:

a) hydrogen-type functions whose radial part forms a Laguerre polynomial (cf. wave functions for the hydrogen atom), *b*) Slater functions whose radial part is a function of the type

$$r^{n-1} \exp(-\xi r),$$

where n is the principal quantum number of the orbital and ξ is a constant specific for the atom and its electronic shell characterized by n (cf. Section 8.1).

c) Gaussian-type functions whose exponential factor is of the type

$$\exp(-\eta r^2)$$

When performing calculations in molecules, fixed combinations of Gaussian functions are frequently used and we then speak of a contracted basis set. Such a function often simulates an atomic orbital. Clementi and Davis optimized contracted basis sets using calculations on atoms.

Each of the described types of atomic orbitals has its advantages and disadvantages. The first two types of orbitals satisfactorily describe the electron density near the atomic nuclei but lead to very complicated many-centre integrals of type (10-6). Gaussian functions in matrix elements can be much more easily integrated but they provide a poor description of the electron distribution in the close vicinity of and at a great distance from the atomic nuclei. This disadvantage is often compensated by increasing the number of atomic orbitals.

The treatments which have been mentioned here are called "*ab initio*", sometimes also "*absolute*", as there is only a single step between the quantum mechanical formulation of the problem and the result of the calculation, consisting of the choice of the analytical form and the number of functions undergoing the optimization process. There is a number of difficulties hindering the extensive application of these methods to large molecules:

a) The number of two-electron integrals of type $(\mu v \mid \varrho \sigma)$ is proportional to the fourth power of m, where m denotes the number of atomic orbitals, so that calculations performed with a relatively small atomic orbital basis set are the only practicable ones.

b) The calculation of some integrals of type $(\mu v \mid \varrho \sigma)$ was, until recently, rather difficult.

c) Most serious is that the error in the calculated binding energies of molecules (the binding energy is defined as the energy of the molecule less the energy of the atoms forming it) amounts to about 1%, even if the Hartree-Fock equations are solved accurately. For chemical purposes – estimation of the equilibrium and the rate constants – an accuracy of one to two orders greater is needed. In the one-electron model each particle moves in the average field of all the other particles, i.e., correlation of particle movements is not taken into consideration. Physically, it is evident that Coulomb repulsion between the electrons does not allow two particles to occur simultaneously at the same point in space; it is customary to describe this situation by saying that each electron is surrounded by a "Coulomb hole". In the Hartree-Fock scheme only the "Fermi hole" appears, so that in a given region the existence of two electrons with parallel spins is excluded [being against the Pauli principle, cf. Eq. (5-42) and below], where the same (space) molecular orbital may be occupied by electrons with different spins. This shortcoming in the model is termed the correlation error and a quantity called *the correlation energy*, E_{corr}, is introduced for its quantitative estimation. E_{corr} is defined for a given system as the difference between the exact value of its total energy, E_{exact}, and Hartree-Fock energy limit, E_{HF}:

$$E_{corr} = E_{exact} - E_{HF}, \qquad (10\text{-}7)$$

where the relativistic contributions are not included in the exact energy value. For example, the SCF treatment[1] for the total energy yields a value of $-2\,722.65$ eV for the hydrogen fluoride molecule, while the experimental value is equal to $-2\,734.16$ eV. This appears to be satisfactory agreement. The dissociation energy, D_E, of a molecule is defined as the difference between the energy of the molecule, D_M, and the energy D_{AT} of the individual atoms formed on dissociation of the molecule,

$$D_E = D_M - D_{AT}, \qquad (10\text{-}8)$$

and for the SCF treatment it amounts to

$$D_E = -2\,722.65 + 2\,718.54 = -4.11 \text{ eV}$$

The experimental D_E value is -6.08 eV; it therefore follows that, for chemically and physically important quantities, obtained as the difference of two large numbers, the error amounts to several tens of percent of the correct value and that "absolute" calculation on the level of the one-particle model is quite insufficient. The correlation energy then assumes values of the same order as the calculated quantities themselves; consequently, calculation of the correlation energy is of particular interest in quantum chemistry today.

Fortunately, semiempirical methods present a way of avoiding these difficulties. The use of these methods is accompanied by two characteristics. On the one hand, approximations are introduced which lead to a substantial decrease in the number of electron repulsion integrals (and complete elimination of the most difficult ones); on the other hand, certain integrals are replaced by numerical values chosen so that the semiempirical theory describes the experimental characteristics of the substance (or of a small group of substances) chosen for adjusting the parameters.

In the reduction of the number of repulsion integrals a very important part is played by *the zero differential overlap approximation* (denoted by the abbreviation ZDO, see below) which Pople in England and Pariser and Parr in the USA employed independently in 1953 in two important semiempirical methods. The Pople approximation is a direct continuation of the Roothaan SCF method and represents a simplified self-consistent field method, whereas, in the Pariser and Parr method, the individual molecular states (the ground state and the excited states) are described using a linear combination of a certain number of Slater determinants. This is not an iteration method. Essentially the approximations employed and the evaluation of the integrals are very similar in these two methods. Both methods were elaborated in the form of π-electron approximations.

Table 10-1

Survey of Semiempirical Methods Employed in Quantum Chemistry

I. All electrons (or valence electrons) are included		II. π-electron approximation	
		1. Closed electron shell	2. Open electron shell
Semiempirical methods	CI Jaffé[2] Dewar[3] SCF Klopman[4] Pople, Santry, Segal[5] Katagiri, Sandorfy[6] Jungen, Labhart[7]	CI Craig[12]; Pariser, Parr[13] SCF Roothaan[14] SCF Pople[15]	CI Ishitani, Nagakura[19] SCF Roothaan[20] SCF Longuet-Higgins, Pople[21]
Empirical methods	Hoffmann[8] Sandorfy[9] Brown[10] Del Re[11]	Improved HMO (e.g. Coulson – Gołebiewski)[16] HMO[17] Perturbation treatment[18] (within HMO formation)	McLachlan[22] HMO[17]

The Roothaan method forms a basis for various semiempirical methods, those considering all the valence electrons and the π-electron methods. The following survey summarizes these methods, starting with the more generally applicable ones (considering the valence electrons), followed by the more special methods.

Before considering the evaluation of the $F_{\mu\nu}$ expressions (matrix elements of the Hartree-Fock operator) it will be expedient to classify these methods into several groups:

I. Methods which explicitly consider either all the electrons or at least all the valence electrons (as well as methods limited to the σ-electrons alone).

II. Methods in which only π-electrons are explicitly considered.

All the methods are further divided into two groups according to the type of electron configuration in the electronic ground state:

1. methods suitable for systems with closed electron shells,
2. methods suitable for systems with open electron shells.

A more detailed survey is given in Table 10-1, where the level of sophistication of the individual methods is also taken into consideration.

Tables 10-2 and 10-3 contain detailed data on the Hamiltonian, wave functions, matrix elements and the regions where the individual versions of the MO-LCAO method can be best employed.

10.2 All-valence electron MO-LCAO methods

10.2.1 Methods explicitly considering electron repulsion

Semiempirical methods based on the SCF theory, suitable for studying large systems in which all the electrons in the valence shells are explicitly considered, were developed in 1965 to 1967. They are particularly attractive because they can be applied to a great variety of types of inorganic and organic systems.

Pople, Santry and Segal[5] studied the nature of these methods in detail and published a general analysis of semiempirical methods considering valence electrons. Among the proposed schemes, the CNDO (complete neglect of differential overlap) method was the first to be developed and employed for the calculation of charge distributions in some large organic molecules. Among further methods various modifications of this semiempirical method should be mentioned, such as the INDO[27] (intermediate neglect of differential overlap) method, the MINDO[28,29] (modified intermediate neglect of differential overlap) method, the PNDO[3] (partial neglect of differential overlap) method and related methods[30], which were mostly intended for the calculation of particular physical properties of molecules.

Table 10-2

Hamiltonian, Wave Functions and Matrix Elements of Various Versions of the MO-LCAO Method (methods considering all valence electrons; systems with a closed electron shell in the ground state)[a]

Method	$\mathscr{H}$	Ψ	Matrix elements of the Hamiltonian	Application range and limitations
1 CNDO/1 (Pople, Santry, Segal)[5,b]	$\sum_i \mathscr{H}_i^c + \sum_i\sum_{i<j} g(i,j)$	Δ_0	$F_{\mu\mu} = U_\mu + \left(P_M - \frac{1}{2}P_{\mu\mu}\right)\gamma_{MM} +$ $+ \sum_{B\neq M}(P_B\gamma_{MB} - v_{MB})$ $F_{\mu\nu} = \beta_{MN}^\circ S_{\mu\nu} - \frac{1}{2}P_{\mu\nu}\gamma_{MN}$ $(\mu \neq \nu)$	Ground state properties of a great variety of substances (inorganic as well as organic): valence angles, deformation vibrations, dipole moments, internal rotation barriers and NMR chemical shifts. So far unsatisfactory for bond lengths and dissociation energies.
2 CNDO/2 (Pople, Segal)[5]	$\sum_i \mathscr{H}_i^c + \sum_i\sum_{i<j} g(i,j)$	Δ_0	$F_{\mu\mu} = -\frac{1}{2}(I_\mu + A_\mu) +$ $+ \left[(P_M - Z_M) - \frac{1}{2}(P_{\mu\mu} - 1)\right]\gamma_{MM} +$ $+ \sum_{B\neq M}(P_B - Z_B)\gamma_{MB}$ $F_{\mu\nu} = \beta_{MN}^\circ S_{\mu\nu} - \frac{1}{2}P_{\mu\nu}\gamma_{MN}$ $(\mu \neq \nu)$	See 1; improved parametrization.

3 INDO (Pople, Beveridge, Dobosh)[26,27]	$\sum_i \mathcal{H}_i^c + \sum_i \sum_{i<j} \mathscr{g}(i,j)$	Δ_0	$F_{\mu\mu} = H_{\mu\mu}^c + \sum_{v\in(A)} P_{vv}\left[(\mu\mu\mid vv) - \frac{1}{2}(\mu v\mid\mu v)\right] + \sum_{B\neq A} P_B \gamma_{AB}$ $F_{\mu v}^{AA} = P_{\mu v}\left[\frac{3}{2}(\mu v\mid\mu v) - \frac{1}{2}(\mu\mu\mid vv)\right]$ $F_{\mu v}^{AB} = H_{\mu v}^c - \frac{1}{2}P_{\mu v}\gamma_{AB}$ $(\mu \neq v)$	See 1; in contrast to the CNDO/2 method, it is capable of distinguishing states of different multiplicity belonging to the same configuration. In addition, it is more suitable for spin density calculation. (INDO = *Intermediate Neglect of Differential Overlap*; expressions for $H_{\mu\mu}^c$ elements are rather complex and have different forms for individual elements.)
4 MINDO/2 (Dewar, Baird, Haselbach)[28,29]	$\sum_i \mathcal{H}_i^c + \sum_i \sum_{i<j} \mathscr{g}(i,j)$	Δ_0	See 3, however, two-centre repulsion integrals, core matrix elements ($H_{\mu v}^c$), and nuclear repulsion energies are calculated using other expressions than in the INDO method [see Ref. 28, 29]	Semiempirical parameters optimized using experimental data. Particularly suitable for calculation of heats of formation, geometry, force constants, and ionization potentials.
5 Extended HMO, EHT (Hoffmann)[8]	$\sum_i \mathcal{H}_i^{\text{ref}}$	Δ_0	$H_{\mu\mu} = I_\mu$ (valence state ionization potential) $H_{\mu v} = 0.5K(H_{\mu\mu} + H_{vv})S_{\mu v}$ $(\mu \neq v)$ (mostly $K = 1.75$)	Ground state properties of aliphatic, aliphatic-aromatic and aromatic systems: conformation of rings, geometrical isomers, internal rotation barriers, deformation vibrations. Fails for bond lengths and for valence vibrations.
6 MO – very simple version[11]	$\sum_i \mathcal{H}_i^{\text{ref}}$	Δ_0	$H_{\mu\mu} = \alpha + \delta_\mu\beta$ $H_{\mu v} = \varepsilon_{\mu v}\beta$ $(\mu \neq v)$ $(\alpha, \beta$: parameters of the method)	An extremely simple method (not requiring a computer) applicable for a rough estimate of electron distribution, dipole moments and equilibrium constants in aliphatic compounds.

[a] The symbols have the usual meaning. Subscript μ (v) denotes the AO located on atom M (N), $P_{\mu v}$ is the usual bond order, P_A is the total electron density on atom A; γ_{MN} and v_{MN} denote the interaction energies, β_{MN}^0 are empirically determined quantities. Δ_0 is the Slater determinant of the ground state.

[b] This version of the CNDO method is presented only for the sake of completeness and for pedagogical reasons; it is not used for numerical calculations.

Table 10-3

Hamiltonian, Wave Functions and Matrix Elements of Various Versions of the MO-LCAO Method (π-electron systems with closed shell in the ground state)[a]

Method	$\mathscr{H}$	Ψ	Matrix elements of the Hamiltonian	Application range and limitations
1 HMO (Hückel)[17]	$\sum_i \mathscr{H}_i^{\text{ref}}$	Δ_0	$H_{\mu\mu} = \alpha + h_\mu \beta$ $H_{\mu\nu} = k_{\mu\nu}\beta$ $(\mu \neq \nu)$	Qualitative discussion of properties of the ground and excited states. In addition, HMO provides relative values of various characteristics for correlation of experimental data.
2 HMO-SC (Wheland, Mann[16]; the ω-technique)	$\sum_i \mathscr{H}_i^{\text{ref}}$	Δ_0	$H_{\mu\mu} = \alpha + (1 - q_\mu)\,\omega\beta$ (for hydrocarbons $\omega = 1.4$) $H_{\mu\nu} = k_{\mu\nu}\beta$ $(\mu \neq \nu)$	See 1. Particularly suitable for calculation of dipole moments, μ, and of ionization potentials.
3 HMO-SC (Coulson, Golebiewski)[16]	$\sum_i \mathscr{H}_i^{\text{ref}}$	Δ_0	$H_{\mu\mu} = \alpha + \xi_{\mu\mu}\beta$ $H_{\mu\nu} = k_{\mu\nu}\exp[-2.683(1.517 - 0.180P_{\mu\nu})]\beta$ $(\mu \neq \nu)$	See 1. Useful for correlation of bond lengths and bond orders.
4 HMO-SC (Janssen, Sandström)[16]	$\sum_i \mathscr{H}_i^{\text{ref}}$	Δ_0	$H_{\mu\mu} = \alpha + [\xi_{\mu\mu} + \omega(n - q_\mu)]\beta$ $H_{\mu\nu} = k_{\mu\nu}(1 + 0.5P_{\mu\nu})\beta$ $(\mu \neq \nu)$	See 1, 2 and 3.

5 SCF (Pople)[15]	$\sum_i \mathcal{H}_i^c + \sum\sum_{i<j} g(i,j)$	Δ_0	$F_{\mu\nu} = \beta_{\mu\nu}^c - \frac{1}{2} P_{\mu\nu}\gamma_{\mu\nu} + [\sum_{\sigma\neq\mu}(P_{\sigma\sigma} - Z_\sigma)\gamma_{\mu\sigma} + \gamma_{\mu\mu}]\delta_{\mu\nu}$	Ground state properties: electron energy, ionization potential, electron affinity, q (μ, ESR), P (geometry). Unsuitable for systems with strong alternation of bond lengths (polyenes, fulvenes).
6 SCF-CI (Brown, Heffernan[23]; VESCF)	$\sum_i \mathcal{H}_i^c + \sum\sum_{i<j} g(i,j)$	Δ_0	$F_{\mu\nu}$, see 5 and further: $\gamma_{\mu\mu} = f(Z)$ $I_\mu = f'(Z_\mu)$ $(Z_\mu = N_\mu + C_1 - 0.35q_\mu)$	Ground state properties, see 5. Particularly suitable for systems with significant charge alternation (cf. 3).
7 SCF-SC (β^c-variation)[24]	$\sum_i \mathcal{H}_i^c + \sum\sum_{i<j} g(i,j)$	Δ_0	$F_{\mu\nu}$, see 5 and further: $\beta_{\mu\nu}^c = f(l_{\mu\nu})$	Ground state properties, see 5. Significant for systems with strong bond alternation.
8 SCF-SC (α^c and β^c variations)[25]	$\sum_i \mathcal{H}_i^c + \sum\sum_{i<j} g(i,j)$	Δ_0	$F_{\mu\nu}$, see 5 and further: $\gamma_{\mu\mu} = f(Z_\mu)$ $I_\mu = f'(Z_\mu)$ $\beta_{\mu\nu} = f(l_{\mu\nu})$	Ground state properties, see 5. This version should be generally applicable.
9 LCI (Pariser, Parr)[13]	$\sum_i \mathcal{H}_i^c + \sum\sum_{i<j} g(i,j)$	$C_0\Delta_0 + \sum_{i\to j} C_{i\to j}\Delta_{i\to j}$	$F_{jl}\delta_{ik} - F_{ik}\delta_{jl} + 2(kj\vert g\vert li) - (kj\vert g\vert il)$ $F_{jl} = \sum_{\mu\neq\nu} c_{\mu j}F_{\mu\nu}c_{\nu l}$ $(kj\vert g\vert li) = \sum_\mu\sum_\nu c_{\mu k}c_{\nu j}c_{\mu k}c_{\nu l}\gamma_{\mu\nu}$	Properties of electronically excited states: π-electron energies of S and T states (electronic spectra), q (μ), P (geometry) in excited states. LCI SCF-MO calculations necessary for heterocyclic systems.

a Symbols have the usual meaning; α and β denote the Coulomb and resonance integrals of the HMO theory, $P_{\mu\nu}$ is the (Coulson) bond order, Δ_0 ($\Delta_{i\to j}$) is the Slater determinant of the ground ($i\to j$ excited) state. $\gamma_{\mu\nu}$ is the electronic repulsion integral.

Among the large number of these highly related approximations, the CNDO method will be discussed in greater detail. Recognition and understanding of the sequence of approximations leading to the CNDO scheme will provide a basis for orientation amongst the other methods when the expressions for the respective matrix elements are known.

Equations for the CNDO scheme can be derived from the expressions characterizing the general MO-LCAO version of the Hartree-Fock method and from Eqs. (10-1) to (10-6) by the following series of approximations:

a) Of the total electron system of the molecule, only the electrons in the valence shells of all the participating atoms are explicitly considered. Therefore, for example, the hydrogen atom contributes an electron in the 1s state to the total electronic system, first-row elements contribute electrons in the 2s and 2p states, etc. This assumption allows us to exclude electrons in the inner shells, which are assumed to electrostatically shield the atomic nuclei, from consideration. The remainder of the atom, by which we understand, for example, atom A deprived of n_A valence electrons, is generally represented physically as a positive point charge $n_A e$, where e is the proton charge; it is usual to call it the core.

b) Each atom in the molecule contributes only atomic orbitals corresponding to the principal quantum number, which is related to the highest occupied orbital of the isolated atom, to the molecular orbitals of type (10-1). Thus a hydrogen atom supplies a 1s orbital, first-row elements supply 2s and 2p orbitals, etc. It is assumed that the atomic orbitals form a set of orthonormalized functions.

c) In order to decrease the number of two-electron integrals of type $(\mu v \mid \varrho \sigma)$, the zero differential overlap assumption is introduced, according to which

$$\chi_\mu^*(1)\,\chi_v(1) = 0 \tag{10-9}$$

for $\mu \neq v$. Relationship (10-9) is a stricter condition than the requirement of orthogonality of functions χ_μ and χ_v and permits rearrangement of the double sum on the right-hand side of Eq. (10-4) in the form

$$\sum_{\varrho=1}^{m} \sum_{\sigma=1}^{m} P_{\varrho\sigma}(\mu v \mid \varrho\sigma) \rightarrow \{ \sum_{\varrho=1}^{m} P_{\varrho\varrho}(\mu\mu \mid \varrho\varrho)\}\,\delta_{\mu v} \tag{10-10a}$$

$$-\frac{1}{2} \sum_{\varrho=1}^{m} \sum_{\sigma=1}^{m} P_{\varrho\sigma}(\mu\varrho \mid v\sigma) \rightarrow -\frac{1}{2} P_{\mu v}(\mu\mu \mid vv) \tag{10-10b}$$

Although neglecting the differential overlap might seem to be too great an interference with the general equations, its usefulness must be judged from the quality of the numerical calculations, which will be discussed below. It is sufficient here to note that a partial correction is introduced by the NDDO[5] (neglect of diatomic differential overlap) scheme, according to which relation (10-9) applies only to orbitals located on different atoms.

d) It is evident that the solution of the physical problem should not depend upon the choice of the coordinate system, which is arbitrary. Pople, Santry and Segal[5] showed that the zero differential overlap approximation calls for additional approximations which are contingent on the requirement that the results be independent of the choice of the coordinate system (similarly as with the exact Hartree-Fock solution). An example will be useful here.

Coulomb interaction will be assumed to exist between two electrons, of which one occupies the $(2p)_A$ orbital on atom A and the second, the $(2s)_B$ orbital on atom B; the orientation of the $(2p_x)_A$ orbital is determined by the unprimed coordinate system depicted in Fig. 6-1, where the origin of the coordinate system lies in the nucleus of atom A. We would have been equally justified in choosing the primed system of Cartesian coordinates, which has a common origin with the original system but differs by rotation through angle α about the z-axis. Because the p orbitals have the same transformation properties as the axes of the Cartesian system, for the p$'$ orbitals expressed with respect to the primed system and for the p orbitals expressed with respect to the non-primed system, the relationship

$$\left\| \begin{matrix} (2p_x)' \\ (2p_y)' \end{matrix} \right\| = \left\| \begin{matrix} \cos\alpha, & \sin\alpha \\ -\sin\alpha, & \cos\alpha \end{matrix} \right\| \left\| \begin{matrix} (2p_x) \\ (2p_y) \end{matrix} \right\| \tag{10-11}$$

is valid. If the angle of rotation is $\alpha = \pi/4$, for which

$$\cos\frac{\pi}{4} = \sin\frac{\pi}{4} = \frac{1}{\sqrt{2}}, \tag{10-12}$$

then the product is given by

$$(2p_x)'\,(2p_y)' = \frac{1}{2}\left[(2p_y) + (2p_x)\right]\left[(2p_y) - (2p_x)\right] =$$

$$= \frac{1}{2}\left[(2p_y)(2p_y) - (2p_x)(2p_x)\right] \tag{10-13}$$

It is evident, furthermore, that because of its spherical symmetry the $(2s)_B$ orbital remains invariant to rotation through any angle.

If zero differential overlap is introduced for the primed system, then

$$((2p_x)'_A\,(2p_y)'_A\,|\,(2s)_B\,(2s)_B) = 0 \tag{10-14}$$

and therefore, considering Eq. (10-13), it follows that

$$((2p_x)_A\,(2p_x)_A\,|\,(2s)_B\,(2s)_B) = ((2p_y)_A\,(2p_y)_A\,|\,(2s)_B\,(2s)_B) \tag{10-15}$$

This result indicates that the Coulomb integrals must not depend on the orientation of the p orbitals. For this reason Pople and co-workers introduced the assumption that the Coulomb two-electron integrals $(\mu\mu\,|\,vv)$ depend only on the atoms on which the atomic orbitals χ_μ and χ_v are located, thus being independent of the specific type of orbital. They denoted these integrals as

$$\gamma_{MN} = (\mu\mu\,|\,vv) = ((2s)_M\,(2s)_M\,|\,(2s)_N\,(2s)_N) \qquad (10\text{-}16)$$

where χ_μ is the orbital located on atom M and χ_v is the orbital located on atom N. This approximation corresponds to "averaging" of the interaction of the electrons located in valence states of different atoms of the molecule and also fulfills the requirement that the solution be invariant to a transformation which leads to combination of the 2s and 2p orbitals on one centre. The second kind of invariance called the "hybridization invariance" is far less important and a number of semiempirical schemes do not even require its fulfilment (for example, the EHT method, see below).

e) The approximations concerning one-electron integrals will be analyzed by discussing the diagonal and nondiagonal elements separately. If $\mu = v$, then, from Eq. (5-19a), it follows for the one-electron part of the matrix element $F_{\mu\mu}$ [cf. Eq. (10-4)] that

$$\langle\chi_\mu\,|\,\hat{h}\,|\,\chi_\mu\rangle = \left\langle\chi_\mu\,\left|\,-\frac{\hbar^2}{2m}\Delta - \frac{Z'_M e^2}{4\pi\varepsilon_0\,|\,r - R_M\,|}\,\right|\,\chi_\mu\right\rangle -$$

$$- \sum_{I(\neq M)}\left\langle\chi_\mu\,\left|\,\frac{Z'_I e^2}{4\pi\varepsilon_0\,|\,r - R_I\,|}\,\right|\,\chi_\mu\right\rangle, \qquad (10\text{-}17)$$

where M denotes the atom on which the χ_μ orbital is located and Z'_I is the core charge expressed in multiples of the proton charge. If we assume that atomic orbital χ_μ, in accordance with the Goeppert-Mayer and Sklar[31] approximation, is an eigenfunction of the one-electron atomic Hamiltonian

$$\left(-\frac{\hbar^2}{2m}\Delta - \frac{Z'_M e^2}{4\pi\varepsilon_0\,|\,r - R_M\,|}\right)\chi_\mu = U_\mu\chi_\mu, \qquad (10\text{-}18)$$

then quantity U_μ can be considered to be the energy of an electron which is in the atomic valence state χ_μ. The value of U_μ can either be determined by calculation or can be taken from the experimentally determined atomic energy levels. Pople and co-workers chose the latter and set U_μ equal to the negative value of the ionization potential of the electron occurring in valence state χ_μ. This second kind of parameter was used in the version of the method known as CNDO/1. It appears, however, that the method yields better numerical results if U_μ is determined from the average of the ionization potential I_μ and the electron affinity A_μ, where both quantities are related to valence state χ_μ; this is obviously a better way of describing the energy conditions when loss of an electron and acceptance of an electron into the orbital χ_μ are equally probable. This alternative is used[5] in the CNDO/2 method.

The expressions

$$\left\langle \chi_\mu \left| \frac{Z'_I e^2}{4\pi\varepsilon_0 \left| r - R_I \right|} \right| \chi_\mu \right\rangle, \qquad I \neq M,$$

correspond to the electrostatic interaction between the electron, whose probability density is determined by function χ_μ located on atom M, and the remainder of the atom (the core) I. If the analytical form of χ_μ is given (for example, as a Slater orbital) it is relatively easy to calculate these integrals, as they are of the two-centre type (the coordinates of χ_μ are related to the nucleus of atom M, and the point charge of core I occurs at the nucleus of atom I). Similarly as in Coulomb two-electron integrals, the individual orbitals on the atom are also not differentiated in these integrals and the following average interaction is always introduced:

$$\left\langle (2s)_M \left| \frac{Z'_I e^2}{4\pi\varepsilon_0 \left| r - R_I \right|} \right| (2s)_M \right\rangle = v_{MI}, \qquad (10\text{-}19)$$

which again depends only on the type of participating atoms, M and I. This way of expressing the interaction is replaced in the CNDO/2 version by introducing the modified Goeppert-Mayer and Sklar potential. The core is then represented by a superposition of the neutral atom and the electron "holes", so that it is assumed that

$$-\left\langle \chi_\mu \left| \frac{Z'_I e^2}{4\pi\varepsilon_0 \left| r - R_I \right|} \right| \chi_\mu \right\rangle = \langle \chi_\mu | \mathscr{V}'_I | \chi_\mu \rangle - Z'_I \gamma_{IM} \qquad (10\text{-}20)$$

The first expression on the right-hand side of the equation represents the penetration integral corresponding to the interaction between the electron located (on atom M) in orbital χ_μ and neutral atom I, which is in the valence state. It can be assumed that the integral has a small value, which can therefore be neglected, so that

$$\langle \chi_\mu | \mathscr{V}'_I | \chi_\mu \rangle = 0 \qquad (10\text{-}21)$$

Because of approximation (10-9), the non-diagonal elements of h should be zero. It appears, however, that a semiempirical method using such an approximation would not yield physically reasonable results.

It is therefore assumed in the CNDO method that

$$\langle \chi_\mu | \hat{h} | \chi_\nu \rangle = \beta^\circ_{MN} \langle \chi'_\mu | \chi'_\nu \rangle, \tag{10-22}$$

where $\langle \chi'_\mu | \chi'_\nu \rangle$ is the overlap integral and β°_{MN} is a parameter depending on atoms M and N on which the atomic orbitals are located. It is necessary to investigate only cases when $M \neq N$, because the atomic orbitals of the valence shell located on the same atom are always orthogonal, so that the one-centre matrix elements (10-22) vanish. The actual form of the atomic orbitals on the right-hand side of Eq. (10-22) expresses an additional assumption for calculation of the matrix elements and is consequently not directly related to the basis set of the atomic orbitals [cf. approximation b)] in terms of which molecular orbitals are expressed. The introduction of primed orbitals into expression (10-22) permits the geometry of the molecule to be considered when calculating the matrix elements of $\hat{h}$ and ensures space and "hybridization" invariance of the solution.

The introduction of approximations a) to e), i.e. Eqs. (10-10a), (10-10b), (10-16), (10-18), (10-19) and (10-22), causes the secular determinant (10-3) and the matrix elements of the Hartree-Fock operator (10-4) to assume the following form in the CNDO/1 method:

$$\det \| F_{\mu\nu} - \varepsilon\delta_{\mu\nu} \| = 0 \tag{10-23}$$

$$F_{\mu\mu} = U_\mu + \left(P_M - \frac{1}{2} P_{\mu\mu} \right) \gamma_{MM} + \sum_{N(\neq M)} (P_N \gamma_{MN} - v_{MN}) \tag{10-24}$$

$$F_{\mu\nu} = \beta^\circ_{MN} S_{\mu\nu} - \frac{1}{2} P_{\mu\nu} \gamma_{MN}, \qquad \mu \neq \nu \tag{10-25}$$

P_M denotes the total electron charge on atom M:

$$P_M = \sum_{\mu \in (M)} P_{\mu\mu}, \tag{10-26}$$

where the sum is carried out over atomic orbitals located on atom M. $S_{\mu\nu}$ is an abbreviated notation for the overlap integral:

$$S_{\mu\nu} = \langle \chi'_\mu | \chi'_\nu \rangle \tag{10-27}$$

Numerical calculation according to this scheme also requires specification of the necessary integrals or matrix elements. Integrals $S_{\mu\nu}$, v_{MN} and γ_{MN} are calculated using the Slater orbitals (with the Slater value of the exponential factor for all atoms except hydrogen for which the value 1.2 was chosen). The experimentally determined values of the ionization potentials U_μ for the first and second row elements are given in Table 10-4.

Table 10-4

Values of Parameter $(-U_\mu)$ in the CNDO/1 Method (in eV) [from Ref. 5]

Atom	H	Li	Be	B	C	N	O	F
1s	13.06	–	–	–	–	–	–	–
2s	–	5.39	9.32	14.05	19.44	25.58	32.38	40.20
2p	–	3.54	5.96	8.30	10.67	13.19	15.85	18.66

The arithmetical mean of the corresponding atomic parameters is used for the value of β°_{MN}:

$$\beta^\circ_{MN} = \frac{1}{2}(\beta^\circ_M + \beta^\circ_N) \tag{10-28}$$

The values of the atomic parameters β°_I have been determined by comparison with the "ab initio" type of calculation for a number of small molecules to achieve optimum agreement of the electron charge distribution with the distribution resulting from semiempirical calculations. The choice of β°_{MN} in form (10-28) keeps the number of semiempirical parameters used in the calculation scheme at an acceptable level. The β°_M values for various atoms are given in Table 10-5.

Table 10-5

Values of Parameter β°_M in the CNDO/1 Method (in eV) [from Ref. 5]

Atom	H	Li	Be	B	C	N	O	F
$-\beta^\circ_M$	9	9	13	17	21	25	31	39

The SCF treatment based on the definition of the matrix elements (10-24) and (10-25) can be numerically solved in a standard manner as described in Section 5.5. Pople and Segal[5] proposed suitable expressions for construction of the zero approximation to the Hartree-Fock operator elements:

$$F^\circ_{\mu\mu} = U_\mu \tag{10-29}$$

$$F^\circ_{\mu\nu} = \beta^\circ_{MN}S_{\mu\nu} \tag{10-30}$$

Calculations at the SCF level are usually carried out to obtain theoretical information on the ground state of the electronic system. It was shown in Section 5.5 that the total electronic energy of a system can be simply expressed in terms of the eigenvalues of the SCF operator and of the interelectronic interaction energy [cf. Eq. (5-62)]. It is preferable for some purposes to express the total energy E_{tot} of a molecule (differing from the total electronic energy by a constant contribution corresponding to the core repulsion) in terms of one-atom and two-atom contributions:

$$E_{tot} = \sum_M E_M + \sum_M \sum_N E_{MN}, \qquad (10\text{-}31)$$

where

$$E_M = \sum_{\mu \in (M)} P_{\mu\mu} U_\mu + \frac{1}{2} \sum_{\mu \in (M)} \sum_{\mu' \in (M)} \left(P_{\mu\mu} P_{\mu'\mu'} - \frac{1}{2} P_{\mu\mu'}^2 \right) \gamma_{MM} \qquad (10\text{-}32)$$

$$E_{MN} = \sum_{\mu \in (M)} \sum_{\nu \in (N)} \left[2 P_{\mu\nu} \beta_{MN}^\circ S_{\mu\nu} - \frac{1}{2} P_{\mu\nu}^2 \gamma_{MN} \right] +$$

$$+ \left[\frac{Z_M' Z_N' e^2}{4 \pi \varepsilon_0 R_{MN}} - P_M v_{MN} - P_N v_{NM} + P_M P_N \gamma_{MN} \right] \qquad (10\text{-}33)$$

In expression (10-33) it is assumed that the core interaction terms can be expressed in the form of Coulomb repulsion between point charges at a distance of R_{MN}.

As mentioned in Section 10.1, the semiempirical methods are accompanied by a considerable decrease in the number of integrals involved in the calculation. Table 10-6, taken from the paper by Klopman and O'Leary[32], demonstrates this situation by comparing the CNDO method with the SCF "ab initio" treatment on the example of the propane molecule.

Table 10-6

Number of Molecular Integrals Necessary for Calculation of the Propane Molecule [from Ref. 32]

Integrals	SCF "ab initio" minimum basis set of atomic orbitals	CNDO
one-centre	368	11
two-centre	6 652	55
three- and four-centre	31 206	0
total	38 226	66

The results originally achieved using the CNDO/2 method for more than twenty molecules (with 4 (BeH_2) to 26 (NF_3) valence electrons) were very encouraging. The calculated quantities included dipole moments, valence angles and deformation vibrations. In recent years, methods of the CNDO type and related methods (INDO, MINDO) have been widely applied and proved satisfactory in calculation of various physical properties for a variety of compounds. It might be added that Del Bene and Jaffé[2] introduced new parameters into the CNDO scheme and thus succeeded in interpreting the electronic spectra of some hydrocarbons and their heteroanalogues. They utilized the limited configuration interaction in their method, which will be discussed in detail in Section 10.3.2 on the theoretical treatment of π-electron systems.

A semiempirical method involving valence electrons was also developed by Klopman[33]. He applied it to more than 100 diatomic molecules and to a number of triatomic molecules. A number of papers[34-39] are also concerned with these subjects.

10.2.2 Methods using an effective Hamiltonian

Hoffmann[8] introduced a method using the effective Hamiltonian – similarly as in the HMO method – but involving all the valence electrons. Formally, this is the Hückel method with an extended atomic orbital basis set, referred to as the EHT method ("extended Hückel theory") in the literature. The basis set for a hydrocarbon consists of carbon 2s and 2p orbitals and hydrogen 1s orbitals; the method takes into account overlap and non-neighbouring interactions. The calculations are performed using the following values of the ionization potentials:

$$H_{\mu\mu}(\text{C 2p}) = -11.4 \text{ eV}$$

$$H_{\mu\mu}(\text{C 2s}) = -21.4 \text{ eV}$$

$$H_{\mu\mu}(\text{H 1s}) = -13.6 \text{ eV}$$

For the hydrocarbon $C_n H_m$ the following atomic orbitals basis set is used:

m hydrogen Slater orbitals,   exponent 1.0

n 2s Slater orbitals for carbon $\Big\}$ exponent 1.625
$3n$ 2p Slater orbitals for carbon

These data are essential for calculation of the overlap integrals between Slater orbitals.

The method is applicable to a great variety of organic and inorganic molecules. It has proven satisfactory in studies concerning conformation of cycles, internal rotation, geometric isomerism and distribution of σ and π electrons. It fails in the calculation of bond lengths (and valence vibrations) and it overestimates steric factors.

Several methods[10,11,40-44] operating on a similar level were introduced earlier. They were mainly intended for studies of saturated hydrocarbons. Hoffmann's method has been described in greater detail because it is the most universal. It has also contributed to the formulation of a series of empirical rules (the Woodward-Hoffmann rules), which are used e.g. to predict the stereospecific course of various reactions of organic and inorganic compounds, and was particularly welcomed by chemists.

The method proposed by Del Re[11] is rather interesting for its simplicity. It solves the secular determinant for all localized bonds of the studied molecule (A):

$$\begin{vmatrix} H_{\mu\mu} - E, & H_{\mu\nu} - ES_{\mu\nu} \\ H_{\mu\nu} - ES_{\mu\nu}, & H_{\nu\nu} - E \end{vmatrix} = 0 \qquad (10\text{-}34)$$

$$- - -\underset{\mu}{\bullet}\!\!-\!\!-\!\!-\!\!\underset{\nu}{\bullet}\, - - -$$

(A)

The overlap is neglected. The evaluation of $H_{\mu\mu}$, $H_{\nu\nu}$ and $H_{\mu\nu}$ (see Table 10-2), which depend on the nature of atoms μ and ν forming the bond, proceeds in the following way: it is assumed that quantities $\varepsilon_{\mu\nu}$ (Table 10-2) depend only on atoms μ and ν and not on the surroundings and that δ_μ depends solely on the nature of the atoms directly bound to atom μ, so that

$$\delta_\mu = \delta_\mu^\circ + \sum_{\substack{\lambda\,(\text{neighbours of} \\ \text{orbital } \mu)}} \gamma_{\mu(\lambda)}\delta_\lambda \qquad (10\text{-}35)$$

Table 10-7

Parameters of the Method According to Del Re[11]

Bond	C–H	C–C	C–N	C–O	C–F	N–H	O–H	C–Cl
ε_{AB}	1.00	1.00	1.00	0.95	0.85	0.45	0.45	0.65
$\gamma_{A(B)}$	0.3	0.1	0.1	0.1	0.1	0.3	0.3	0.2
$\gamma_{B(A)}$	0.4	0.1	0.1	0.1	0.1	0.4	0.4	0.4
δ_A°	0.07	0.07	0.07	0.07	0.07	0.24	0.40	0.07
δ_B°	0.00	0.07	0.24	0.40	0.57	0.00	0.00	0.35

Equations of type (10-35) are set up in a number corresponding to the number of non-equivalent atoms in the system, giving a set of n equations for n unknown values, δ_μ. The values of δ_μ°, $\gamma_{\mu(\nu)}$ and $\varepsilon_{\mu\nu}$ are given in Table 10-7. The secular equation can then be solved and the electron density and bond orders can be calculated in the usual manner.

10.3 π-Electron theory

10.3.1 π−σ-Electron separation

The π-electron approximation is based on the assumption that, in unsaturated and aromatic compounds, only the π-electron system is considered explicitly in quantum mechanical calculations. The remaining electrons of the molecule, including the σ electrons, are considered to be a rigid skeleton, in the electrostatic field of which the π electrons move, and it is assumed that they are independent of changes in the π-electron system. Their effect is described semiempirically either by the values of parameters or by means of a potential, which is, for example, of a purely electrostatic nature in the Goeppert-Mayer and Sklar approximation. Within the framework of the π-electron theory, interpretations have been made with remarkable accuracy and a number of physical properties of aromatic and conjugated compounds, for example, their heats of formation and electronic spectra, have been predicted.

Strictly speaking, as the electrons are indistinguishable we should speak of π states and σ states described by wave functions of suitable symmetry, instead of π electrons or σ electrons. However, concepts such as the π electron, etc., are already in common use and have an established place in quantum chemical terminology.

The definition of states of π and σ symmetry is based on the fact that, in planar polyatomic molecules, it is possible to divide the atomic orbitals forming the basis set for the expansion of the molecular orbitals into two distinct groups. One group contains π orbitals, which are antisymmetric with respect to reflection in the molecular plane; the other group consists of σ orbitals, which are symmetric. For example, in the ethylene molecule, whose atoms lie in the xy-plane, it is possible to separate $(2p_z)_1$ and $(2p_z)_2$ orbitals located on the $(1, 2)$ carbon atoms of the molecule, which are of the π type, from the minimum atomic orbital basis set (cf. Section 6.6). The other orbitals of this basis set are σ orbitals. The π atomic orbitals form the basis for the construction of the molecular orbitals of π symmetry, on which the description of π bonds is founded. In ethylene there are two electrons in the π bond (one double bond), benzene has six π electrons, etc.

π-Electrons differ from σ electrons not only in their symmetry properties. First, each group of electrons occurs in a different part of the molecule. A π electron has zero probability density in the molecular plane, whereas a σ electron occurs in this plane with maximum probability. Generally, π electrons are more weakly bound to the molecule than σ electrons so that they are more easily ionized and more reactive and are therefore generally responsible for the chemical and physical properties of π-electron-containing compounds (electronic spectra, ionization potentials etc.). A further difference between the two types of electrons consists in the fact that the σ states can be localized in space. In this way we obtain a description for electrons in bonds between a pair of adjacent atoms, or for lone electron pairs on individual atoms, whereas the π electrons form a delocalized system over the entire conjugated molecular skeleton.

McWeeny[45] and Lykos and Parr[46] systematically studied the π-electron approximation and the region of validity of $\pi - \sigma$ separability. They concluded that, under certain conditions, the π-electron Hamiltonian (i.e. the Hamiltonian which depends solely on the coordinates of the π electrons) can be defined as

$$\mathcal{H}_\pi = \sum_{i=1}^{n_\pi} \mathcal{H}^c(i) + \sum_{i<j}^{n_\pi} \mathcal{G}(i, j), \tag{10-36}$$

which, after substituting into the expression for the mean energy value,

$$E_\pi = \frac{\langle \Phi_\pi | \mathcal{H}_\pi | \Phi_\pi \rangle}{\langle \Phi_\pi | \Phi_\pi \rangle}, \tag{10-37}$$

where $\Phi_\pi(1, 2, \ldots, n_\pi)$ is the wave function describing the π electrons, yields the π-electron contribution to the total energy of the system including interaction with the other electrons and nuclei of the molecular system. $\mathcal{H}^c$ in Eq. (10-36) is the one-electron operator involving the kinetic energy of the π electrons and their interaction with the nuclei of the atoms and with all the σ electrons. The conditions ensuring the validity of $\sigma - \pi$ separability restrict the form of the wave function and can be summarized as follows:

a) The normalized wave function Φ for the entire electron system can be written in the form of an antisymmetrized product,

$$\Phi = \mathcal{A}_{\pi\sigma} \Phi_\pi(1, 2, \ldots, n_\pi) \, \Phi_\sigma(n_{\pi+1}, \ldots, n_\pi + n_\sigma), \tag{10-38}$$

where Φ_π is an antisymmetric function with respect to permutation of the π electrons, Φ_σ is antisymmetric with respect to permutation of the σ electrons and operator $\mathcal{A}_{\pi\sigma}$ carries out permutations of σ and π electrons in such a way that the total wave function also fulfils the Pauli principle [cf. Eqs. (4-91) and (4-93)].

b) Each of partial functions Φ_π and Φ_σ is, in itself, normalized, i.e.,

$$\langle \Phi_\pi | \Phi_\pi \rangle = \langle \Phi_\sigma | \Phi_\sigma \rangle = 1 \tag{10-39}$$

c) Each of functions Φ_π and Φ_σ can be expanded [cf. Eq. (5-28)] in a series of orthonormal Slater determinants,

$$\Phi_\pi = \sum_i c_i^\pi \Delta_i^\pi \tag{10-40}$$

$$\Phi_\sigma = \sum_i c_i^\sigma \Delta_i^\sigma, \tag{10-41}$$

in which determinants Δ_i^π and Δ_i^σ are built by means of π and σ orbitals, respectively. It is assumed that the subsets of π and σ orbitals possess no common elements — atomic orbitals.

d) Wave function Φ_σ is identical for the ground state and for all the excited states of the molecule. In other words, all changes in the molecule (excitation, ionization) occur only in the π-electron system and the other electrons are not affected.

On fulfillment of conditions *a*) to *d*), the total electronic energy of a molecule can be written in the form

$$E = E_\sigma^\circ + E_\pi, \tag{10-42}$$

where E_σ° denotes a constant energy value (common for the electronic ground state and for the electronically excited states of the molecule) which is contributed by electrons in σ states (including all the electrons in the inner, i.e. non-valence-atomic shells), and the π-electron energy, E_π, is defined by Eq. (10-37). In agreement with this conclusion, the variation principle can be applied directly to expression (10-37) without considering the σ electrons, as they contribute only a constant value to the total energy.

10.3.2 The Pople version of the SCF method for π-electron systems

Initially, it should be noted that similar approximations are employed in the calculation scheme of the Pople version[15] of the SCF method for π-electron systems as those encountered when deriving the equations characterizing the CNDO method. This is essentially because the SCF-method for π-electron systems had already been elaborated (in the early fifties), and had proven satisfactory; consequently, an attempt was later made to apply similar calculation schemes to the description of systems involving all the valence electrons.

From the general LCAO-SCF expressions and Eqs. (10-1) to (10-6) it is possible to obtain the Pople version of the SCF method by introducing the following approximations:

a) The π electrons in the studied aromatic or conjugated molecule can be treated independently of the other electrons. Molecular orbitals (10-1) are expressed in the form of a linear combination of atomic orbitals of π symmetry, for example, in atoms with atomic numbers 3 to 10, the 2p orbitals perpendicular to the molecular plane are considered.

b) The zero differential overlap approximation expressed by Eq. (10-9) is used, enabling a considerable reduction of the number of two-electron integrals [cf. Eqs. (10-10a) and (10-10b)], as well as neglect of all overlap integrals.

c) The one-centre Coulomb integrals are expressed using atomic spectroscopic data for the ionization potential I_μ and electron affinity A_μ of an electron occurring in the $\pi-$orbital χ_μ of the atom M:

$$(\mu\mu \mid \mu\mu) = \gamma_{\mu\mu} = I_\mu - A_\mu, \tag{10-43}$$

where a similar notation as in Eq. (10-16) was introduced for the Coulomb integrals except that indices which simultaneously denote the atom are retained, which is possible because only one atomic orbital located on atom M can occur in molecular orbital (10-1). Relation (10-43) was introduced by Pariser[47] and is based on the energy balance for the model reaction of a simple electron transfer process; the respective electron is assumed to occupy atomic valence state p. For two carbon atoms we can write

$$\dot{C} + \dot{C} \;\; \rightarrow \;\; \ddot{C} + C^+ \tag{10-44}$$

This type of charge transfer can be represented by the superposition of two processes:

$$\dot{C} \;\; \rightarrow \;\; C^+ + e \;\;\; (I_C)$$
$$\dot{C} + e \;\; \rightarrow \;\; \ddot{C} \;\;\; (-A_C),$$

whence

$$I_C - A_C = \Delta E = \gamma_{\mu\mu} \tag{10-45}$$

On the left-hand side of Eq. (10-45) are quantities related to the carbon atom. Their numerical values are evident from Table 10-8. Analogous relations are also used for the determination of one-centre Coulomb integrals for other atoms (cf. Table 10-8).

d) The matrix elements of operator $\mathscr{H}^c$ are expressed similarly as in the CNDO scheme.

The diagonal element $\langle \chi_\mu \mid \mathscr{H}^c \mid \chi_\mu \rangle$ can be written as given by Eq. (10-17). Eqs. (10-18) to (10-21) yield

$$\langle \chi_\mu \mid \mathscr{H}^c \mid \chi_\mu \rangle = U_\mu - \sum_{\nu(\neq\mu)} Z_\nu \gamma_{\mu\nu}, \tag{10-46}$$

Table 10-8

Parameters I_μ, A_μ, $\beta^c_{\mu C}$ (in eV) of the Pople SCF Method[a]

Atom (μ)	Type of compound	I_μ	A_μ	$\beta^c_{\mu C}$
C	conjugated planar	11.22	0.69	−2.318
N	pyridine	14.1	1.8	−2.318
	aniline, pyrrole	27.3	9.3	−1.854
O	ketone, quinone	13.6	2.3	−2.318
	phenol, furan	32.9	10.0	−2.318
O⁻	phenolate	21.0	9.5	−2.318
S	thiophene	20.0	9.16	−1.623

[a] The same parameters are also used in the common PPP method (PPP is an abbreviation of Pariser, Parr and Pople), *a limited configuration interaction method* (LCI) utilizing the SCF molecular orbitals (i.e. LCI-SCF method, see below).

where Z_v denotes the number of electrons which the atom v (previously denoted by N) contributes to the π-electron system. The Goeppert-Mayer and Sklar assumptions[31], represented by Eqs. (10-18) and (10-20), were originally introduced in a treatment of π-electron systems. However, assumption (10-21) of negligibility of the penetration integrals, although used frequently in various versions of parametrization, is not always employed.

The non-diagonal matrix elements are considered to be empirical parameters, which are usually chosen so that the calculation optimally reproduces the experimental data for one molecule or a group of molecules.* The "tight binding" approximation is very often introduced, according to which

$$\langle \chi_\mu | \mathscr{H}^c | \chi_v \rangle = \begin{cases} \beta^c_{\mu v} & \text{if } \mu \text{ and } v \text{ correspond to neighbouring atoms} \\ 0 & \text{in all other cases.} \end{cases} \quad (10\text{-}47)$$

The rearrangement of the matrix elements of the Hartree-Fock operator (10-4) in the sense of approximations *a*) to *d*), i.e., Eqs. (10-10a), (10-10b), (10-43), (10-46) and (10-47), leads to the expressions

$$F_{\mu\mu} = U_\mu + \frac{1}{2} P_{\mu\mu} \gamma_{\mu\mu} + \sum_{\sigma(\neq\mu)} (P_{\sigma\sigma} - Z_\sigma) \gamma_{\mu\sigma} \quad (10\text{-}48)$$

$$F_{\mu v} = \beta^c_{\mu v} - \frac{1}{2} P_{\mu v} \gamma_{\mu v}, \qquad (\mu \neq v) \quad (10\text{-}49)$$

* We speak of the adjustment of parameters to experimental data. The values thus obtained (for example, using heats of formation, and spectral transitions in ethylene and benzene) are then employed in the entire region of structurally related (in this case conjugated) compounds.

The secular determinant has the form given by expression (10-23). The charge-and bond-order matrix elements $P_{\mu\nu}$ were defined by Eq. (10-5) and refer to molecular orbitals of π symmetry. Table 10-8 summarizes the values of I_μ, A_μ, as well as the core resonance integral $\beta^c_{\mu\nu}$ for a number of important atoms and bonds. It should be added when considering the numerical solution of the SCF equations, discussed in detail in Section 5.5, that the secular determinant of the zeroth iteration step is usually constructed using expansion coefficients obtained by the simple molecular orbital method (HMO, see below).

The Pople method appears to be the most convenient semi-empirical method for description of the properties of aromatic and conjugated organic molecules in the electronic ground state (heats of formation, dipole moments, bond lengths, chemical reactivity). The total electron energy at the SCF level can best be expressed in the form of Eq. (5-62). To determine the total energy of the molecule, E_{tot}, it is necessary to add the core repulsion terms [cf. Eq. (10-33)].

The repulsion of electrons corresponding to atoms μ and ν in a neutral molecule with uniform electron charge distribution is approximately the same as the μ and ν core repulsion, so that the Coulomb term (the second term) in Eq. (5-62) is roughly compensated by the core repulsion and Eq. (5-62) therefore assumes the form

$$E_{tot} \approx 2\sum_i \varepsilon_i + \sum_i\sum_j \langle \varphi_i\varphi_j | \mathscr{g} | \varphi_j\varphi_i \rangle \tag{10-50}$$

If the eigenvalues ε_i and the exchange integrals in Eq. (10-50) are expressed in terms of the expansion coefficients of the molecular orbitals, the relationship

$$E_{tot} = \sum_\mu P_{\mu\mu}\left(U_\mu + \frac{1}{2}P_{\mu\mu}\gamma_{\mu\mu}\right) + \sum\sum_{\mu\neq\nu} P_{\mu\nu}\left(\beta^c_{\mu\nu} - \frac{1}{4}P_{\mu\nu}\gamma_{\mu\nu}\right) \tag{10-51}$$

is obtained. If the expressions in parentheses are replaced by the Coulomb (α_μ) and resonance $(\beta_{\mu\nu})$ integrals from the HMO method, the HMO expression for the total energy is obtained (Section 10.3.5).

10.3.3 The Pariser-Parr method of limited configuration interaction

Approximation of the wave function as a single-determinant function, is, as a rule, insufficient for the calculation of the electronic structure of molecules in excited states. A remedy for this situation can be found by describing the electronic states of the molecule in terms of a linear combination of Slater determinants (cf. Section 5.4).

Slater determinants can be constructed on the basis of either HMO (see below) or SCF molecular orbitals. In some systems (for example, benzenoid hydrocarbons) the HMO and SCF orbitals lead to practically the same results. In systems with heteroatoms, however, the application of SCF molecular orbitals is preferable; in addition, it is possible to use the fact that configurations of singly excited states do not interact with the ground state configurations [the Brillouin theorem (cf. Section 5.5)]. The integrals which must be evaluated when employing the configuration interaction method are analogous to the integrals appearing in treatments using the SCF method.

In the semiempirical method of Pariser and Parr[13], the expansion of the wave function involves, in addition to the determinant of the ground state, only the determinants of singly excited configurations, obtained from the ground state configuration (cf. Sections 5.4 and 5.5) by replacing the i-th occupied molecular orbital by the j-th unoccupied molecular orbital. Such a configuration will be denoted by the symbol (i, j).

The wave function describing state a can then be written in the form

$$\Psi_a = C_{0a}\Delta_0 + \sum_{(i,j)} C_{(i,j)a}\Delta_{ij}, \tag{10-52}$$

where expansion coefficients C are variation parameters. Expansion of the wave function using all the singly excited configurations is sometimes too tedious from the aspect of computation and then only some of them are considered. The general algorithm for the calculation has been described in Section 5.4.

The expression for the charge distribution derived from the wave function described above warrants particular attention. Whereas the electron densities on atoms (q_μ) and bond orders $(P_{\mu\nu})$ can be simply expressed within the single-determinant approximation (SCF or HMO; closed shell system) in the form

$$P_{\mu\nu} = 2 \sum_{i(occ.)} c_{\mu i}^* c_{\nu i} \tag{10-53a}$$

$$q_\mu = P_{\mu\mu}, \tag{10-53b}$$

the expression for the bond orders in the limited configuration interaction (LCI) method involving singly excited configurations alone is defined in a far more complicated way[24]:

$$^a p_{\mu\nu}^{(LCI)} = P_{\mu\nu} + \sum_{(i,j)} [C_{(i,j)a}]^2 (c_{\mu j} c_{\nu j} - c_{\mu i} c_{\nu i}) +$$

$$+ \sqrt{(2)} \sum_{(i,j)} (C_{0,a})(C_{(i,j)a})(c_{\mu i} c_{\nu j} + c_{\mu j} c_{\nu i}) +$$

$$+ \sum_{(i,j)<(k,l)} (C_{(i,j)a})(C_{(k,l)a})[(c_{\mu j} c_{\nu l} + c_{\mu l} c_{\nu j})\delta_{ik} - (c_{\mu i} c_{\nu k} + c_{\mu k} c_{\nu i})\delta_{jl}] \tag{10-54}$$

$P_{\mu\nu}$ is the bond-order matrix element of the ground state configuration [cf. Eq. (10-53a)], indices i and k denote occupied molecular orbitals, j and l are unoccupied molecular orbitals and it is assumed that expansion coefficients are real numbers. The summation $\sum\limits_{(i,j)<(k,l)}$ is equivalent to the summation $\frac{1}{2}\sum\limits_{(i,j)\neq(k,l)}$ over all the considered singly excited configurations. The expression for the generalized bond order $^a p_{\mu\nu}^{(LCI)}$ follows from the expression for the first-order density matrix corresponding to wave function Ψ_a,

$$\tilde{\gamma}_a(\boldsymbol{r};\boldsymbol{r}') = \sum_{\mu\nu} {}^a p_{\mu\nu}^{(LCI)} \chi_\mu(\boldsymbol{r}) \chi_\nu(\boldsymbol{r}') \tag{10-55}$$

(cf. Section 11.2.2). In this equation, the atomic orbitals are assumed to be real functions.

10.3.4 A survey of semiempirical π-electron methods

Having become acquainted with the most important types of semiempirical methods used for studies of conjugated systems, the most important approximations which have so far been employed will be surveyed (Table 10-9). In this connection, we will mention various possibilities for the approximation of integrals and describe the most important expressions.

The neglecting of individual terms and approximations will be discussed systematically; the scheme given in Table 10-9 applies for the further discussion.

Group A.

The very numerous theoretical characteristics of planar (or almost planar) conjugated systems, which are interesting for chemists and physicists, are not significantly influenced by this group of approximations.

Group B.

Subgroup B.1. Within the Goeppert-Mayer and Sklar approximation, in instance (*i*), α_μ^c (c denotes core) will be approximated by the corresponding ionization potential of the atom in the valence state. In instance (*ii*) the penetration integrals are explicitly considered, but this is rather rare[24,48−50]. Their inclusion is manifested in two chief ways:

a) in a non-uniform charge distribution in alternant hydrocarbons,

b) in the fact that the theoretical transitions corresponding to α bands in the electronic spectra of benzenoid hydrocarbons (see below) become allowed.

Table 10-9

Survey of Neglected Terms and Approximations in π-Electron Semiempirical Methods Used for the Study of Conjugated Systems

---

A.1. Neglect of relativistic corrections

2. Born-Oppenheimer approximation

3. Electron correlation is included only in empirical parameters

4. π-electron approximation

$$E = \frac{\int \Delta_0 \mathcal{H} \Delta_0 \, d\tau}{\int \Delta_0^2 \, d\tau}$$

$$\mathcal{H} = \sum_i \mathcal{H}_i^c + \sum\sum_{i<j} g(i,j)$$

Δ_0...normalized Slater determinant of the ground state (LCAO approximation, χ denotes AO's)

B.1. $\mu = \nu$

(i) approximation through effective ionization potentials I (neglect of penetration integrals)

(ii) calculation of penetration integrals

2. $\mu \neq \nu$

(i) μ and ν are neighbours

α) studies with constant empirical values (2 to 3 eV)

β) values depend on the bond length $[\beta_{\mu\nu}^c = f(r_{\mu\nu})$ or $f(P_{\mu\nu})]$

(ii) μ and ν are not neighbours

α) $\beta_{\mu\nu}^c = 0$ (tight binding approximation)

β) $\beta_{\mu\nu}^c \sim S_{\mu\nu}$

a) $\int \chi_\mu(1) \, \mathcal{H}^c(1) \, \chi_\nu(1) \, d\tau(1)$...one- and two-centre core integrals

b) $\int \chi_\mu(1) \chi_\nu(1) \, g(1,2) \chi_\varrho(2) \chi_\sigma(2) \, d\tau(1) \, d\tau(2) \equiv (\mu\nu \,|\, \varrho\sigma)$...many-centre electron repulsion integrals

C.1. "m^4 catastrophe" [ZDO: zero differential overlap...$(\mu\nu \,|\, \varrho\sigma) \, \delta_{\mu\nu}\delta_{\varrho\sigma}]$

2. One-centre electron repulsion integrals: $(\mu\mu \,|\, \mu\mu) \equiv \gamma_{\mu\mu} = I_\mu - A_\mu$

3. Two-centre electron repulsion integrals: $(\mu\mu \,|\, \nu\nu) \equiv \gamma_{\mu\nu}$ approximated using various formulas (see the text)

Subgroup B.2.

Case *I*: μ and ν are neighbouring atoms

a) If constant quantities are employed, it is possible to use the values given in Table 10-8.

b) In general it is necessary to include the dependence of these integrals on the bond lengths[51,52]. Several empirical formulas have been proposed for this purpose*, 1 and 6 are particularly important:

1. $$\beta_{\mu\nu}^c = \beta_0^c \exp\{a(1.397 - r_{\mu\nu})\} \qquad (10\text{-}56)$$

a is equal to either 4.5988 (reference 53), or

3.2196 (reference 54)

---

* The formulae were taken directly from the quoted papers. The distances are mostly in 10^{-10} m and the calculated quantities are in eV. Before using these formulae, it is recommended that the reader consult the original literature.

and further $r_{\mu\nu} = b - cP_{\mu\nu}$
with $b = 1.52$, $c = 0.19$, or
$\qquad b = 1.50$, $c = 0.15$ (cf. cited papers)

β_0^c denotes the core resonance integral of the standard C−C bond; the value $\beta_0^c = -2.318$ eV is frequently used.

2.
$$\beta_{\mu\nu}^c = k/r_{\mu\nu}^6 \quad \text{(reference 55)} \tag{10-57}$$

$$k_{C=C} = -17.464$$

$$k_{C=N} = -13.983$$

$$k_{C=O} = -8.8086$$

3.
$$\beta_{\mu\nu}^c = -6442 \exp(-5.6864 r_{\mu\nu}) \quad \text{(reference 13)} \tag{10-58}$$

4. $\beta_{\mu\nu}^c = -2524 \exp\left\{-5.047\left(\dfrac{\xi_\mu - \xi_\nu}{\xi_c} - 2\right)^2 - 5r_{\mu\nu}\right\}$ (reference 56)

$\qquad\qquad\qquad\qquad$ (10-59)

(ξ's are the effective nuclear charges of the orbitals)

5.
$$\beta_{\mu\nu}^c = -1.60 + aP_{\mu\nu} + bP_{\mu\nu}^2 \quad \text{(reference 57)} \tag{10-60}$$

6. Finally, Mulliken's relation between the resonance and overlap integrals must be mentioned:

$$\beta_{\mu\nu}^c = \beta_0^c \frac{S_{\mu\nu}}{S_0} \tag{10-61}$$

(the quantities with index 0 refer to the reference bond)

7.
$$\beta_{\mu\nu}^c = \frac{1}{r_{\mu\nu}} \frac{dS_{\mu\nu}}{dr_{\mu\nu}} \quad \text{(reference 58)}$$

Case *II*: μ and ν are not neighbouring atoms

a) $\beta_{\mu\nu}^c = 0$ ("tight binding" approximation; this approximation is used very frequently).

b) If all β's are considered, then the greatest difficulty lies in finding a suitable function to correctly describe the interaction between more distant centres. Flurry and Bell[59] tested several approximations:

$$\beta_{\mu\nu}^c \sim S_{\mu\nu}$$
$$\beta_{\mu\nu}^c = f[a \exp(-hr_{\mu\nu})]$$

The following relation appeared to them to be useful (although not optimal):

$$\beta_{\mu\nu}^c = \frac{(2 - S_{\mu\nu}) S_{\mu\nu}}{2 - S_{\mu\nu}^2} (H_{\mu\mu} H_{\nu\nu})^{1/2} \tag{10-62}$$

($H_{\mu\mu}$ is the ionization potential for the valence state.)

Flurry, Stout and Bell[56] employed the formula of Katagiri and Sandorfy[6] in studies of phenols and related compounds.

$$\beta_{\mu\nu}^c = - \frac{S_{\mu\nu}}{4} \{b_\mu + (\mu\mu \,|\, \mu\mu) + b_\nu + (\nu\nu \,|\, \nu\nu) + 2(\mu\mu \,|\, \nu\nu)\}, \quad (10\text{-}63)$$

where b_μ and b_ν are empirical constants:

Atom	Type	b_μ
C	–	7.56
N	pyridine	11.15
N	pyrrole	20.0
O	carbonyl	9.0
O	furan	38.0

The following expression appeared to be particularly useful:

$$\beta_{\mu\nu}^c = K \frac{2Z_\mu Z_\nu}{Z_\mu + Z_\nu} S_{\mu\nu}(2 - S_{\mu\nu}), \quad (10\text{-}64)$$

where K is a numerical constant (0.5246), Z_μ (Z_ν) are the core charges and $S_{\mu\nu}$ is the overlap integral between Slater orbitals.

Group C.

Subgroup C.1. The zero differential overlap approximation (ZDO) reduces "catastrophe m^4" to "unpleasantness m^2". It is almost universally applied.

Subgroup C.2. Approximation $I - A$ [cf. Eq. (10-45)] of one-centre electron repulsion integrals, introduced by Pariser, has proven very satisfactory.

Subgroup C.3. Two-centre electron repulsion integrals for π electrons located in the χ_μ and χ_ν AO's

$$\gamma_{\mu\nu} \equiv (\mu\mu \,|\, \nu\nu) = \int \chi_\mu^*(1)\, \chi_\mu(1)\, [e^2/(4\pi\varepsilon_0 r_{12})]\, \chi_\nu^*(2)\, \chi_\nu(2)\, d\tau(1)\, d\tau(2) \quad (10\text{-}65)$$

must fulfill two conditions:

$$\lim_{r \to 0} \gamma(r) = \gamma_{11} \quad (10\text{-}66)$$

$$\lim_{r \to \infty} \gamma(r) = e^2/4\pi\varepsilon_0 r$$

Many formulae are used in the literature for approximating these integrals; several of them are given here for illustration; formula 4.* below has proven very useful.

---

* The distances are substituted in multiples of 10^{-10} m; $\gamma_{\mu\nu}$ is in eV.

1. The expression given by Parr[60] based on a model in which the distribution of the π charges is approximated by charged spheres:

$$\gamma_{\mu\nu} = \frac{7.1975}{r_{\mu\nu}} \left\{ \left[1 + \left(\frac{R_\mu - R_\nu}{2r_{\mu\nu}} \right)^2 \right]^{-1/2} + \left[1 + \left(\frac{R_\mu + R_\nu}{2r_{\mu\nu}} \right)^2 \right]^{-1/2} \right\}$$

$$(r_{\mu\nu} \geqq 2.8 \times 10^{-10} \text{ m}) \qquad (10\text{-}67)$$

R_μ (R_ν) denotes the diameter of a homogeneously charged sphere. For R_μ it holds

$$R_\mu = \frac{4.597}{Z_\mu} 10^{-8} \text{ cm}$$

In addition,

$$\gamma_{\mu\nu} = \tfrac{1}{2}[\gamma_{\mu\mu} + \gamma_{\nu\nu}] - ar_{\mu\nu} - br_{\mu\nu}^2 \qquad (r_{\mu\nu} < 2.8 \times 10^{-10} \text{ m}) \quad (10\text{-}68)$$

2. The Pople approximation[15]

$$\gamma_{\mu\nu} = 14.399/r_{\mu\nu} \qquad (10\text{-}69)$$

3. The Löwdin-Ohno approximation[61]

$$\gamma_{\mu\nu} = \frac{e^2}{\sqrt{(r_{\mu\nu}^2 + c^2)}}, \qquad (10\text{-}70)$$

where $c = \dfrac{14.399}{0.5[\gamma_{\mu\mu} + \gamma_{\nu\nu}]}$.

4. The Mataga-Nishimoto approximation for hydrocarbons[48]

$$\gamma_{\mu\nu} = \frac{e^2}{a + r_{\mu\nu}} = \frac{14.399}{1.328 + r_{\mu\nu}} \qquad (10\text{-}71)$$

This completes the detailed discussion of parametrization in π-electron methods. There are several reasons for the extent of this information. First of all, when using semiempirical methods, the chemist frequently encounters many of the given formulae in the literature. Furthermore, the π-electron methods are not obsolete, as some authors believed in the early seventies. In spite of the development of theoretically more exact methods and the use of computers, it cannot be expected that, in the near future, these methods will yield better numerical results for planar conjugated systems than those based on the π-electron approximation; in addition, financial outlay connected with calculations on systems containing 20 to 50 atoms (50 to 120 electrons) would be unjustifiable. Furthermore, the parametrization of π-electron methods is sufficiently developed so that these methods are very suitable, not only for the interpretation of experimental data, but even for relatively safe predictions; they can also contribute to the solution of structural problems.

At present there is, however, a certain characteristic of semiempirical quantum chemical methods which should be mentioned; this could be termed a method and parameter explosion. The reader has probably noticed that the number of combinations of proposed approximations to the individual integrals (Table 10-9) amounts to several hundred. For the chemist who is not a quantum chemist, selection of a method for the interpretation of experimental data is not an easy task. It is obviously necessary to choose the optimum combination (i.e., the one best describing the experimental results). This choice is almost impossible. Confusion can be overcome by applying a version which has proven satisfactory in a well established laboratory. There is, however, also a second possibility: to investigate systematically, after sufficient theoretical consideration, the very numerous available possibilities. Though this is rather thankless work, it has led[62] to useful results. It would be foolish to believe that it yields the best results for conjugated compounds of all known types and for all important physical characteristics. But we can safely claim that it yields good results for the characteristics of the ground state (heats of atomization, dipole moments, bond lengths) as well as for the characteristics of electronically excited states (excitation energy, transition and dipole moments). This is a very positive result, because in the literature it is widely believed that description of the characteristics in ground and excited states can be achieved only by using two different sets of parameters.

Because of its universal applicability, because of the possibility of considering the σ-core polarization and because of its "objectiveness" (in the sense of "independence" of the person doing the calculation as far as the parametrization is concerned), the proposed procedure deserves more detailed description. The diagonal elements $\mathscr{H}^{c}_{\mu\mu}$ are approximated by the formula

$$\mathscr{H}^{c}_{\mu\mu} = \alpha_{\mu} - \sum_{\nu \neq \mu} Z_{\nu} \gamma_{\mu\nu} - \sum_{\sigma} Z_{\sigma} \gamma_{\mu\sigma}, \qquad (10\text{-}72)$$

where the individual terms are defined as follows:

$$\alpha_{\mu} = b_{\mu} Z_{\mu}^{2} + c_{\mu} Z_{\mu} + d_{\mu} \qquad (10\text{-}73)$$

$$\gamma_{\mu\mu} = c_{\mu} Z_{\mu} + f_{\mu} \qquad (10\text{-}74)$$

$$\vartheta_{\mu} = \vartheta_{\mu}^{\circ} + \Delta\vartheta_{\mu}(Z_{\mu} - Z_{\mu}^{\circ}) \qquad (10\text{-}75)$$

The second summation in Eq. (10-72) is carried out only over atoms bound by σ bonds.

The change in the core charge, ΔZ_μ, is calculated using the formula

$$\Delta Z_\mu = -c_\mu \sum_{\nu \neq \mu} \frac{\chi_\nu - \chi_\mu}{\chi_\mu + \chi_\nu} r_{\mu\nu}^{-2}, \tag{10-76}$$

where the summation is carried out over all the neighbours of position μ, i.e., not only over all the atoms of the conjugated skeleton. χ_μ is the Mulliken electronegativity of atom μ; in references 62 are given tables of the optimal values of constants b_μ, c_μ, d_μ, f_μ, ϑ_μ° and $\Delta\vartheta_\mu$ for elements of the first two periods of the periodic table of the elements and also for As, Se, Br, Sb, Te and I. The ϑ_μ terms appear in the expressions for the bond length

$$r_{\mu\nu} = \sqrt{\left[\frac{2.9 + 0.175(Z_\mu^\circ - Z_\nu^\circ)}{\vartheta_\mu + \vartheta_\nu}\right]}(1.523 - 0.19 P_{\mu\nu}) \frac{n_\mu n_\nu}{n_\mu + n_\nu}, \tag{10-77}$$

where n_μ and n_ν are the principal quantum numbers of the Slater orbitals of atoms μ and ν [cf. Eqs. (8-6) and (8-7)].

Finally, the resonance integrals (all of them, the "tight binding" approximation is not used here) are approximated by the expression

$$H_{\mu\nu}^c = \beta_{\mu\nu} = 0.542 \frac{2\alpha_\mu \alpha_\nu}{\alpha_\mu + \alpha_\nu} S_{\mu\nu}(2 - S_{\mu\nu}), \tag{10-78}$$

where $S_{\mu\nu}$ is the overlap integral between the Slater orbitals of atoms μ and ν. It remains to be added that the two-centre repulsion integrals have been evaluated using the formula introduced by Mataga and Nishimoto [Eq. (10-71)].

10.3.5 Very simple π-electron version of the MO method

One of the oldest versions of the MO method, the Hückel method[17,64-66] (HMO), belongs in this group. It is characteristic for this method that various simplifications were taken to extremes: the introduction of any further simplification would result in the collapse of the whole method. Similarly as in the EHT method, here the electron repulsion is also not explicitly considered and it is assumed that the total Hamiltonian can be expressed as the sum of effective Hamiltonians, each of which depends on the coordinates of a single electron (cf. Section 5.5):

$$\mathscr{H}(1, 2, ..., n_\pi) = \sum_{\mu=1}^{n_\pi} \mathscr{H}_\mu^{ef} \tag{10-79}$$

Consequently, the effective operator $\mathscr{H}_1^{ef}$ corresponds to the first electron, etc. This need not be specified because all matrix elements in which these operators appear are – as mentioned below – considered to be empirical parameters. Therefore, in practice, these integrals are not calculated but are replaced by suitable numerical values (adjustment to the experimental data).

The difference between the HMO method and the EHT method lies in that a) in the HMO method the overlap integrals $\langle \chi_\mu | \chi_\nu \rangle \equiv S_{\mu\nu}$ ($\mu \neq \nu$) are neglected and b) while the p_z atomic orbitals of the carbon atoms alone establish the basis set for the MO's in the HMO method, all atomic orbitals corresponding to the valence electrons are considered in the EHT method.

The optimum values of $c_{\mu i}$ (μ is the AO index, i is the MO index in the molecular orbitals $\varphi_i = \sum_{\mu=1}^{m} c_{\mu i} \chi_\mu$) are determined as usual by the variation method. The system of linear equations for their determination has the form

$$\sum_{\nu=1}^{m} c_\nu (H_{\mu\nu} - E S_{\mu\nu}) = 0; \qquad \mu = 1, 2, ..., m \qquad (10\text{-}80)$$

and it must hold that

$$\det \| H_{\mu\nu} - E S_{\mu\nu} \| = 0 \qquad (10\text{-}81)$$

The following types of matrix elements occur in the energy calculation

$$H_{\mu\nu} = \int \chi_\mu \mathcal{H}^{\text{ef}} \chi_\nu \, d\tau = \langle \chi_\mu | \mathcal{H}^{\text{ef}} | \chi_\nu \rangle, \qquad (10\text{-}82)$$

denoted by α_μ (Coulomb integral: $\mu = \nu$) and by $\beta_{\mu\nu}$ (resonance integral: $\mu \neq \nu$; the atomic orbitals are considered to be real functions).

In the Hückel method, the following simplifying assumptions are made concerning these integrals:

a) The α_μ for all centres (corresponding to all conjugated C atoms) have the same value, α.

b) $\beta_{\mu\nu}$ is considered to equal zero if the carbon atoms in the μ and ν positions are not bound by a σ bond and to equal the uniform value β if the C atoms are bound to each other (the "tight binding" approximation).

c) It is assumed that the atomic orbitals form an orthonormal basis, so that

$$S_{\mu\nu} = \delta_{\mu\nu},$$

where $\delta_{\mu\nu}$ is the Kronecker delta.

The α and β integrals are considered to be empirical parameters of the HMO method, so that numerical calculations can be carried out within the framework of this method without specifying the expressions for the AO's and for the effective Hamiltonian. The total π-electron energy [cf. Eqs. (5-40b) and (10-115)] obtained within the HMO scheme will be denoted by W.

Several refinements have been introduced into the HMO method; a modification enables extension of the method to systems containing heteroatoms (Pauling).

The first refinement (Wheland) concerns the introduction of non-vanishing values for the overlap integrals between adjacent orbitals; it is usually assumed in cyclic systems that $S_{\mu\nu} = 0.25$, where μ and ν designate the $2p_z$ AO's on adjacent carbons connected by a σ bond. The introduction of this refinement does not lead to numerical complications and can be made after completion of the standard HMO calculation. It appears, however, that its introduction is not connected with increased quality of the theoretical data. Considerably more important are modifications which do not employ constant values of α_μ and $\beta_{\mu\nu}$ but rather assume functional dependences either

$$\alpha_\mu = \alpha_\mu(q_\mu)$$

or

$$\beta_{\mu\nu} = \beta_{\mu\nu}(P_{\mu\nu}),$$

or make both assumptions simultaneously. For more detailed information on the respective methods as well as their range of application, see Table 10-3.

Many studies have been devoted to the selection of empirical parameters for heteroatoms and for heteroatom−carbon and hetero-atom−heteroatom bonds. Streitwieser proposed a very useful series of values; we have also used similar values in our laboratory; Table 10-10 indicates a set of values which proved satisfactory in various applications. These parameters are generally stated in the form

$$\alpha_\mu = \alpha + h_\mu\beta \tag{10-83}$$

$$\beta_{\mu C} = k_{\mu C}\beta, \tag{10-84}$$

where μ denotes the heteroatom, μC is the heteroatom−carbon bond, α is the Coulomb integral of the $2p_z$ carbon orbital and β is the resonance integral of the π carbon−carbon bond.

It can be shown that, for some simple systems (for example polyenes, cyclopolyenes, polyacenes), expressions for the calculation of orbital energies and other quantities can be given in closed form (Table 10-11). For polyenes, the general expression of the orbital energy can be obtained in a manner which will be outlined briefly here using a polyene of m

Table 10-10

HMO Parameters h_μ and $k_{\mu\nu}$ [Eqs. (10-83) and (10-84)]

Atom	Example	h_μ	Bond	$k_{\mu C}$
B	(borazole)	-1	B⋯N, B⋯C	0.7
C	(naphthalene, azulene)	0	C⋯C	1
C	(hexatriene)	0	C=C	1.1
			C−C	0.9
N	(pyridine)	0.5	C⋯$\overline{N}$	1
$\overset{+}{N}$	(pyridine cation)	2.0	C⋯$\overset{+}{N}$	1
N	(pyrrole, aniline)	1.5	C−$\overline{N}$	0.8
N	(nitrile)	0.5	C≡N|	1.4
N	(nitrobenzene)	2.0	C⋯N	0.8
			$N\overset{\nearrow O}{\searrow O}$	0.7
O	(ketone, phenolate)	1.0	C=O, C−O$^-$	1.0
O	(furan, phenol)	2.0	C−O	0.8
S	(thioketone)	0.5	C=S	0.9
S	(thiophene, thiophenol)	1	C−S	0.7
F	(fluorobenzene)	3	C−F	0.7
Cl	(chlorobenzene)	2	C−Cl	0.4
Br	(bromobenzene)	1.5	C−Br	0.3
I	(iodobenzene)	1.3	C−I	0.25

Table 10-11

General Formulas for Calculation of HMO Orbital Energies in Several Types of Systems

Formula in Fig. 10-1	System	k_j	
I	linear polyenes	$2\cos\dfrac{j\pi}{m+1}$;	$j = 1, 2, ..., m$
II	cyclic Hückel polyenes	$2\cos\dfrac{2j\pi}{m}$;	$j = 1, 2, ..., m$
III	cyclic Möbius polyenes	$2\cos\dfrac{j\pi}{m}$;	$j = \pm 1, \pm 3, ..., \begin{cases}\pm(m-1), \ m \text{ even} \\ +m, \ m \text{ odd}\end{cases}$

carbon atoms as an example. The construction of the determinant according to Eq. (10-81) including the HMO approximation is straightforward; for the sake of simplicity, the equation obtained will be divided by β and the substitution $(\alpha - E)/\beta = k$ will be introduced, to give

$$\begin{vmatrix} k & 1 & 0 & 0 & \dots & 0 \\ 1 & k & 1 & 0 & \dots & 0 \\ 0 & 1 & k & 1 & \dots & 0 \\ & \vdots & & & & \\ 0 & 0 & 0 & 0 & \dots & k \end{vmatrix} = 0 \qquad (10\text{-}85)$$

This determinant of order m will be denoted D_m; expansion by cofactors of the first row elements leads directly to the recurrence formula

$$D_m = kD_{m-1} - D_{m-2} \qquad (10\text{-}86)$$

It will be convenient to find a pair of numbers r and s which possess the following properties:

$$r + s = k \qquad (10\text{-}87a)$$

$$rs = 1 \qquad (10\text{-}87b)$$

Solving this system of equations gives

$$r = k/2 + \sqrt{(k^2/4 - 1)} \qquad (10\text{-}88a)$$

$$s = k/2 - \sqrt{(k^2/4 - 1)} \qquad (10\text{-}88b)$$

The recurrence formula can be expressed in terms of r and s:

$$D_m - sD_{m-1} = r(D_{m-1} - sD_{m-2}) \qquad (10\text{-}89a)$$

$$D_m - rD_{m-1} = s(D_{m-1} - rD_{m-2}) \qquad (10\text{-}89b)$$

The left-hand sides of these equations are terms of geometric series with quotients r and s; it is therefore possible to write

$$D_m - sD_{m-1} = r^{m-2}(D_2 - sD_1) \qquad (10\text{-}90a)$$

$$D_m - rD_{m-1} = s^{m-2}(D_2 - rD_1) \qquad (10\text{-}90b)$$

D_{m-1} can easily be eliminated from these equations and the following relation is then obtained:

$$D_m = \frac{r^{m-1}(D_2 - sD_1) - s^{m-1}(D_2 - rD_1)}{r - s} \qquad (10\text{-}91)$$

Eq. (10-91) can be changed into a more convenient form by the substitution

$$k = 2 \cos \varphi \qquad (10\text{-}92)$$

Then for r and s it holds that

$$r = \cos \varphi + i \sin \varphi \qquad (10\text{-}93a)$$

$$s = \cos \varphi - i \sin \varphi \qquad (10\text{-}93b)$$

Using the Moivre theorem and considering the expressions

$$D_1 = k$$

$$D_2 = k^2 - 1$$

then

$$D_m = \frac{\sin (m + 1) \varphi}{\sin \varphi} \tag{10-94}$$

Now the equation $D_m = 0$ can be solved. From Eq. (10-94)

$$\sin (m + 1) \varphi = 0, \tag{10-95a}$$

so that

$$\varphi = \frac{j\pi}{m + 1} \tag{10-95b}$$

and therefore

$$k_j = 2 \cos \frac{j\pi}{m + 1} \tag{10-96}$$

It is evident from the course of the cosine that all the required solutions will be found by considering a total of m values for j: 1, 2, 3, ..., m. It is obvious that the expression obtained for k_j is identical with that given in Table 10-11. It remains to be added that expressions are available in the literature[64] which allow direct calculation of orbital energies for further systems: for cyclopolyenes in which conjugation is caused by overlap of the p and d orbitals (IV), for radialenes (V), dendralenes (VI) and polyacenes (VII) (Fig. 10-1).

In systems I to VII (and similarly in all further periodic conjugated molecules), expressions for the expansion coefficients and for quantities derived from them can be stated in closed form. This fact can be illustrated by the expression for the bond orders in a polyene with m carbon atoms (where m is an even number),

$$P_{\mu,\mu+1} = \frac{1}{m + 1} \left[\operatorname{cosec} \frac{\pi}{2m + 2} + (-1)^{\mu - 1} \operatorname{cosec} \frac{(2\mu + 1)\pi}{2m + 2} \right], \tag{10-97}$$

and by the expression for the π-electron energy of a cyclopolyene with m carbon atoms:

$$W = m\alpha + 2\beta \left(\operatorname{cosec} \frac{\pi}{2m + 2} - 1 \right), \tag{10-98}$$

where α and β denote the Coulomb and resonance integrals, respectively.

It is desirable to briefly discuss the numerical part of the calculations. The HMO method is an ideal example because it provides very lucid results. It has, however, some important features in common with more

228

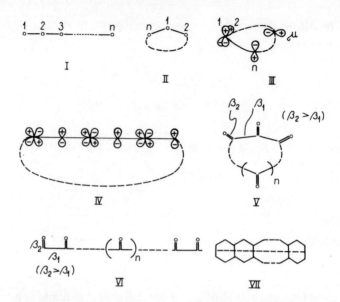

Fig. 10-1. Schematic representation of some π-electron systems. I – linear polyenes, II – cyclic Hückel polyenes, III – cyclic Möbius polyenes, IV – cyclic p–d polyenes, V – radialenes, VI – dendralenes, VII – polyacenes.

complicated methods. The solution of the secular determinant is, for example, met in the HMO, SCF and the CI methods. However, the difficulty of obtaining the matrix elements in the individual methods differs a great deal.

For illustration, methylenecyclopropene (VIII) can be chosen. The following system of equations is then valid:

$$
\begin{aligned}
(\alpha - E)c_1 + c_2\beta &= 0 \\
(\alpha - E)c_2 + c_1\beta + c_3\beta + c_4\beta &= 0 \\
(\alpha - E)c_3 + c_2\beta + c_4\beta &= 0 \\
(\alpha - E)c_4 + c_2\beta + c_3\beta &= 0
\end{aligned}
\tag{10-99}
$$

The substitution

$$
-(\alpha - E)/\beta = k
\tag{10-100}
$$

yields

$$
\begin{aligned}
-kc_1 + c_2 &= 0 \\
-kc_2 + c_1 + c_3 + c_4 &= 0 \\
-kc_3 + c_2 + c_4 &= 0 \\
-kc_4 + c_2 + c_3 &= 0
\end{aligned}
\tag{10-101}
$$

These equations have a non-trivial solution only if the determinant of the multiples of c is zero, i.e.

$$\begin{vmatrix} -k & 1 & 0 & 0 \\ 1 & -k & 1 & 1 \\ 0 & 1 & -k & 1 \\ 0 & 1 & 1 & -k \end{vmatrix} = 0 \qquad (10\text{-}102)$$

Expansion of the determinant leads to a quartic algebraic equation, the solutions of which are

$$k_1 = 2.170; \quad k_2 = 0.311; \quad k_3 = -1.000; \quad k_4 = -1.481$$

Because of the relationship between k and the orbital energy E, it then holds that

$$E = \alpha + k\beta \qquad (10\text{-}103)$$

Since the Coulomb (α) and resonance (β) integrals are negative quantities, the lowest of the four values of the orbital energy can be written in the form

$$E_1 = \alpha + 2.170\beta \qquad (10\text{-}104)$$

The energy of the least favourable level is evidently

$$E_4 = \alpha - 1.481\beta \qquad (10\text{-}105)$$

By stepwise solution of system of Eqs. (10-101) for k_1, k_2, k_3 and k_4, the expansion coefficients of all four MO's are obtained. Their normalized values are given in Table 10-12. The procedure for the calculation is shown in detail, for example, in the collection of examples[65]. Here it is sufficient to give the expression for φ_1 corresponding to ε_1:

$$\varphi_1 = 0.282\chi_1 + 0.612\chi_2 + 0.523\chi_3 + 0.523\chi_4 \qquad (10\text{-}106)$$

Because errors occur very easily in the numerical computations, it is desirable to discuss the checking of the results in somewhat greater detail.

First, the orbital energies $(E_i = \alpha + k_i\beta)$ and molecular orbitals will be discussed; it can be shown that the sum of the eigenvalues of the HMO matrix (quantities k_i) must equal the trace of the HMO matrix [cf. Eq. (4-130)] and that the sum of the squares of the eigenvalues equals the sum of the squares of all elements of the HMO matrix, i.e.:

$$\sum_{i=1}^{m} k_i = \sum_{\mu=1}^{m} a_{\mu\mu} \qquad (10\text{-}107)$$

$$\sum_{i=1}^{m} k_i^2 = \sum_{\mu=1}^{m} \sum_{\nu=1}^{m} a_{\mu\nu}^2 \tag{10-108}$$

Because of the orthonormality of the molecular orbitals it holds that

$$\int \left(\sum_{\mu} c_{\mu i} \chi_{\mu}\right)^2 d\tau = \sum_{\mu} c_{\mu i}^2 \langle \chi_{\mu} | \chi_{\mu} \rangle + \sum_{\mu(\neq\nu)} \sum_{\nu} c_{\mu i} c_{\nu i} \langle \chi_{\mu} | \chi_{\nu} \rangle = 1 \tag{10-109}$$

Because the AO's employed form an orthonormal set, the expansion coefficients evidently fulfil the condition

$$\sum_{\mu=1}^{m} c_{\mu i}^2 = 1 \tag{10-110}$$

In the EHT method, where the overlap is not neglected, a different condition is, of course, valid. It can similarly be shown that, for orthogonal MO's (cf. Table 10-12)

$$\sum_{\mu=1}^{m} c_{\mu i} c_{\mu j} = 0; \quad i \neq j \tag{10-111}$$

In order to simplify checking of the correctness of the results given in Table 10-12, a table of the squares of values k_i and $c_{\mu i}$ is drawn up.

Table 10-12

HMO Orbital Energies and Expansion Coefficients of Methylenecyclopropene (VIII)

i	k_i	$\mu =$ 1	2	3	4	$\sum_{\mu} c_{\mu i} c_{\mu j}$
		Expansion coefficients ($c_{\mu i}$)				
1	2.170	0.282	0.612	0.523	0.523 $\}$	0
2	0.311	−0.815	−0.254	0.368	0.368 $\}$	0
3	−1.000	0	0	−0.707	0.707 $\}$	0
4	−1.481	−0.506	0.749	0.302	0.302 $\}$	0
$\Sigma k_i =$	0.000					

i	k_i^2	$\mu =$ 1	2	3	4	$\sum_{\mu} c_{\mu i}^2$
		Squares of expansion coefficients ($c_{\mu i}^2$)				
1	4.709	0.080	0.374	0.273	0.273	1.000
2	0.097	0.664	0.065	0.135	0.135	0.999
3	1.000	0	0	0.500	0.500	1.000
4	2.193	0.256	0.561	0.091	0.091	0.999
$\Sigma k_i^2 =$	7.999	$\sum_i c_{\mu i}^2$: 1.000	1.000	0.999	0.999	

From the HMO matrix* of the investigated system

$$\begin{Vmatrix} 0 & 1 & 0 & 0 \\ 1 & 0 & 1 & 1 \\ 0 & 1 & 0 & 1 \\ 0 & 1 & 1 & 0 \end{Vmatrix}$$

it follows that

$$\sum_{\mu=1}^{4} a_{\mu\mu} = 0 \quad \text{and} \quad \sum_{\mu=1}^{4} \sum_{\nu=1}^{4} a_{\mu\nu}^2 = 8$$

The values of k_i and $c_{\mu i}$ are apparently correct.

Furthermore, in all the conjugated hydrocarbons, the sum of the squares of k_i equals double the number of C$-$C bonds; this can easily be verified.

It is also simple to check the correctness of the calculation of the electron charge densities. The sum of the electron charge densities in the individual positions $(q_\mu \equiv P_{\mu\mu})$ equals the total number of π electrons (n_π):

$$\sum_{\mu=1}^{m} q_\mu = n_\pi \tag{10-112}$$

The check-up on the correctness of the bond orders is more complicated. It can be shown that the orbital energies (the total energy) of the system are related to the expansion coefficients (bond orders). This relationship is obtained using the general expression for the energy. For the orbital energy,

$$E_i = \frac{\langle \varphi_i | \mathscr{H}^{ef} | \varphi_i \rangle}{\langle \varphi_i | \varphi_i \rangle}, \tag{10-113}$$

where $\varphi_i = \sum_{\mu=1}^{m} c_{\mu i} \chi_\mu$. Because of the orthonormality properties of the φ_i and the AO's it holds that

$$E_i = \sum_{\mu=1}^{m} c_{\mu i}^2 \alpha_\mu + 2 \sum_{\mu<\nu}^{m} \sum c_{\mu i} c_{\nu i} \beta_{\mu\nu} \tag{10-114}$$

Considering the definition of the π-electron energy, W (as the sum of the MO energies multiplied by the occupation numbers), valid in empirical methods, a transition from the expression for E_i to the expression for W can be made by summing over all the doubly occupied MO's:

$$W = 2 \sum_{i(occ)} E_i = 2 \sum_{i(occ)} \sum_\mu c_{\mu i}^2 \alpha_\mu + 2 \sum_{i(occ)} 2 \sum_{\mu<\nu} \sum c_{\mu i} c_{\nu i} \beta_{\mu\nu} \tag{10-115}$$

---

* The HMO matrix is identical with the topological matrix, which has ones in the positions corresponding to C$-$C bonds and zeros in all the other positions.

232

The summation over the MO's can easily be performed employing the definition of the electron charge densities and bond orders, giving the relationship

$$W = \sum_{\mu} q_{\mu}\alpha_{\mu} + 2\sum_{\mu < \nu}\sum P_{\mu\nu}\beta_{\mu\nu} \qquad (10\text{-}116)$$

It follows from Eqs. (10-114) and (10-116) that the orbital and total π-electron energy can be calculated using expansion coefficients; because W is determined simply by summing the corresponding orbital energies and because the correctness of q can be verified using relation (10-112), it follows that Eq. (10-116) can be used for checking of the bond order values. The expressions for E_i and W have an important role in the perturbation treatment.

First it will be verified whether $E_1 = \alpha + 2.170\beta$ corresponds to φ_1 [Eq. (10-106)] in methylenecyclopropene. According to Eq. (10-113)

$$E_1 = \int (0.282\chi_1 + 0.612\chi_2 + 0.523\chi_3 + 0.523\chi_4) \times$$
$$\times \mathscr{H}^{\text{ef}}(0.282\chi_1 + 0.612\chi_2 + 0.523\chi_3 + 0.523\chi_4)\, d\tau = \alpha + 2.170\beta$$

Thus the previously quoted value of E_1 is correct. The data will now be verified in the molecular diagram:

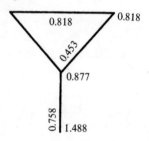

First it holds that

$$\sum_{\mu=1}^{4} q_{\mu} = 1.488 + 0.877 + 0.818 + 0.818 = 4.001$$

From the given orbital energy values it follows that

$$W = \sum_{i=1}^{2} 2E_i = 4\alpha + 4.962\beta$$

W can be calculated using Eq. (10-116). Because uniform values of α and β are employed in the HMO method the relation

$$W = 1.478\alpha + 0.882\alpha + 0.820\alpha + 0.820\alpha +$$
$$+ 2(0.758\beta + 0.453\beta + 0.453\beta + 0.818\beta) = 4\alpha + 4.962\beta$$

is valid. The correctness of the data in the molecular diagram has therefore been verified.

10.3.6 Perturbation methods within the framework of the simple MO method

Now a group of methods yielding values which can be considered as approximations to HMO values will be considered. Let us assume that the quantum-chemical (HMO) solution for a particular system is known (i.e., all the E_i's and $c_{\mu i}$'s are known) and that the solution for a system differing only slightly from the initial system is required; the new system can differ in the value of the Coulomb integral in the μ position or in the value of the resonance integral of the $\varrho\sigma$ bond. This situation can be symbolically described as follows:

Original system   New (perturbed) system

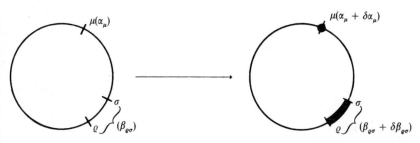

$$\text{Known values} \begin{cases} E_i \;\ldots\; E_i' \;= E_i + \delta E_i \\ W \;\ldots\; W' = W + \delta W \\ q \;\;\ldots\; q' \;\; = q + \delta q \\ \mathbf{P} \;\;\ldots\; \mathbf{P}' \;\; = \mathbf{P} + \delta\mathbf{P} \\ c_{\mu i} \;\ldots\; c_{\mu i}' \end{cases} \text{Values to be calculated}$$

The characteristics of the new system could, of course, be obtained by the usual HMO procedure, i.e. by arranging and solving the respective secular equation. It is typical for the perturbation treatment that the approximate values of the required characteristics (of the new system) can be obtained from (known) characteristics of the original system. The function allowing calculation of the orbital energy E will be approximated by another function, this substitution being meaningful within a certain interval. A Taylor series is a suitable function for this purpose:

$$E(\alpha_x) = E(\alpha) + \frac{\partial E(\alpha)}{\partial \alpha}(\alpha_x - \alpha) + \frac{1}{2!}\frac{\partial^2 E(\alpha)}{\partial \alpha^2}(\alpha_x - \alpha)^2 +$$

$$+ \frac{1}{3!}\frac{\partial^3 E(\alpha)}{\partial \alpha^3}(\alpha_x - \alpha)^3 + \ldots \tag{10-117}$$

The expression for the total π-electron energy can, of course, be expanded quite similarly. A rough estimate of the extent of the perturbation, which is expressed by the values $\delta\alpha_\mu$ and $\delta\beta_{\mu\nu}$, is generally not difficult. For illustration, the perturbation can be imagined to represent the substitution of the =CH− group by a nitrogen atom, i.e. the formal formation of pyridine from benzene:

The change in the value of the Coulomb integral, $\delta\alpha$, is then given by the expression

$$\delta\alpha = \underbrace{\alpha + 0.5\beta}_{\substack{\text{new} \\ \text{Coulomb integral}}} - \underset{\substack{\downarrow \\ \text{original}}}{\alpha} = 0.5\beta$$

It remains to calculate the values of the partial derivatives. Eq. (10-114) leads to the result*

$$\frac{\partial E_i}{\partial\alpha_\mu} = c_{\mu i}^2 \qquad \frac{\partial E_i}{\partial\beta_{\mu\nu}} = 2c_{\mu i}c_{\nu i} \qquad (10\text{-}118)$$

From Eq. (10-116)* it follows that

$$\frac{\partial W}{\partial\alpha_\mu} = q_\mu \qquad \frac{\partial W}{\partial\beta_{\mu\nu}} = 2P_{\mu\nu} \qquad (10\text{-}119)$$

Thus, the first differential coefficient in expansion (10-117), corresponding to the change in the Coulomb and resonance integrals, is estimated for both the orbital energy and the total energy. The calculation of further differential coefficients will be discussed later. First, however, the number of terms in expansion (10-117) to be considered must be decided: it is necessary to compromise between accuracy (which demands as many terms as possible) and ease of calculation (which, of course, requires as few terms as possible). According to the number of derivatives considered, we speak of first-order, second-order, or higher-order perturbation calculations:

---

* The terms $\dfrac{\partial c_{\mu i}^2}{\partial\alpha_\mu}\alpha_\mu$ and $\dfrac{\partial q_\mu}{\partial\alpha_\mu}\alpha_\mu$ are negligible compared to $c_{\mu i}^{2\,\prime}$ and q_μ.

$$\delta E = \text{1st derivative} + \text{2nd derivative} + \text{higher derivatives}$$

Perturbation calculation of

> first order $\longrightarrow$
>
> second order $\longrightarrow$
>
> higher order $\longrightarrow$

Fortunately, it appears that the first-order derivative often suffices for qualitative or semiquantitative solution of problems. The energy expressions for the perturbed system are given in Table 10-13.

Table 10-13

Expressions for the Energy of a Perturbed System

Energy	Perturbation	
	Coulomb integral	resonance integral
orbital	$E_i + c_{\mu i}^2 \delta\alpha_\mu$	$E_i + 2c_{\mu i}c_{\nu i}\delta\beta_{\mu\nu}$
total π	$W + q_\mu\delta\alpha_\mu$	$W + 2P_{\mu\nu}\delta\beta_{\mu\nu}$

If changes in α and β occur simultaneously, then the following relation holds for the total energy:

$$W' = W + q_\mu\delta\alpha_\mu + 2P_{\mu\nu}\delta\beta_{\mu\nu} \qquad (10\text{-}120)$$

The perturbation can, of course, occur in several centres simultaneously; in the general case

$$W' = W + \sum_\mu q_\mu\delta\alpha_\mu + 2\sum_{\mu\nu} P_{\mu\nu}\delta\beta_{\mu\nu}, \qquad (10\text{-}121)$$

where the summation is carried out over all "perturbed" atoms (μ) and bonds ($\mu\nu$). Calculation of orbital energies is usually confined to the first-order perturbation theory (although higher terms can, of course, be included). In the calculation of the total energy, it is interesting to consider the second-order derivatives, as they enable calculation of electron densities and bond orders. The expressions for the second-order derivatives can be written in the form

$$\frac{\partial^2 W}{\partial\alpha_\mu^2} = \frac{\partial}{\partial\alpha_\mu}\left(\frac{\partial W}{\partial\alpha_\mu}\right) = \frac{\partial q_\mu}{\partial\alpha_\mu} \qquad (10\text{-}122)$$

$$\frac{\partial^2 W}{\partial\beta_{\mu\nu}^2} = \frac{\partial}{\partial\beta_{\mu\nu}}\left(\frac{\partial W}{\partial\beta_{\mu\nu}}\right) = 2\frac{\partial P_{\mu\nu}}{\partial\beta_{\mu\nu}}, \qquad (10\text{-}123)$$

where relations (10-119) were employed.

The physical significance of quantity $\partial q_\mu / \partial \alpha_\mu$ is clear; it corresponds to the change in the electron density induced by a change in the Coulomb integral; quantity $\partial P_{\mu\nu} / \partial \beta_{\mu\nu}$ has analogous meaning. These quantities are called[66] polarizabilities and they are denoted by the symbol Π. It is possible to generalize and to extend the definition of polarizability:

$$
\text{polarizability} \begin{cases}
\text{atom} - \text{atom} \ldots \Pi_{\mu,\nu} &= \dfrac{\partial q_\mu}{\partial \alpha_\nu} \\[2mm]
\text{atom} - \text{bond} \ldots \Pi_{\mu,\varrho\sigma} &= \dfrac{\partial q_\mu}{\partial \beta_{\varrho\sigma}} \\[2mm]
\text{bond} - \text{atom} \ldots \Pi_{\varrho\sigma,\mu} &= \dfrac{\partial P_{\varrho\sigma}}{\partial \alpha_\mu} \\[2mm]
\text{bond} - \text{bond} \ldots \Pi_{\varrho\sigma,\mu\nu} &= \dfrac{\partial P_{\varrho\sigma}}{\partial \beta_{\mu\nu}}
\end{cases}
\tag{10-124}
$$

The polarizabilities discussed so far in connection with the perturbation calculation are special cases and are called self-polarizabilities (atom $-$ atom, $\Pi_{\mu\mu}$; bond $-$ bond $\Pi_{\mu\nu,\mu\nu}$). For the π-electron energy of the perturbed system,

$$
W' = W + q_\mu \delta\alpha_\mu + \frac{1}{2} \Pi_{\mu\mu} (\delta\alpha_\mu)^2
\tag{10-125}
$$

$$
W' = W + 2P_{\mu\nu} \delta\beta_{\mu\nu} + \Pi_{\mu\nu,\mu\nu} (\delta\beta_{\mu\nu})^2
\tag{19-126}
$$

It should again be borne in mind that the energy W' of the new (perturbed) system is calculated solely in terms of characteristics of the original system $(q, \mathbf{P}, \Pi)$; it remains, for general information, to give the expression for the $\Pi_{\mu\nu}$ polarizabilities:

$$
\Pi_{\mu\nu} = 4 \sum_{j=1}^{occ} \sum_{k}^{unocc} \frac{c_{\mu j} c_{\nu j} c_{\mu k} c_{\nu k}}{k_j - k_k},
\tag{10-127}
$$

where the k_j values are defined by Eq. (10-100). The first summation is carried out over the occupied MO's, the second over the unoccupied MO's.

Before giving a few examples, it should be noted that the polarizabilities are used for perturbation calculations of indices occurring in molecular diagrams; by rearranging the definitions, the relationships

$$
\delta q_\mu = \Pi_{\mu\nu} \delta\alpha_\nu \qquad \delta P_{\varrho\sigma} = \Pi_{\varrho\sigma,\mu\nu} \delta\beta_{\mu\nu}
\tag{10-128}
$$

are obtained. The numerical calculation of the polarizabilities in more extensive systems is rather lengthy; however, for hundreds of systems these data are available in the literature. It is also true, however, that polarizabilities are no longer used as much as they once were.

As an example, the orbital energies, the total energy and the π-electron density distribution for cyclopropenone will be calculated by the perturbation method from the data for methylenecyclopropenone. In order to determine how closely the perturbation data approach the data obtained by solving the secular equation, the results obtained from the perturbation method will be compared with the accurate (HMO) calculation:

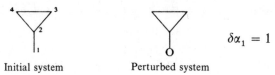

Initial system   Perturbed system

Table 10-12 gives the HMO data for the initial system, the orbital energies and the expansion coefficients; the polarizability values are as follows:

μ \ ν	1	2	3	4
1	0.402			
2	−0.247	0.311		
3	−0.078	−0.032	0.434	
4	−0.078	−0.032	−0.324	0.435

μν \ ρσ	1 2	2 3	2 4	3 4
1 2	0.265			
2 3	−0.201	0.329		
2 4	−0.201	−0.005	0.329	
3 4	0.137	−0.123	−0.123	0.110

The polarizability matrices are symmetrical and the self-polarizabilities are the diagonal elements of these matrices.

The values of the orbital energies [cf. Eqs. (10-100), (10-117) and (10-118)] can now be calculated:

$$k'_1 = k_1 + (c_{11})^2\,\delta\alpha_1 = \quad 2.170 + (0.282)^2 \quad = \quad 2.249$$
$$k'_2 = k_2 + (c_{12})^2\,\delta\alpha_1 = \quad 0.311 + (-0.815)^2 = \quad 0.976$$
$$k'_3 = k_3 + (c_{13})^2\,\delta\alpha_1 = -1.000 + 0 \quad\quad\quad = -1.000$$
$$k'_4 = k_4 + (c_{14})^2\,\delta\alpha_1 = -1.481 + (0.506)^2 \quad = -1.225$$

For the π-electron energy

$$W' = W + q_1\delta\alpha_1 = 4\alpha + 4.962\beta + 1.488\beta = 4\alpha + 6.450\beta$$

so that

$$q'_1 = q_1 + \Pi_{11}\delta\alpha_1 = 1.488 + 0.402 = 1.890$$
$$q'_2 = q_2 + \Pi_{12}\delta\alpha_1 = 0.877 - 0.247 = 0.630$$
$$q'_3 = q_3 + \Pi_{13}\delta\alpha_1 = 0.818 - 0.078 = 0.740$$
$$q'_4 = q_4 + \Pi_{14}\delta\alpha_1 = 0.818 - 0.078 = 0.740$$

The characteristics of the original system (A) and those of the perturbed system (B) (perturbation model of cyclopropenone) can be compared with the HMO data for cyclopropenone (C) (Table 10-14). It is then interesting to investigate the extent to which the W' value approaches the W (HMO) value if the second-order perturbation calculation is used:

$$W' = W + q_1\delta\alpha_1 + \Pi_{11}(\delta\alpha_1)^2 \tag{10-129}$$

$$(= 4\alpha + 4.962\beta + 1.488\beta + 0.402\beta = 4\alpha + 6.852\beta)$$

Table 10-14

Energy Characteristics of Cyclopropenone
(Expressed in Multiples of the β Value) and π-Electron Densities

	A	B	C
k_1	2.170	2.249	2.303
k_2	0.311	0.976	1.000
k_3	-1.000	-1.000	-1.000
k_4	-1.481	-1.225	-1.303
$W - 4\alpha$	4.962	6.450	6.606
q_1	1.488	1.890	1.759
q_2	0.877	0.630	0.723
q_3	0.818	0.740	0.759
q_4	0.818	0.740	0.759

In the chapter on electronic spectroscopy, the very useful application of the perturbation method for estimation of the effect of substituents on the position of the longest wavelength bands is discussed.

10.4 The FE-MO method[67-70]

The free electron method is now used very little in practice. It seems unable to compete with the very flexible and universally applicable MO-LCAO method. For study of the theory of the chemical bond, however, it has several interesting features. The basic idea is very simple: the conditions for the π-electrons in polyene (free mobility in the vicinity of the skeleton of $C-C$ bonds and the "impossibility" of leaving this skeleton) are reminiscent of the conditions for a particle in a potential box. This model is undoubtedly very primitive; its advantage lies in that it leads to such a simple form for the Schrödinger equation that the solution can be achieved in a closed form (by direct integration). Except for the hydrogen atom, this is one of the few such cases.

Transition to more complicated systems requires the following modification of the simple model: (*i*) transition to a two-dimensional box (this allows the treatment of planar conjugated systems), (*ii*) the possibility of introducing positions with changed potential values allows description of systems containing heteroatoms. In spite of these possibilities, the FE-MO method is unable to seriously compete with the MO-LCAO method.

10.5 Valence bond theory (VB method)

The valence bond (VB) theory[71] and the molecular orbital theory are the two basic methods for construction of an approximate wave function, describing the electronic states in a molecule. Historically, the valence bond method evolved from the Heitler−London theory of the hydrogen molecule, published in 1927, i.e. one year after the appearance of the fundamental papers on quantum mechanics. The Heitler−London work was, in principle, the first successful attempt at quantum-mechanical interpretation of covalent bonding in a molecule. Because the VB theory represents an extension of the Heitler−London model of the hydrogen molecule, it will be useful to demonstrate the physical meaning of assumptions used in the VB method on this example.

Let us start by considering two hydrogen atoms which are initially so far apart that interaction between them is impossible. The electronic states of the atoms are described by the wave functions $\chi_\mu(1)$ and $\chi_\nu(2)$, where the indices of the functions (of atomic orbitals) denote the nuclei and the numbers denote the individual electrons (their coordinates). The system as a whole is described by the product of these functions,

$$\chi_\mu(1)\,\chi_\nu(2),\qquad\qquad(10\text{-}130)$$

as it is assumed that, for non-interacting atoms, the relationships

$$\mathscr{H}(1)\,\chi_\mu(1) = E_\mu\chi_\mu(1)\qquad\qquad(10\text{-}131a)$$

$$\mathscr{H}(2)\,\chi_\nu(2) = E_\nu\chi_\nu(2)\qquad\qquad(10\text{-}131b)$$

240

are valid, where $\mathcal{H}(1)$ and $\mathcal{H}(2)$ are the Hamiltonians for electron 1 and electron 2, respectively. The Hamiltonian of the entire system is then

$$\mathcal{H}^0 = \sum_{i=1}^{2} \mathcal{H}(i);$$ (10-132)

the wave function in the Schrödinger equation

$$\mathcal{H}^0\Psi^\circ = E^\circ\Psi^\circ$$ (10-133)

can therefore be sought in the form

$$\Psi^\circ = \chi_\mu(1)\,\chi_\nu(2)$$ (10-134)

and the total energy E of the system is given by

$$E^\circ = E_\mu^\circ + E_\nu^\circ$$ (10-135)

[cf. Eq. (5-40b)]. This leads to the obvious result that the total electronic energy in the absence of mutual interaction equals the sum of the energies of the subsystems.

When the hydrogen atoms gradually approach each other, interactions begin to occur between them, which are of three types depending on the kind of participating particles:

    a) electron – electron,

    b) electron – nucleus of the second atom,

    c) nucleus – nucleus.

These interactions must be included in the total Hamiltonian of the system and their resultant leads to the formation of a stable hydrogen molecule when the hydrogen atoms reach the equilibrium distance. The construction of the complete Hamiltonian $\mathcal{H}$ encounters no difficulties [cf. Eq. (5-18)] and therefore the principal problem lies in determination of the approximate wave function Ψ for the system. It is plausible to demand that at large internuclear distances the wave function takes the form (10-134), describing two separate atoms. As the electrons are indistinguishable, functions $\chi_\mu(1)\,\chi_\nu(2)$ and $\chi_\mu(2)\,\chi_\nu(1)$ are equally probable and therefore wave function Ψ describing the hydrogen molecule can be written in the form (omitting the normalization factor)

$$\Psi \sim \chi_\mu(1)\,\chi_\nu(2) \pm \chi_\mu(2)\,\chi_\nu(1)$$ (10-136)

Function Ψ is, of course, still incomplete because it does not include the spin states of the electrons. The two-electron problem was discussed in Section 6.7, where it was found that the total wave function satisfies the Pauli principle in two instances:

a) if the part of the wave function that depends on the spatial coordinates is symmetrical and the spin part describes a singlet state,

b) if the part of the wave function depending on the spatial coordinates is antisymmetrical and the spin part describes a triplet state.

Thus the total electronic wave function can be written in the form [cf. Eq. (6-113)]

$$\Psi_g = N_g[\chi_\mu(1)\chi_\nu(2) + \chi_\mu(2)\chi_\nu(1)]\frac{1}{\sqrt{2}}[\alpha(1)\beta(2) - \alpha(2)\beta(1)], \qquad (10\text{-}137)$$

where the index g (gerade) indicates that the spatial part of the wave functions is symmetrical with respect to permutation of the electrons. A function of the following type is, of course, also admissible:

$$\Psi_u = N_u[\chi_\mu(1)\chi_\nu(2) - \chi_\mu(2)\chi_\nu(1)]\frac{1}{\sqrt{2}}[\alpha(1)\beta(2) + \alpha(2)\beta(1)], \qquad (10\text{-}138)$$

where the index u (ungerade) indicates that the spatial part of the wave function is antisymmetrical. N_g and N_u denote the normalization constants in the given equations. In Eq. (10-138), only one of the three possible spin functions of the triplet state is considered (cf. Section 6.7), as the remaining two correspond to the same (degenerate) energy level.

We can determine which of the two functions, (10-137) or (10-138), describes a stable bond between hydrogen atoms. Calculation of the energy expectation value for Ψ_g yields

$$E_g = \frac{\langle\Psi_g|\mathcal{H}|\Psi_g\rangle}{\langle\Psi_g|\Psi_g\rangle} =$$

$$= \frac{N_g^2\langle\chi_\mu(1)\chi_\nu(2) + \chi_\mu(2)\chi_\nu(1)|\mathcal{H}|\chi_\mu(1)\chi_\nu(2) + \chi_\mu(2)\chi_\nu(1)\rangle}{2N_g^2[1 + \langle\chi_\mu(2)|\chi_\nu(2)\rangle\langle\chi_\nu(1)|\chi_\mu(1)\rangle]} =$$

$$= \frac{J' + K'}{1 + S_{\mu\nu}^2} \qquad (10\text{-}139)$$

and similarly for Ψ_u,

$$E_u = \frac{J' - K'}{1 - S_{\mu\nu}^2}, \qquad (10\text{-}140)$$

where the normality of the spin functions was used and the following symbols were introduced; for the Coulomb integral

$$J' = \langle\chi_\mu\chi_\nu|\mathcal{H}|\chi_\mu\chi_\nu\rangle, \qquad (10\text{-}141a)$$

for the exchange integral

$$K' = \langle\chi_\mu\chi_\nu|\mathcal{H}|\chi_\nu\chi_\mu\rangle \qquad (10\text{-}141b)$$

and for the overlap integral

$$S_{\mu v} = \langle \chi_\mu \mid \chi_v \rangle \tag{10-141c}$$

The notation for the integrals has already been defined by Eqs. (5-31a) and (5-31b).

The qualitative discussion of energy expressions (10-139) and (10-140) is simplified by neglecting the overlap integrals. Then the total energy of the hydrogen molecule, E_{tot}, including the nuclear repulsion energy, E_{rep}, can be written in the form

$$E_{(tot)g} = J' + E_{rep} + K' \tag{10-142}$$

$$E_{(tot)u} = J' + E_{rep} - K' \tag{10-143}$$

At large internuclear distances the value of K' is negligible and the expression $(J' + E_{rep})$ represents the energy of the atoms and their Coulomb interaction. At smaller distances the value $(J' + E_{rep})$ varies slowly and has a shallow minimum in the region of the equilibrium distance of the nuclei. In this region K' is a relatively large negative number and represents about 90% of the binding energy of the molecule, provided it is in the singlet state, corresponding to the wave function Ψ_g. For the triplet state corresponding to Ψ_u, the K' term leads to repulsion of the hydrogen atoms at all internuclear distances, causing spontaneous dissociation of the molecule into two hydrogen atoms.

It should be pointed out that the overlap integral is particularly important and that this was not considered in the qualitative discussion. Neglect of this factor can lead to serious inconsistencies in the quantitative treatments and, therefore, generalizing the Heitler-London approach to a many-atom system leads to serious difficulties connected with the non-orthogonality of the atomic orbital basis set.

The important conclusions derived from the given solution for the hydrogen molecule form the basis for the logical structure of the valence bond method. The covalent bond between two atoms, depicted in the chemical formula by a dash, is described in the VB method by a function of type (10-137), corresponding to antiparallel spins for the electrons forming the bond. These functions are said to describe a "local singlet" state—local because it corresponds specifically to the bond between two atoms. The form of the wave function for the polyatomic molecule must then satisfy the condition of forming "local singlet" states in the respective bond regions.

Although the Heitler-London theory yields qualitatively correct results, there is a considerable quantitative disagreement with the experimental values of physical quantities. Several modifications of this method have therefore been suggested, which are also suitable for the calculation of other types of covalent bonds. Their use has led to a substantial improvement in the results. One of these procedures consists of using hybrid orbitals instead of atomic orbitals for construction of the spatial part of the function describing the bond. The hybrid orbitals (cf. Sections 6.6 and 7.2) then have the advantage that their orientation can be chosen in the bond direction. In other instances, more accurate results were obtained when "a certain amount" of the ionic structure was included, i.e. functions of the type $\chi_\mu(1)\,\chi_\mu(2)$ or $\chi_\nu(1)\,\chi_\nu(2)$, so that the required VB function (without the spin part) would have the form

$$[\chi_\mu(1)\,\chi_\nu(2) + \chi_\mu(2)\,\chi_\nu(1)] + \lambda[\chi_\mu(1)\,\chi_\mu(2) + \chi_\nu(1)\,\chi_\nu(2)], \quad (10\text{-}144)$$

where coefficient λ specifies the extent of inclusion of the ionic form and is usually considered to be a variation parameter.

The expressions for the VB functions can be compared with the wave functions constructed on the basis of molecular orbitals (expressed as a linear combination of atomic orbitals). It is sufficient to investigate the part of the wave function which depends on the spatial coordinates of the electrons. The molecular orbital for the given problem, corresponding to the lowest occupied one-electron state, has the form

$$\varphi \sim [\chi_\mu + \chi_\nu], \quad (10\text{-}145)$$

where it is assumed that φ is expanded in terms of the AO minimum basis set. Two electrons with different spins can occupy this molecular orbital and the spatial part of the product wave function is then given by the expression

$$\varphi(1)\,\varphi(2) \sim [\chi_\mu(1)\,\chi_\nu(2) + \chi_\mu(2)\,\chi_\nu(1) + \chi_\mu(1)\,\chi_\mu(2) + \chi_\nu(1)\,\chi_\nu(2)], \quad (10\text{-}146)$$

allowing direct comparison with Eqs. (10-137) and (10-144). Comparison with the uncorrected VB function indicates that function (10-146) contains additional ionic contributions which can be physically interpreted as corresponding to the extreme electron density distribution when both electrons occur on the same nucleus, representing the H^+H^- ionic state. These states are known to be considerably less stable than the states corresponding to a uniform electron distribution and if this difference is not considered, the implications connected with the use of MO's in form (10-145) cannot be properly understood. At large interatomic distances, the single-determinant MO theory fails completely, because a wave function of form (10-146) predicts the formation of ions (which is a process requiring energy consumption) with the same probability as dissociation into two neutral atoms. The single-determinant MO theory then yields excessively high values for the total energy of the system.

The whole problem can also be interpreted in terms of electron correlation, i.e. by the concept introduced in Section 10.1. Evidently the single-determinant MO theory underestimates electron correlation: the distribution of electrons 1 and 2 in a certain molecular orbital is quite independent, leading, for example, to the same probability for structures $\chi_\mu \chi_\nu$ and $\chi_\mu \chi_\mu$. The VB method, on the other hand, overestimates correlation, because it admits only the possibility of complete separation of the electrons on the two atoms. It is therefore evident that (*i*) correct description would lead to results lying between the VB and MO data; when the two methods yield similar results, then the results can be considered to be reliable, (*ii*) in order to improve the VB description, it is necessary to consider ionic forms, i.e. wave functions of type (10-144).

For a wave function constructed on a molecular orbital basis, improvement is possible using the configuration interaction method (cf. Section 5.4). If two electrons are placed in the antibonding orbital

$$\varphi' \sim [\chi_\mu - \chi_\nu] \qquad (10\text{-}147)$$

(i.e. a doubly excited configuration; the symmetry of the singly excited configuration is unsuitable and therefore does not interact with the ground state configuration, cf. Section 6.7), a determinant function is obtained, which interacts with the ground state determinant. The spatial part of this wave function can be written in the (unnormalized) form

$$\varphi(1)\, \varphi(2) + k\varphi'(1)\, \varphi'(2), \qquad (10\text{-}148)$$

whence, after substituting Eqs. (10-145) and (10-147), multiplying and comparing with expression (10-144), it follows that the two functions (VB and CI) are equivalent, as long as

$$\lambda = \frac{1 + k}{1 - k} \qquad (10\text{-}149)$$

Generally, the two methods yield similar results in a broad region.

To describe the general procedure for constructing the VB wave functions, it is advantageous to investigate a suitable model molecule. We shall choose the water molecule. It is preferable to choose a coordinate system with the nucleus of the oxygen atom lying in the origin and both $O - H$ bonds in the xy-plane, forming equal angles with the x- and y-axes. It will be satisfactory in a rough model to assume that the OH bonds form a $90°$ angle, i.e. that the nuclei of the hydrogen atoms are located on the x- and y-axes (the experimental value of this angle is 105° and refinement of the model could be achieved, for example, by introducing suitable hybrid orbitals). The atomic orbitals required here are the $(1s)$, $(2s)$, $(2p_x)$, $(2p_y)$ and $(2p_z)$ orbitals located on the nucleus of the oxygen atom and the $(1s)_\mu$ and $(1s)_\nu$ orbitals corresponding to the hydrogen atoms, where the μ atom lies on the x-axis and the ν atom on the y-axis. To a first approximation it can be assumed that the bond is formed through the electron pairs described by the $(2p_x)$ and $(1s)_\mu$ orbitals and the $(2p_y)$ and $(1s)_\nu$ orbitals, and that the remaining orbitals are occupied by two electrons so that they are unable to contribute to the bonding in the molecule. The electron configurations of the participating atoms can therefore be written as follows:

O: $(1s)^2, (2s)^2, (2p_z)^2, (2p_x), (2p_y)$
H: $(1s)_\mu$
H: $(1s)_\nu$

Because electrons with opposite spins are responsible for the bond formation, the spin functions multiplying spatial $(2p_x)$ and $(1s)_\mu$ orbitals must differ; the same must also be true for the $(2p_y)$ and $(1s)_\nu$ orbitals. This condition is satisfied by the four combinations of functions indicated in Table 10-15 as cases 2 to 5, of which case 2 will be discussed in greater detail as an example:

$$(2p_x)\,\alpha, \quad (1s)_\mu\,\beta, \quad (2p_y)\,\alpha, \quad (1s)_\nu\,\beta$$

Although the closed shells of the oxygen atom also play a certain part in the total wave function of the electron system of the water molecule, they are not given in Table 10-15, because their occupation is fixed and identical for all six possibilities indicated in the table. The wave functions must be antisymmetric with respect to the permutation of the electron coordinates — this property can easily be achieved by replacing the product functions with Slater determinants (cf. Sections 5.4 and 5.5).

Table 10-15

Valence Structures of the Electron System of the H_2O Molecule

Case	$(2p_x)$	$(1s)_\mu$	$(2p_y)$	$(1s)_\nu$	Function
1	α	α	β	β	Ψ_1
2	α	β	α	β	Ψ_2
3	β	α	α	β	Ψ_3
4	α	β	β	α	Ψ_4
5	β	α	β	α	Ψ_5
6	β	β	α	α	Ψ_6

We can therefore assign a Slater determinant Ψ to each case given in Table 10-15, which will be illustrated again with case 2:

$$\Psi_2 = |\,(1s)\,\alpha, (1s)\,\beta, (2s)\,\alpha, (2s)\,\beta, (2p_z)\,\alpha, (2p_z)\,\beta, (2p_x)\,\alpha, (1s)_\mu\,\beta,$$
$$(2p_y)\,\alpha, (1s)_v\,\beta\,| \tag{10-150}$$

So far, four determinants which are all eigenfunctions of the operator of the z-component of the total spin momentum $\mathscr{S}_z$ have been formed; in addition, the spin functions of the bonding orbitals are suitable for bond formation. A VB wave function can be formed from these four determinants if there exists a linear combination such that the resultant function Ψ is antisymmetric with respect to interchange of the spins assigned to the $(2p_x)$ and $(1s)_\mu$ orbitals, as well as to the $(2p_y)$ and $(1s)_v$ orbital pair; the electrons in the bond regions then form "local singlet" states. In general,

$$\Psi \sim \sum_{i=2}^{5} a_i \Psi_i \tag{10-151a}$$

Since Ψ must be antisymmetric with respect to interchange of the spin functions assigned to orbitals $(2p_x)$ and $(1s)_\mu$, then

$$\Psi \sim [-a_2\Psi_3 - a_3\Psi_2 - a_4\Psi_5 - a_5\Psi_4] \tag{10-151b}$$

and, because of the second condition also

$$\Psi \sim [-a_2\Psi_4 - a_3\Psi_5 - a_4\Psi_2 - a_5\Psi_3] \tag{10-151c}$$

Equations (10-151a) to (10-151c) are fulfilled when

$$a_2 = a_5,$$
$$a_3 = a_4 = -a_2,$$

giving for the unnormalized function describing the bonding situation in the water molecule,

$$\Psi \sim [\Psi_2 - \Psi_3 - \Psi_4 + \Psi_5] \tag{10-152}$$

This function is sometimes referred to as an eigenfunction of the $(2p_x) - (1s)_\mu$ and $(2p_y) - (1s)_v$ bonds corresponding to a certain valence structure. It can be shown that this function is an eigenfunction of the operator of the square of the total spin momentum $\mathscr{S}^2$ with the eigenvalue $S = 0$.

Sometimes, a larger number of valence structures can be attributed to a certain molecule. It appears, for example, that in a system of $2n$ π-electrons described by $2n$ atomic functions there are $\dfrac{(2n)!}{n!(n+1)!}$

independent structures with covalent bonds of zero total spin. Consequently, for benzene, five structures form a complete set of VB functions. It is obviously not sufficient to consider only the two Kekulé structures, I and II, but it is necessary also to consider structures with long bonds (III − V), called Dewar structures.

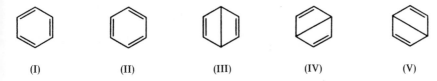

(I) (II) (III) (IV) (V)

The total VB wave function is considered in the form

$$\Psi = C_1(\Psi^I + \Psi^{II}) + C_2(\Psi^{III} + \Psi^{IV} + \Psi^V) \qquad (10\text{-}153)$$

The symmetry of the molecule is also taken into account here. The set of functions I − V is complete because any further structure can be expressed as a combination of these five structures.

The calculation of VB functions and their energies is formally quite simple. In general, the VB function is assumed to have the form

$$\Psi = \sum_{i=1}^{m} C_i \Psi^i, \qquad (10\text{-}154)$$

where m is the number of valence structures.

The variation method leads to the usual system of equations

$$\sum_{i=1}^{m} C_i(H_{ji} - ES_{ji}) = 0; \qquad j = 1, 2, \ldots, m \qquad (10\text{-}155)$$

Similarly as in the MO-LCAO method, the values of the expansion coefficients can be calculated by solving system of equations (10-155); the allowed energy values are determined from the condition (10-156):

$$\det \| H_{ij} - ES_{ij} \| = 0 \qquad (10\text{-}156)$$

The squares of coefficients (C_i^2) represent the weight of the i-th structure in the VB wave function. Of course, Ψ appearing in Eq. (10-154) is a many-electron wave function. The matrix elements have the usual meaning

$$H_{kl} = \int (\Psi^k)^* \mathscr{H} \Psi^l \, d\tau \qquad (10\text{-}157)$$

$$S_{kl} = \int (\Psi^k)^* \Psi^l \, d\tau, \qquad (10\text{-}158)$$

where $\mathscr{H}$ is the Hamiltonian of the studied electron system. The diagonal element H_{kk} represents the energy of the k-th structure.

248

The following expression has proven useful for calculation of non-diagonal elements:

$$H_{kl} - ES_{kl} = \frac{1}{2^{n-i}} [J' - E + bK'], \qquad (10\text{-}159)$$

where J' is Coulomb integral; K' is the exchange integral; i is the number of cyclic (closed) formations (called "islands") formed on superposition of the k-th and l-th structures, the matrix elements of which are calculated*, $b = c - \frac{1}{2}d$; c is the number of pairs of neighbouring centres (chemically bonded) in the islands; and d is the number of neighbouring pairs of centers on neighbouring islands (bonded in the respective compound).

As an example, the expression for the matrix element between the Kekulé (k) and Dewar (l) structures of naphthalene can be derived:

$$k \qquad\qquad l$$

The π bonds of these structures:

Superposition:

1st   2nd   3rd island, consequently, $i = 3$

Determination of c Determination of d

$$c = 6 \qquad\qquad\qquad d = 4$$

---

* Two π bonds connecting two centres are also considered to be an island. The number of islands is determined by investigating the pattern formed by superposition of the diagrams of the respective structure in which only π-bonds and long bonds are depicted.

Substitution into Eq. (10-159) gives ($2n = 10$)

$$H_{kl} - ES_{kl} = \frac{1}{2^{5-3}}\left[J' - E + \left(6 - \frac{1}{2}4 \right)K' \right] =$$

$$= \frac{1}{4}[J' - E + 4K']　　　　(10\text{-}160)$$

The non-diagonal elements can be determined by several different procedures.

The lowest energy value obtained by solving the secular equation corresponds to the ground state; the other values correspond to the electronically excited states. The difference between the ground state energy and the energy of the Kekulé structure equals the resonance energy.

10.6 The crystal field and ligand field theories[72-81]

10.6.1 Introductory comments

These theories originated from the necessity of interpreting properties, mainly optic and magnetic, of numerous series of compounds in which the ion of an element (mostly of a transition element) is surrounded by a certain number of other molecules or ions which are called ligands. These compounds are termed complex compounds. The arrangement of the ligands around the central atom is regular and mostly corresponds to one of the three arrangements depicted in Fig. 10-2. In cases (a) and (b) the arrangements are of octahedral and tetrahedral symmetry, respectively. Case (c) represents a square complex.

The crystal field theory was established[80] as early as 51 years ago, shortly after quantum mechanics was introduced. It has the particularly

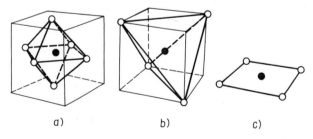

　　　　a)　　　　　　　b)　　　　　　　c)

Fig. 10-2. Octahedral, tetrahedral, and square complexes: central ion (●), ligands (○). In (a) and (b) the tetrahedron and the octahedron are, for lucidity, drawn in a cube.

interesting feature that, from mere knowledge of the symmetry of the ligand arrangement around the central ion, it is, for example, possible to predict that the (five-fold) degeneracy of the d orbitals will be removed (cf. Sections 6.6 and 6.8). In addition it is also possible to determine how the d orbitals will split and to find the degeneracy of the new levels. This is a valuable and interesting result. However, these considerations tell us nothing about the energy sequence of the individual orbitals or groups of degenerate orbitals. This is not particularly surprising and is a common feature of all descriptions based on symmetry considerations of molecular configurations, in other words on group theory.

Now when various types of all valence electron methods (EHT, CNDO, INDO) are available, nothing, in principle, prevents quantitative solution of these problems.

The fact that symmetry considerations alone can explain the splitting of degenerate d orbitals is particularly noteworthy and permits explanation of the absorbance by complexes of the first series of transition elements (as in higher series) in the visible region, i.e. in the electromagnetic radiation region with wavenumbers of 12000 to 25000 cm^{-1}. Experience with conjugated compounds (as far as the relationship between structure and colour is concerned) does little to explain why "small" (and unconjugated) formations, such as different complexes in which the ion of the transition element (e.g. V^{2+}, Fe^{2+} or Cu^{2+}) is the central atom, are coloured (i.e. have absorption maxima in the visible region); yet similar compounds containing Ca^{2+} and Zn^{2+} are colourless. This observation can easily be explained qualitatively because a "suitable" energy gap between the d orbitals is formed on removal of the degeneracy. As the incomplete occupation of d orbitals is typical for complexes of transition elements, the reason for their colour is obvious. In Ca^{2+},

Table 10-16

Ground State Terms of Free Atoms and the Corresponding Spectroscopic Notation [Ref. 77]

4s	K	Ca											
	$^2S_{1/2}$	$^1S_0$											
3d			Sc	Ti	V	Cr	Mn	Fe	Co	Ni	Cu	Zn	
			$^2D_{3/2}$	$^3F_2$	$^4F_{3/2}$	$^7S_3$	$^6S_{5/2}$	$^5D_4$	$^4F_{9/2}$	$^3F_4$	$^2S_{1/2}$	$^1S_0$	
4p		Ga	Ge	As	Se	Br	Kr						
		$^2P_{1/2}$	$^3P_0$	$^4S_{3/2}$	$^3P_2$	$^2P_{3/2}$	$^1S_0$						

Table 10-17

Number of d Electrons in Atoms and Ions of the Transition Elements

	Sc	Ti	V	Cr	Mn	Fe	Co	Ni	Cu	Zn
Neutral atom (Me)	1	2	3	5^a	5	6	7	8	10^a	10
Me^{2+}	1	2	3	4	5	6	7	8	9	10
Me^{3+}	0	1	2	3	4	5	6	7	8	9

[a] One electron only in the 4s atomic orbital.

on the other hand, the d orbitals are unoccupied and in Zn^{2+} they are fully occupied.

These phenomena can be considered more specifically, first, by finding which elements are involved (Table 10-16) and how many d electrons are available in the various ionic states (Table 10-17). This section will deal with the electron configurations of dipositive and tripositive ions of the transition elements. It has already been shown that these elements are typified by their incompletely filled d electron shells. The rare earth elements have similar properties as the transition metals and have incompletely filled f electron shells. There are three series of transition metals among the stable elements. To the first series belong elements of atomic numbers 21 to 30 with electron configurations

$$(1s)^2 (2s)^2 (2p)^6 (3s)^2 (3p)^6 (3d)^n (4s)^k,$$

where n has values from 1 to 10 and k the value 2, except for elements Cr and Cu, where $k = 1$. It is noteworthy that the (1s) to (3p) orbitals (inclusive) are fully occupied in these elements and that this partial configuration corresponds to the electron configuration of argon. These orbitals are only slightly influenced by the ligand environment and are usually not considered explicitly. Since the ionization of electrons, when divalent and trivalent ions of the transition elements are formed, occurs chiefly in the 4s atomic orbital, the ion of the transition metal in a complex compound is considered to be a d^n ion perturbed by its immediate environment, the ligands. The data in Table 10-17 can be understood on this basis. The elements with atomic numbers 39 to 48 belong to the second series of transition metals and the elements with atomic numbers 72 to 80 to the third series.

10.6.2 The electrostatic model (crystal field)

The crystal field theory considers ligands replaced by point charges (or dipoles) and assumes solely electrostatic interaction between the central atom and these point charges. It appears that this very simple model often leads to the same sequence of energy levels as more complicated models. Dubious results are, however, sometimes obtained, for example, for square complexes, if the formation of the covalent bond is completely ignored. The combination of the crystal field theory with the molecular orbital theory, known as the ligand field theory, does not have this drawback. It will be discussed later.

From the aspect of the electron configuration of the central ion, the described model of the complex ion corresponds to the intramolecular Stark effect, with the metal ion located in an electrostatic field induced by the ligands, which are considered to be fixed and unpolarizable. It is expedient to assume that the nucleus of the central ion is located at the origin of the coordinate system. The ligand environment is assumed to have a certain symmetry corresponding to point group G. The electrostatic potential caused by the ligands has the same symmetry (cf. Section 6.2). The Hamiltonian of an ion located in the ligand field then has the form

$$\mathcal{H} = \sum_{i=1}^{n} \left\{ -\frac{\hbar^2}{2m} \Delta_i - \frac{Ze^2}{4\pi\varepsilon_0 r_i} + \xi(r_i) \mathcal{L}_i \cdot \mathcal{S}_i + \mathcal{V}(r_i) \right\} +$$
$$+ \sum_{i<j} g(i,j), \qquad (10\text{-}161)$$

where, in addition to known symbols [cf. Eqs. (5-18), (5-19) and (4-88)], the term $\mathcal{V}(r_i)$, denoting the potential energy operator of the i-th electron in the ligand field, also appears. With the exception of $\mathcal{V}(r_i)$, $\mathcal{H}$ therefore corresponds to the Hamiltonian of the free ion. Because the perturbation method will be used for solution of the problem, it will be important to know the ratio of the magnitudes of the different terms occurring in the Hamiltonian. The following cases are of practical importance:

a) $\mathcal{L}\mathcal{S} < \mathcal{V} \leqq g$:   weak field scheme
b) $\mathcal{V} > g \geqq \mathcal{L}\mathcal{S}$:   strong field scheme
c) $\mathcal{V} < \mathcal{L}\mathcal{S} \leqq g$:   rare earth scheme

Notation has been introduced here for the total effects of the spin-orbit interaction, $\mathcal{L}\mathcal{S}$, of the crystal field, $\mathcal{V}$, and of the intereletronic repulsion, g. In the "accurate" solution of the problem (by means of a variation or perturbation calculation, taking into account a sufficient number of perturbation contributions) all three cases coincide and the final result is always the same. If only first-order perturbation contributions are considered, the results differ. A further reason for differentiating individual cases lies in an attempt to render the calculation more convenient and also appears in the specification of the initial functions representing the unperturbed functions in the perturbation method. In case a), where, the effect of the electron interaction exceeds the electrostatic influence of the ligands, the calculation is begun with wave functions describing the terms of the ion to which perturbation $\mathscr{V}$ is applied and, to include the spin-orbit interaction, a further perturbation in the form of the respective operator is introduced. It should be noted that, in metals of the first series of transition elements, the effect of the spin-orbit interaction on the energy spectrum of the ions is comparatively small, and therefore it is not usually taken into consideration. In case b) it is assumed that the crystal field is so strong that it perturbs the $l-l$ coupling and therefore the one-electron orbitals are taken directly as the basis for the perturbation treatment. To factorize the secular problem (cf. Section 6.6), suitable linear combinations of the atomic orbitals are usually used which simultaneously form the bases for the irreducible representations of symmetry group G. In case c) $\mathscr{V}$ is applied as the perturbation to the free ion energy levels in which the effect of the $l-s$ coupling is included.

As will be seen below, qualitative changes in the term system of the free ion can be determined on the basis of the crystal field theory, which also provides a method for their quantitative determination. For this reason the procedure is carried out in two stages.

Table 10-18

Relationship between Terms of "Free" and "Complex" Ions in a Field of Octahedral Symmetry

Term of free ion	S	P	D	F	G
Terms in complex ion	A_1	T_1	E, T_2	A_2, T_1, T_2	A_1, E, T_1, T_2

Using the representation theory of finite groups, first the splitting of the terms of the free ion caused by the electrostatic field is determined. The paper by Bethe[80] is of fundamental importance in this respect; it presents a method for decomposition of the irreducible representations of the full three-dimensional rotation group into irreducible representations of point groups of lower symmetry, especially for octahedral, hexagonal, tetragonal, and rhombic groups. The paper also demonstrates the derivation of the characters of the individual irreducible representations of the given symmetry groups. Thus, for example, in a field of octahedral symmetry, the term of the free ion is split in dependence on quantum number L in the way indicated in Table 10-18 (cf. Section 6.7).

In the further stage of the calculation, the extent of splitting or the shift of unsplit terms is determined by the perturbation method. The general procedure in the quantitative treatment of the crystal field problem was described in Section 6.8. It remains to describe the best way of adapting the perturbation operator (limited to cases when the spin-orbit interaction is negligible). If the ligands can be represented by charge density $\varrho(\mathbf{R})$, the following expression can be written for the potential energy $V(\mathbf{r})$ of the electron occuring in position $\mathbf{r}$:

$$V(\mathbf{r}) = -\frac{1}{4\pi\varepsilon_0} \int \frac{\varrho(\mathbf{R})}{|\mathbf{R} - \mathbf{r}|} \, d\mathbf{R}, \tag{10-162}$$

where $\mathbf{R}$ is the position vector of a general point in the charge cloud and the integration is performed over the entire charge distribution. If the ligands are approximated by point charges, as is frequently done, then the integration is reduced to summation over these point charges. An important step in the modification of expression (10-162) is formulation of the denominator of the integrand as follows:

$$\frac{1}{|\mathbf{R} - \mathbf{r}|} = \frac{1}{(R^2 + r^2 - 2Rr\cos\omega)} =$$

$$= \frac{1}{R}\left[1 + \left(\frac{r}{R}\right)^2 - 2\frac{r}{R}\cos\omega\right]^{-1/2} \tag{10-163}$$

Here r and R are the magnitudes of the corresponding vectors and ω is the angle lying between vectors $\mathbf{r}$ and $\mathbf{R}$; it is assumed that $r < R$. It can be expected that the relationship

$$\left|\left(\frac{r}{R}\right)^2 - 2\frac{r}{R}\cos\omega\right| < 1, \tag{10-164}$$

is valid for this case, and therefore expression (10-163) can be rearranged by expansion in a binomial series of the type

$$[1 + x]^{-1/2} = 1 - \frac{1}{2}x + \frac{3}{8}x^2 - \ldots \tag{10-165}$$

If the terms are arranged according to powers of r/R, it then follows that

$$\frac{1}{|\mathbf{R} - \mathbf{r}|} = \sum_{k=0}^{\infty} \frac{r^k}{R^{k+1}} P_k(\cos\omega), \tag{10-166}$$

where $P_k(\cos\omega)$ are Legendre polynomials which are sometimes defined as coefficients of the respective power series. It is preferable to write Eq. (10-166) in the form

$$\frac{1}{|\mathbf{R} - \mathbf{r}|} = \sum_{k=0}^{\infty} \frac{r_<^k}{r_>^{k+1}} P_k(\cos\omega), \tag{10-167}$$

where, because of convergence of the series, the denominator on the right-hand side of the equation contains the magnitude of the larger of vectors r and R and the numerator contains the smaller of the two. A series of type (10-167) is known as the multipole expansion of a potential in the given point.

As follows from the form of Hamiltonian (10-161), the operator of the potential energy of the electrons in the electrostatic field of the ligands has the form

$$\sum_{i=1}^{n} \mathscr{V}(r_i) \tag{10-168}$$

As every wave function can be expressed in the form of an expansion in terms of Slater determinants (cf. Section 5.4), the calculation of matrix elements

$$\langle \Delta_k | \sum_{i=1}^{n} \mathscr{V}(r_i) | \Delta_j \rangle \tag{10-169}$$

becomes of principal importance.

In Eq. (10-169) Δ_k and Δ_j are determinant functions [cf. Eq. (5-29)] constructed on the basis of one-electron orbitals, here atomic orbitals χ_μ. It has been shown [cf. Table 5-2 and Eqs. (5-33) and (5-34)] that matrix elements (10-169) can be expressed in terms of the integrals

$$\langle \chi_\mu(1) | \mathscr{V}(r_1) | \chi_\nu(1) \rangle, \tag{10-170}$$

where (specifically for this case) χ_μ and χ_ν are two atomic orbitals (for example, d orbitals) centered on the nucleus of the transition metal ion. Atomic orbitals can be expressed as the product of the radial and angular parts, the latter of which is, in principle, an associated Legendre function. The $\mathscr{V}(r_1)$ operator is invariant under all symmetry operations of point group G, which corresponds to the ligand environment. It also appears[81] that if multipole expansion (10-167) is employed for expressing $\mathscr{V}(r_1)$, the Legendre polynomials $P_k(\cos \omega)$ can be represented as an expansion of products of two associated Legendre polynomials, where one coefficient depends solely on the angular electron coordinates and the second on the angular coordinates related to the charge distribution of the ligands. The integrand of matrix element (10-170), which depends on the angular coordinates, therefore has, in principle, the form of the product of three associated Legendre polynomials and it is evident that, in view of the orthogonality relations between these types of functions, the infinite series is reduced to the sum of a few terms. Moreover, it is evident that the calculation will also be simplified by application of the

256

selection rules derived in Section 6.5. More details on this subject are provided in monographs devoted to the theory of the crystal and ligand fields (e.g. Ref. 81).

It has already been noted that a number of interesting conclusions on the bonding conditions in complex compounds can be drawn on the basis of qualitative considerations and taking into account the symmetry of the problem. A single d electron located in an octahedral ligand field is an example. This model can be used to represent the aquo complex of trivalent titanium. Such a system, denoted as d^1, is characteristic in that the procedure for the solution is the same irrespective of whether a weak or a strong field scheme is employed. In Section 6.6 it was concluded that in a field of six ligands of octahedral symmetry (group O), the five originally degenerate d orbitals are separated into two sets. One set consists of the d_{z^2} and $d_{x^2-y^2}$ orbitals, belonging to the irreducible representation E (twofold degeneracy); the second set, consisting of the d_{xy}, d_{yz}, d_{xz} orbitals, spans representation T_2 (threefold degeneracy). Placing the d orbitals in a crystal field is manifested in general by an increase in their energy; this phenomenon is caused by the monopole contribution ($k = 0$) of expansion (10-167). Furthermore, splitting of the levels occurs which is characteristic for the symmetry of the ligand distribution around the central ion. The conditions in octahedral complexes can be understood on the basis of the geometry of the angular parts of the d orbitals; in Fig. 10-3 two of these orbitals (d_{z^2}, $d_{x^2-y^2}$) are depicted in a model of the octahedral complex.

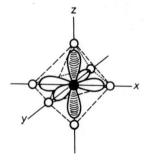

Fig. 10-3. Central ion (●)
and ligands (○)
in an octahedral complex.
The $d_{x^2-y^2}$ and d_{z^2} orbitals are
depicted (the latter hatched).

These two orbitals will be discussed here, as they lie in the direction toward the ligands leading to the greatest interaction due to the repulsion between the electrons in the orbitals and the point charges (representing the ligands). This interaction is manifested by an increase in the energy of orbitals of E symmetry. The centre of gravity of the term must be preserved, however, and the energy of orbitals of T_2 symmetry will consequently decrease. The monopole contribution of the ligand charge

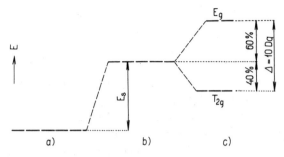

Fig. 10-4. Effect of a crystal field of octahedral symmetry [*b*) and *c*)] formed by six ligands on the fivefold degenerate level corresponding to the d¹ states of the free atom [*a*)].

distribution has already been taken into account in the increase in the energy of all the d orbitals by the value E_s (cf. Fig. 10-4). The second step concerning the removal of the degeneracy depends on the specific geometry of the ligand arrangement. The centre of gravity is preserved – this property is connected with the invariance of the trace of the secular determinant matrix [cf. Eq. (4-130) and (4-160)] toward unitary transformation. The resultant effect of the perturbation is graphically represented in Fig. 10-4. Fig. 10-5 depicts the effect of the intensity of the electrostatic field on the extent of splitting of the E and T_2 levels of the d¹ system; the figure represents the simplest possible Orgel diagram expressing the continuous transition from atomic states to states of the complex ion. It is evident from the figures that the energy difference corresponding to the new terms is important for spectroscopy. This difference, $\Delta = E(E) - E(T_2)$, is a basic parameter for octahedral complexes and is usually denoted by Δ or 10Dq. If the energy of the terms of the complex ion is expressed relative to the centre of gravity of these

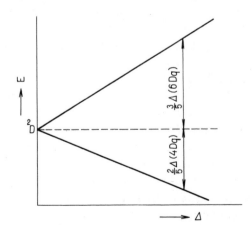

Fig. 10-5. Diagram for the energy levels of the d¹ system.

terms, the value $\frac{3}{5}\Delta$ is obtained for the energy of term E and $-\frac{2}{5}\Delta$ for the energy of term T_2. Transition between these terms (as a d$-$d transition) is forbidden by the selection rules in atomic electron spectroscopy, because it is a transition between states of equal parity. Owing to the interaction of the electronic and the nuclear motion during vibrations in the complex ion, the transition becomes partially allowed. This then explains the existence of the weak absorption band in the visible region of the spectrum of $[Ti(H_2O)_6]^{3+}$ ions.

Many-electron systems are somewhat more complicated, as it becomes necessary to distinguish between weak and strong crystal fields. Consequently, for example, for the d^2 system corresponding to the Ni^{2+} ion, under the influence of a weak crystal field, the terms of the free ion Ni^{2+}, i.e. the functions corresponding to states ^1S, ^3P, ^1D, ^3F and ^1G, are employed as the unperturbed functions for the perturbation treatment.

Use of a strong crystal field scheme allows lucid interpretation of the states of the complex ion, because diagrams similar to that in Fig. 10-5 can also be used for many-electron systems. When carrying out qualitative considerations, it is necessary to resort to *Hund's rule of maximum multiplicity* for the ground states of atomic systems. This rule expresses the empirically determined fact (which has also been theoretically confirmed) that an electron system is in the ground state when the maximum number of electrons have parallel spins. Then the maximum number of electron exchange integrals is nonzero; these integrals appear in the expression for the total energy with a negative sign [cf. Eq. (5-62)], thus decreasing the total value. In the strong crystal field approximation, the ground state of the d^2 system in the octahedral field will therefore correspond to the triplet state of electron configuration $(T_2)^2$ and, similarly, the d^3 system will correspond to quartet $(T_2)^3$. However, the d^k configurations, where k assumes values $k = 4$, 5, 6, 7, are ambiguous because if Δ (the "strength" of the electrostatic field) is small, Hund's rule affects the entire d shell, corresponding to an attempt by the ion to attain maximum multiplicity. The complex ion will then be in the same spin state as the free ion. If Δ is large, the electrons are forced into the energetically more favourable T_2 level, accompanied by a gain in the "orbital energy" and a loss in the exchange energy. There is consequently a change in multiplicity during transition from the free ion to the complex bonded ion. Thus, for example, the Fe^{3+} ion, corresponding to the d^5 configuration, can, according to this model, occur in two electron configurations in complex compounds:

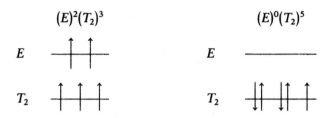

$(E)^2(T_2)^3$ $\qquad\qquad\qquad$ $(E)^0(T_2)^5$

These configurations can be distinguished on the basis of magnetic measurements. The first possibility, in the Pauling nomenclature, corresponds to an ionic complex (or a *"high spin complex"*); the second possibility, where a change in the multiplicity occurs on formation of the complex, corresponds to a covalent complex (*"low spin complex"*).

10.6.3 Ligand field theory

As mentioned in the previous section, the crystal field theory solves the bonding conditions in the complex compound using a model in which the electrostatic field of the ligands influences the electrons of the central ion. This model does not include charge transfer between the ligands and the central atom, in contradiction to a number of experimental results obtained using neutron diffraction, paramagnetic electron resonance and nuclear magnetic resonance. It is a further disadvantage of the crystal field theory that it does not sufficiently explain the relative stability of complexes in dependence on changes in the ligand environment and that it is unable to describe a double bond between a ligand and an ion, which often contributes to the stability of the complex.

It was therefore necessary to include the possibility of charge transfer between the ligands and the central ion, which is fulfilled by a variety of methods in the molecular orbital theory. The ligand field theory is, in essence, a specific version of these methods for calculation of the properties of complex compounds. This theory proceeds from the assumption of the molecular orbital theory and passes into the electrostatic model in the limiting case of zero charge transfer. Jarrett[82], for example, gave a general formulation of the problem.

Using the molecular orbital theory, which explicitly considers the orbital structure of the ligands, one-electron orbitals can be written in the form

$$\varphi = \sum_{\mu} a_{\mu}\chi_{\mu} + \sum_{\lambda} b_{\lambda}\chi_{\lambda}, \qquad (10\text{-}171)$$

where index μ denotes the orbitals located on the central ion and λ

the orbitals located on the ligands; a_μ and b_λ are the corresponding expansion coefficients. It is obvious that any approach using an "ab initio" type of treatment leads to numerical problems in the form of many-centre integrals and high-order secular problems. Consequently, many semi-empirical methods have been developed in this field, resembling those discussed in Section 10.2, involving all the valence electrons. The expression for the calculation of the off-diagonal elements in the EHT method (cf. method 5 in Table 10-2) had already been used ten years prior to its formulation in calculation[83] of the electronic structure of the complex MnO_4^- ion.

It should be mentioned in this connection that a new approach to studies of the electronic structure of inorganic complexes, based on the application of the Slater expression of the exchange energy, has been introduced by Johnson[84] and co-workers within the framework of the scattering model of the SCF theory (the method is denoted briefly SCF-X_α-SW). This treatment does not require the calculation of many-centre integrals and the results of its application to several ionic molecules [e.g. to the $(SO_4)^{2-}$ ion] are rather promising.

REFERENCES

1. Nesbet R. K.: *J. Chem. Phys.* **36**, 1518 (1962).
2. Del Bene J., Jaffé H. H.: *J. Chem. Phys.* **48**, 1807, 4050 (1968).
3. Dewar M. J. S., Klopman G.: *J. Am. Chem. Soc.* **89**, 3089 (1967).
4. Klopman G.: *J. Am. Chem. Soc.* **87**, 3300 (1965).
5. Pople J. A., Santry D. P., Segal G. A.: *J. Chem. Phys.* **43**, S129 (1965); Pople J. A., Segal G. A.: *J. Chem. Phys.* **43**, S136 (1965); **44**, 3289 (1966).
6. Katagiri S., Sandorfy C.: *Theoret. Chim. Acta* **4**, 203 (1966).
7. Jungen M., Labhart H.: *Theoret. Chim. Acta* **9**, 345, 366 (1968).
8. Hoffmann R.: *J. Chem. Phys.* **39**, 1397 (1963); **40**, 2480 (1964).
9. Sandorfy C.: *Canad. J. Chem.* **33**, 1337 (1955).
10. Brown R. D.: *J. Chem. Soc.* **1953**, 2615.
11. Del Re G.: *J. Chem. Soc.* **1958**, 4031.
12. Craig D. P.: *Proc. Roy. Soc.* (London) **A200**, 474 (1950).
13. Pariser R., Parr R. G.: *J. Chem. Phys.* **21**, 466, 767 (1953).
14. Roothaan C. C. J.: *Rev. Mod. Phys.* **23**, 69 (1951).
15. Pople J. A.: *Trans. Faraday Soc.* **49**, 1375 (1953); Brickstock A., Pople J. A.: *Trans. Faraday Soc.* **50**, 901 (1954).
16. Wheland G. W., Mann D. E.: *J. Chem. Phys.* **17**, 264 (1949); Coulson C. A., Gołebiewski A.: *Proc. Phys. Soc.* (London) **78**, 1310 (1961); Janssen M. J., Sandström J.: *Tetrahedron* **20**, 2339 (1964).
17. Hückel E.: *Z. Physik* **70**, 204 (1931); **76**, 628 (1932).
18. Coulson C. A.: *Proc. Phys. Soc.* (London) **A65**, 933 (1952); Longuet-Higgins H. C., Sowden R. G.: *J. Chem. Soc.* **1952**, 1404.

19. Ishitani A., Nagakura S.: *Theoret. Chim. Acta* **4**, 236 (1966); Zahradník R., Čársky P.: *J. Phys. Chem.* **74**, 1235, 1240 (1970).
20. Roothaan C. C. J.: *Rev. Mod. Phys.* **32**, 179 (1960).
21. Longuet-Higgins H. C., Pople J. A.: *Proc. Phys. Soc.* (London) **A68**, 591 (1955).
22. McLachlan A. D.: *Mol. Phys.* **3**, 233 (1960).
23. Brown R. D., Heffernan M. L.: *Trans. Faraday Soc.* **54**, 757 (1958).
24. Hochmann P.: *Thesis, Czechoslovak Acad. Sci.*, Praha 1967.
25. Zahradník R., Kröhn A., Pancíř J., Šnobl D.: *Collect. Czech. Chem. Commun.* **34**, 2553 (1969).
26. Pople J. A. in *Aspects de la chimie quantique contemporaine, Editions du CNRS*, Paris 1971, p. 17.
27. Pople J. A., Beveridge D. L., Dobosh P. A.: *J. Chem. Phys.* **47**, 2026 (1967).
28. Baird, N. C., Dewar M. J. S.: *J. Chem. Phys.* **50**, 1262 (1969).
29. Dewar M. J. S., Haselbach E.: *J. Am. Chem. Soc.* **92**, 590 (1970).
30. Fisher H., Kollmar H.: *Theoret. Chim. Acta* **13**, 213 (1969); Sichel J. M., Whitehead M. A.: *Theoret. Chim. Acta* **7**, 32 (1967); Sichel J. M., Whitehead M. A.: *Theoret. Chim. Acta* **11**, 220 (1968); Klopman G., Polák R.: *Theoret. Chim. Acta* **22**, 130 (1971); Klopman G., Polák R.: *Theoret. Chim. Acta* **25**, 223 (1972).
31. Goeppert-Mayer M., Sklar A. L.: *J. Chem. Phys.* **6**, 645 (1938).
32. Klopman G., O'Leary B.: *Fort. chem. Forschung* **15**, 445 (1970).
33. Klopman G.: *J. Am. Chem. Soc.* **86**, 4550 (1964); **87**, 3300 (1965).
34. Cook D. B., Hollis P. C., McWeeny R.: *Mol. Phys.* **13**, 553 (1967).
35. Kaufman J. J.: *J. Chem. Phys.* **43**, S152 (1965).
36. Pohl H. A., Rein R., Appel K.: *J. Chem. Phys.* **41**, 3385 (1964).
37. Skancke P. N.: *Arkiv f. Fysik* **29**, 573 (1965); **30**, 449 (1965).
38. Zahradník R., Čársky P.: *Collect. Czech. Chem. Commun.* **38**, 1876 (1973).
39. Dixon R. N.: *Mol. Phys.* **12**, 83 (1967).
40. Sandorfy C., Daudel R.: *Compt. rend.* **238**, 93 (1954).
41. Fukui K., Kato H., Yonezawa T.: *Bull. Chem. Soc. Japan* **33**, 1197 (1960); **34**, 442 (1961).
42. Sandorfy C.: *Canad. J. Chem.* **33**, 1337 (1955).
43. Pople J. A., Santry D. P.: *Mol. Phys.* **7**, 269 (1964).
44. Klopman G.: *Tetrahedron* **19** (suppl. 2), 110 (1963).
45. McWeeny R.: *Proc. Roy. Soc.* (London) **A223**, 306 (1954); *Rev. Mod. Phys.* **32**, 335 (1960).
46. Lykos P. G., Parr R. G.: *J. Chem. Phys.* **24**, 1166 (1956); **25**, 1301 (1956).
47. Pariser R.: *J. Chem. Phys.* **21**, 568 (1953).
48. Mataga N., Nishimoto K.: *Z. Phys. Chem.* (Frankfurt) **13**, 140 (1957).
49. Ruedenberg K.: *J. Chem. Phys.* **34**, 1861 (1961).
50. Fischer-Hjalmars I.: *J. Chem. Phys.* **42**, 1962 (1965).
51. Pancíř J., Matoušek I., Zahradník R.: *Collect. Czech. Chem. Commun.* **38**, 3039 (1973).
52. Hosoya H., Nagakura S.: *Theoret. Chim. Acta* **8**, 319 (1967).
53. Nakajima T.: *Bull. Chem. Soc. Japan* **38**, 83 (1965).
54. Longuet-Higgins H. C., Salem L.: *Proc. Roy. Soc.* (London) **A257,** 445 (1960).
55. Kon H.: *Bull. Chem. Soc. Japan* **28,** 275 (1955).
56. Flurry R. L., Jr., Stout E. W., Bell J. J.: *Theoret. Chim. Acta* 8, 203 (967).
57. Straub P. A., Meuche D., Heilbronner E.: *Helv. Chim. Acta* **49**, 517 (1966).
58. Linderberg J.: *Chem. Phys. Letters* **1**, 39 (1967).

262

59. Flurry R. L., Jr., Bell J. J.: *J. Am. Chem. Soc.* **89**, 525 (1967).
60. Parr R. G.: *J. Chem. Phys.* **20**, 1499 (1952).
61. Ohno K.: *Notes on M. O. Calculations of π-electron Systems.* Quantum Chemical Group, Uppsala 1962.
62. Pancíř J., Matoušek I., Zahradník R.: *Collect. Czech. Chem. Commun.* **38**, 3039 (1973).
63. Parr R. G.: *Quantum Theory of Molecular Electronic Structure.* Benjamin, New York 1964.
64. Heilbronner E., Bock H.: *The HMO-Model and its Application.* Wiley, New York and Verlag Chemie, Weinheim 1976. (Three volumes.)
65. Zahradník R., Čársky P.: *Organic Quantum Chemistry Problems.* Plenum Press, New York 1973.
66. Coulson C. A., Longuet-Higgins H. C.: *Proc. Roy. Soc.* (London) **A191**, 39 (1947); **A192**, 16 (1947); **A193**, 447 (1948); **A195**, 188 (1948).
67. Bayliss N. S.: *Quart. Revs.* **6**, 319 (1952).
68. Platt J. R. in *Encyclopedia of Physics, Handbuch der Physik* (ed. Flügge S.) vol. 37/2, Springer, Berlin 1961, pp. 173 – 281.
69. Kuhn H.: Synopsis of lextures given at the Summer School in Theoretical Chemistry, Konstanz, Germany, September 1963.
70. *Free-Electron Theory of Conjugated Molecules* (a source book). Publications of the Chicago group of J. R. Platta from the period 1949 – 1961. Wiley, New York 1964.
71. Coulson C. A.: *Valence.* Clarendon Press. Oxford 1952.
72. V ek A. A.: *Struktura a vlastnosti koordinačních sloučenin.* Academia, Praha 1966.
73. Orgel L. E.: *An Introduction to Transition-Metal Chemistry. Ligand-Field Theory.* Methuen, London 1960.
74. Ballhausen C. J.: *Introduction to Ligand Field Theory.* McGraw-Hill, New York 1962.
75. Phillips L. F.: *Basic Quantum Chemistry*, Chap. 6. Wiley, New York 1965.
76. Dunn T. M., McClure D. S., Pearson R. G.: *Some Aspects of Crystal Field Theory.* Harper & Row, New York 1965.
77. Sutton D.: *Electronic Spectra of Transition Metal Complexes.* McGraw-Hill, London 1968.
78. Hill H. A. O., Day P. (eds.): *Physical Methods in Advanced Inorganic Chemistry.* Wiley-Interscience, New York 1968.
79. Manch W., Fernelius W. C.: *J. Chem. Educ.* **38**, 192 (1961).
80. Bethe H.: *Ann. Physik* **3**, 133 (1929).
81. Griffith J. S.: *The Theory of Transition-Metal Ions.* Cambridge University Press, 1961.
82. Jarrett H. S.: *J. Chem. Phys.* **31**, 1579 (1959).
83. Wolfsberg M., Helmholz L.: *J. Chem. Phys.* **20**, 837 (1952).
84. Johnson K. H.: *Advan. Quant. Chem.* **7**, 143 (1973).

11. USE OF THE SOLUTION
TO THE SCHRÖDINGER EQUATION

11.1 Quantities related to the molecular energy (the total electron energy, ionization potential, electron affinity, excitation energy)[1]

In methods in which electron repulsion is not considered explicitly (HMO, EHT), the relations are simple. For illustration it will be useful to consider a system described, say, by six molecular orbitals $(\varphi_1, \varphi_2, ..., \varphi_6)$ and by the corresponding orbital energies $(E_1, E_2, ..., E_6)$ (Fig. 11-1). The total energy is given by [cf. Eq. (10-115)].

$$W = \sum_i n_i E_i ,\qquad (11\text{-}1)$$

where n_i is the occupation number of the i-th MO (and can assume a value of 0, 1 or 2) and E_i is the orbital energy of the i-th MO for which

$$E_i = \alpha + k_i \beta \qquad (11\text{-}2)$$

The Coulomb (α) and resonance (β) integrals in this formula can be expressed in the usual energy units. This does not imply that the corresponding integrals need be solved, but only that numerical values are assigned to them such that the theoretical quantities in which

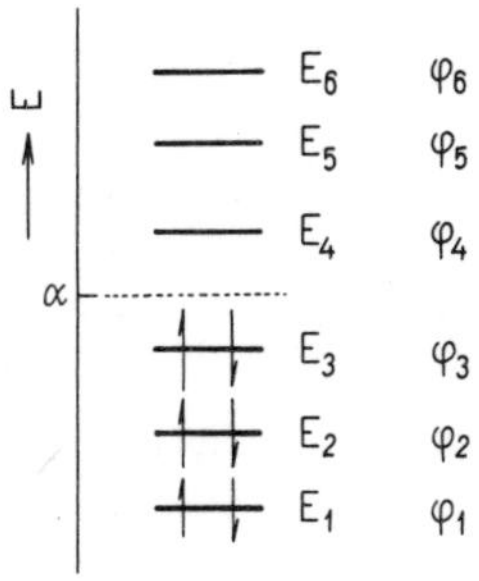

Fig. 11-1. Molecular orbitals φ_i and their energies E_i for the studied system.

264

these integrals appear give a true representation of different experimental characteristics for specimens of various classes of substances. In the HMO method energy characteristics are, as a rule, expressed in terms of quantities α and β. When comparing theoretical (HMO) data with experimental results, an incongruency appears: for the standard resonance integral (the resonance integral corresponding to the carbon $2p_z$ atomic orbitals on neighbouring atoms), a value of $20-200$ kJ/mol is obtained according to the nature of the experimental data (instead of the expected constant value). It follows that the numerical value of β, corresponding to a certain characteristic and to a certain group of substances, cannot be used for another characteristic or another group of substances.

In the considered methods, calculation of the electron energy (π or $\pi + \sigma$) is very simple; it is given by the sum of the occupied one-electron energy levels [cf. Eq. (11-1)]; this energy can also be expressed in terms of electron charge densities and bond orders and of the Coulomb and resonance integrals. The expression valid in the HMO theory has already been given [cf. Eq. (10-116)].

The heats of formation* of non-conjugated organic compounds can be calculated relatively accurately using group contributions. On the other hand, in conjugated (most frequently planar) compounds this is not true and the differences between experimental and calculated values are considerable. This difference is called the resonance energy. Theoretically, a similar quantity, called the delocalization energy (E_D), is defined as the difference between the π-electron energy of the system (W), whose delocalization energy is calculated, and the π-electron energy of the energetically most favourable Kekulé structure (W_K); consequently (cf. Sections 10.5 and 15.2)

$$E_D = W - W_K \tag{11-3}$$

The ionization potential (I) represents the energy that must be added to a system to transfer an electron from the system to a site of zero potential. On the other hand, the energy which is liberated on the addition of one electron to a system is called *the electron affinity* (A). The most important of these quantities is connected with the highest occupied and the lowest unoccupied MO's (the first ionization potential and the first electron affinity). Fig. 11-2 illustrates changes in the occupation of MO's due to these two processes (b, c) as well as to the electron

---

* The heat which is liberated on formation of 1 mol of a substance from the elements, where the reactants and products are in their standard states. This heat can be determined indirectly from the heat of combustion or of hydrogenation (see below).

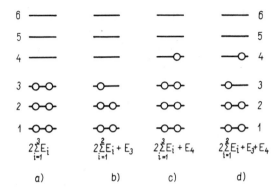

Fig. 11-2. Ground state (*a*), state after ionization (radical cation) (*b*), state after attachment of an electron (radical anion) (*c*), state following the lowest energy excitation (*d*). The electronic energies of individual systems are given.

excitation with the lowest energy requirement (called the $N \rightarrow V_1$ excitation, N and V being symbols for the "normal" and "valence" states). From the energies of these structures and from the energy of a molecule in the initial state very useful theoretical characteristics can be calculated from simple differences (cf. Fig. 11-2):

$$I = W_b - W_a = -E_3 \tag{11-4}$$

$$A = W_c - W_a = E_4 \tag{11-5}$$

$$E(N \rightarrow V_1) = W_d - W_a = E_4 - E_3 \tag{11-6}$$

It follows that the i-th ionization potential (the i-th electron affinity) is, in general, equal to the orbital energy of the i-th MO (except for the sign of the ionization potential). In Fig. 11-3 the first ionization potentials are plotted against the HMO energies of the highest occupied MO's for a series of conjugated hydrocarbons (Table 11-1). Separation of data for hydrocarbons with markedly alternating bonds (polyenes) is not surprising, since in these systems one of the HMO assumptions (assumption of equality of the resonance integrals) is not fulfilled. If various values of β, according to the lengths of the $C-C$ bonds, are introduced into the calculation, quite satisfactory results are obtained.

In methods where electron repulsion is considered explicitly (e.g. the Pople approximation of the Roothaan SCF method, CNDO methods), the relationships are more complex. In the expression for the total electron energy appear not only orbital energies ε_i, but also terms derived from electron repulsion [cf. Eq. (10-31)]. Electron Coulomb and exchange integrals are defined as follows [cf. Eqs. (5-59c) and (5-59d)]:

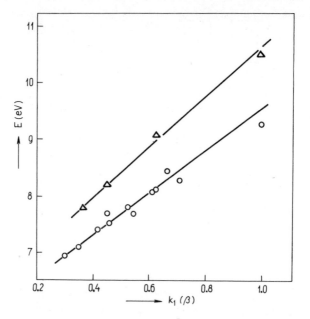

Fig. 11-3. Dependence of adiabatic ionization potentials on the HMO energy of the highest occupied π-molecular orbital for polyenes ($\triangle$) and for benzenoid hydrocarbons ($\bigcirc$).

Table 11-1

Theoretical and Experimental First Ionization Potentials and Electron Affinities

Hydrocarbon	$\dfrac{k_1}{\beta}$	$\dfrac{I^{a}}{eV}$	$\dfrac{A^{b}}{eV}$
Ethylene	1.000	10.52	–
1,3-Butadiene	0.618	9.07	–
1,3,5-Hexatriene	0.445	8.23	–
1,3,5,7-Octatetraene	0.347	7.80	–
Benzene	1.000	9.24	–
Naphthalene	0.618	8.14	0.148
Anthracene	0.414	7.42	0.556
Tetracene	0.295	6.94	–
Phenanthrene	0.605	8.07	0.307
Benz[a]anthracene	0.452	7.52	0.630
Pyrene	0.445	7.70	0.591
Chrysene	0.520	7.82	0.397
Diphenyl	0.705	8.27	–

[a] Ionization potential determined from photoionization and electron spectroscopy measurements (from various sources).
[b] R. S. Becker, E. Chen: *J. Chem. Phys.* **45**, 2403 (1966).

$$J_{ij} = \frac{e^2}{4\pi\varepsilon_0} \iint \varphi_i^*(1)\,\varphi_j^*(2)\,\frac{1}{r_{12}}\,\varphi_i(1)\,\varphi_j(2)\,d\tau_1\,d\tau_2 \tag{11-7}$$

$$K_{ij} = \frac{e^2}{4\pi\varepsilon_0} \iint \varphi_i^*(1)\,\varphi_j^*(1)\,\frac{1}{r_{12}}\,\varphi_i(2)\,\varphi_j(2)\,d\tau_1\,d\tau_2 \tag{11-8}$$

The expressions for the individual energy characteristics are given in Table 11-2. The expression for the ionization energy (and also for the electron affinity) is particularly interesting, because it is formally the same as the expression appearing in the simple methods. This is a consequence of the Koopmans theorem, according to which the SCF orbital energies ε_i of the parent system can also be used for calculation of the total energy of the ion derived (by removal or acceptance of an electron) from this system. At the same time the change in geometry which generally accompanies ionization is not considered, nor is the fact that the system formed is (in contrast to the parent system) a system with an open electron shell. These are undoubtedly rather drastic simplifications for which the use of Koopmans theorem has been repeatedly criticized. The correct procedure requires calculation of the SCF energy for the parent system with a closed shell and for the radical-ion with an open electron shell; in both systems it is necessary to take correct interatomic distances into account. Numerical values of ionization potentials obtained by the two procedures differ only slightly in rigid molecules (e.g. conjugated hydrocarbons), which supports the use of the Koopmans theorem in these systems. However, in small molecules, the difference in the results usually amount to about 1 eV. For radicals the Koopmans theorem leads to more

Table 11-2

MO-Energy Characteristics (Closed Shell Systems)

Characteristics	Methods	
	semiempirical[a]	empirical
ionization potential; ionization of the electron from the i-th MO	ε_i	E_i
excitation energy; excitation of an electron from the i-th MO to the j-th MO		
$(S\rightarrow S')$	$\varepsilon_j - \varepsilon_i - J_{ij} + 2K_{ij}$	$\left.\begin{array}{c}\\\\\end{array}\right\} E_j - E_i$
$(S\rightarrow T)$	$\varepsilon_j - \varepsilon_i - J_{ij}$	
Total electron energy	$2\sum_i \varepsilon_i + \sum_i\sum_j (2J_{ij} - K_{ij})$	$2\sum_i E_i$

[a] For definition of integrals J_{ij} and K_{ij} see Eqs. (11-7) and (11-8).

Table 11-3

MO-Energy Characteristics (Open Shell Systems)[a]

Characteristics	SCF method of Longuet-Higgins and Pople
ionization potential; ionization of electron from the singly occupied m-th MO	$\varepsilon_m - \dfrac{1}{2} J_{mm}$
ionization potential; ionization of electron from the doubly occupied i-th MO	
a) leading to a singlet state	$\varepsilon_i - \dfrac{3}{2} K_{im}$
b) leading to a triplet state	$\varepsilon_i + \dfrac{1}{2} K_{im}$
electron affinity (acceptance of an electron into the m-th MO)	$\varepsilon_m + \dfrac{1}{2} J_{mm}$
excitation energy; excitation of an electron (D→D′)	
a) from a doubly occupied into a singly occupied MO $(i \rightarrow m)$[b]	$\varepsilon_m - \varepsilon_i + \dfrac{1}{2}(K_{im} + J_{mm} - 2J_{im})$
b) from a singly occupied into an unoccupied MO $(m \rightarrow r)$[c]	$\varepsilon_r - \varepsilon_m + \dfrac{1}{2}(K_{mr} + J_{mm} - 2J_{mr})$
Total electron energy	$\displaystyle\sum_\mu \sum_\nu \dfrac{1}{2} P_{\mu\nu}(F_{\mu\nu} - H^c_{\mu\nu}) - \dfrac{1}{4} J_{mm}$

[a] For definition of integrals J_{ij} and K_{ij} see Eqs. (11-7) and (11-8); m is a subscript of the singly occupied MO in the initial system.

[b] Excitation of this type is designated as A-type excitation.

[c] B-type excitation.

complex expressions (Table 11-3). Some interesting information follows from this table, for example, that ionization from an arbitrary doubly occupied level leads to two different values for the ionization potential (according to the multiplicity of the system after ionization). Most important is that Table 11-3 points out the incorrectness of mechanical transfer of expressions from closed shell systems to systems with open shells.

In calculations within the framework of many-electron methods (e.g. methods of configuration interaction), the theoretical intepretation of electronic excitation requires knowledge of the energy difference for two states of the molecule studied (the state before and after excitation); in ionization processes, the energy difference for two-electron systems, differing in the number of electrons which the parent system loses or gains, must be found (cf. Sections 5.4 and 5.5).

It will now be suitable to describe the experimental determination of ionization potentials. For several decades three methods have been used: electron impact, photoionization, and optical spectroscopy. None of the methods is particularly simple, so that the number of experimentally determined ionization potentials was, until recently, not large. A few years ago an interesting and powerful method was developed permitting determination not only of the first, but also of a number of higher ionization potentials in a single experiment using a fairly simple procedure. This is the method of *photoelectron spectroscopy* (PES)[2]. In this method, the studied molecules in the gaseous phase are ionized by photons of a defined energy (photons of 21.21 eV from a helium discharge lamp are generally used), and the kinetic energy of the electrons liberated from the molecule is experimentally determined. A photoelectron spectrum is schematically depicted in Fig. 11-4. The maxima of the bands indicate directly the individual ionization potentials. From the reproduction of a real spectrum of carbon monoxide (Fig. 11-5) it is evident that the situation is more complicated and individual bands have a fine structure. The structure of the spectrum depends on the fact that, when ionized, the molecules are excited to a set of vibrational states of a molecular ion. In Fig. 11-6 this situation is outlined for a diatomic molecule (polyatomic molecules are rather similar).

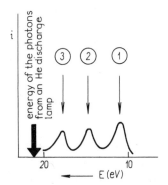

Fig. 11-4. Photoelectron spectrum — schematic representation.

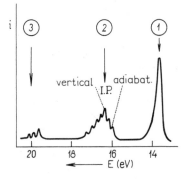

Fig. 11-5. Actual photoelectron spectrum: the 2nd and 3rd bands display vibrational structure.

Ionization begins from the vibrational ground state of the parent system: transition into a set of vibrational states must be considered. The two most important, called *the vertical* and *adiabatic transitions*, are illustrated in Fig. 11-6. It is typical for the first that no change occurs in the interatomic distance; such a transition is called *a Franck-Condon transition* and it is the most intense (cf. the 2nd band in Fig. 11-5). The second important transition is that into the vibrational ground state of

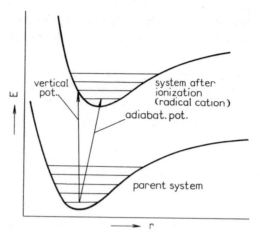

Fig. 11-6. Potential energy curves of the parent system and of the radical cation produced by removing an electron.

the molecular ion. This is obviously the transition with the lowest energy requirement (0−0 transition).

It is necessary to add that the first ionization potentials determined by electron impact are always several tenths of an eV higher than the values determined by photoionization. The difference is very likely due to the fact that, in the first method, vertical potentials are measured, whereas in the second method, adiabatic potentials are found. The reason that vertical transitions (transitions with no change in interatomic distances) are obtained in electron impact measurements is the very short duration of this process − after removal of the electron the molecule simply has not enough time to change its geometry.

While calculation of electron affinities is as simple as calculation of ionization potentials, their experimental determination is more difficult. It is based on the ability of substances to absorb thermal electrons. The relative absorption is expressed by an absorption coefficient which can be shown to be related to the electron affinity of the molecule. Electron affinities have so far been determined in this way for only a few benzenoid hydrocarbons (Table 11-1) and for a number of their derivatives. Correlation of these affinities with HMO energies (of the orbitals which the thermal electron enters) is less satisfactory than correlation of ionization potentials.

With regard to the complexity of direct determination of electron affinities, it is useful to mention that polarographic half-wave potentials of reduction waves can be proportional to electron affinities. This is true of aprotic solvents where the radical-anion formed is not subject to further conversions and where the polarographic process is reversible. The values obtained in this way, however, include the solvation energy of the ion formed.

In the discussion of the possibilities of using one-electron energies, the simple calculation of localization energies is worth mentioning (which, however, must not be mistaken for delocalization energies). These quantities will be used later in the calculation of the π-electron contribution to the activation energy. The atomic localization energy of an atom in a conjugated planar system is defined as the difference between the energy of a system formed after removal of the μ-th AO from a conjugated parent system and the energy of the parent system. This removed orbital (AO) can be occupied by none, one or two electrons. Similarly, the bond localization energy (also called ortholocalization energy) is defined as the energy of a double bond plus that of the remaining part of the original molecule reduced by the energy of the original system. Paralocalization and generally polycentric localization energies can be defined in a similar way. Apparently, the delocalization energy can be considered as a special

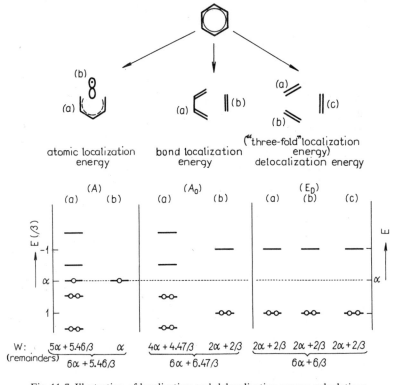

Fig. 11-7 Illustration of localization and delocalization energy calculations.

case of "polybond" localization energy. For illustration*, the atomic localization (A), ortholocalization (A_O) and delocalization (E_D) energies of benzene (cf. Fig. 11-7) can be calculated:

$$A = |6\alpha + 5.46\beta - (6\alpha + 8\beta)| = 2.54\beta$$

$$A_O = |6\alpha + 6.47\beta - (6\alpha + 8\beta)| = 1.53\beta$$

$$E_D = |6\alpha + 6\beta - (6\alpha + 8\beta)| = 2\beta$$

11.2 Quantities derived from the wave function

11.2.1 Introductory comments

The object of any sufficiently general quantum-chemical calculation is to obtain a wave function for the studied molecule or for any other electron system. Knowledge of the wave function is a prerequisite for calculating the expectation values of physical quantities, among which energy assumes an exceptional position – because of its role in the Schrödinger equation and as a universal constant of motion. Among the other measurable physical and chemical quantities are a number of those which are indispensable for characterization of molecules; the theoretical interpretation of these quantities is thus inevitably necessary. Calculation of the dipole moment of molecules in the text below will serve as an example for these quantities.

The use of a wave function for the calculation of physical quantities alone is disadvantageous because a great deal of information contained in the wave function is lost. On the other hand, during the development of modern chemistry, a number of specific concepts were formed, which proved to be very useful for prognosis and interpretation of the chemical and physical properties of electron systems. This applies to terms such as the two-electron bond, lone electron pair and hybridization. In addition, the properties of molecules can be intepreted in terms of atomic properties, such as atomic charges. This fact motivated efforts to elaborate methods which would be capable of analyzing very complex wave functions in terms of these concepts. However, before describing these methods, it would be useful to introduce the density matrices, by means of which these methods can conveniently be classified.

---

* The HMO energies of benzene and butadiene equal $6\alpha + 6\beta$ and $4\alpha + 4.47\beta$, respectively.

11.2.2 Density matrix

Let us start, for the sake of simplicity, from the definition of a spin orbital (5-32):

$$\lambda(x) = \varphi(\mathbf{r})\,\eta(\sigma),\tag{11-9}$$

where the symbols have the same meaning as in Section 5.4. From the statistical interpretation (cf. postulate 2 in Section 4.2) it follows that the expression

$$|\lambda(x)|^2\,dx\tag{11-10}$$

gives the probability of occurrence of an electron in the space-spin element $dx = d\mathbf{r}\,d\sigma$ and thus

$$\gamma(x) = |\lambda(x)|^2 = |\varphi(\mathbf{r})|^2\,|\eta(\sigma)|^2\tag{11-11}$$

is the electron probability density function. If we are not interested in the spin, then integration can be carried out over the spin variable to give

$$\varrho(\mathbf{r})\,d\mathbf{r} = d\mathbf{r}\int \gamma(x)\,d\sigma = |\varphi(\mathbf{r})|^2\,d\mathbf{r},\tag{11-12}$$

where $\varrho(\mathbf{r})$ gives the probability density without reference to the spin.

Generalization of the given relations for the many-electron system (where the number of electrons is n) is relatively easy considering that the expression

$$|\Psi(x_1, x_2, ..., x_n)|^2\,dx_1\,dx_2 ... dx_n\tag{11-13}$$

represents the probability of simultaneously finding electron 1 in element dx_1, electron 2 in element dx_2, ... and finally electron n in element dx_n. The probability of finding electron 1 in a space-spin element dx_1 with an arbitrary distribution of the other electrons is obtained by integration of expression (11-13) over the coordinate of the second to the n-th electron. We are, however, more interested in the probability of finding any of the n electrons in element dx_1, as the electrons in the considered system are indistinguishable. Since the product $\Psi\Psi^*$ is symmetrical in the variables of the n electrons, the desired probability is obtained by multiplication of the given integral by number n:

$$\gamma(x_1) = n\int |\Psi(x_1, x_2, x_3, ..., x_n)|^2\,dx_2 ... dx_n,\tag{11-14}$$

where x_1 denotes "point x_1" at which the probability density of any of the n electrons is evaluated. For the determination of the "spinless" density Eq. (11-12) is again valid.

The probabilities for configurations of any number of particles can also be found. The relationship

$$\Gamma(x_1, x_2) = \frac{n(n-1)}{2}\int |\Psi(x_1, x_2, ..., x_n)|^2\,dx_3 ... dx_n,\tag{11-15}$$

274

giving the probability density of simultaneously finding one electron at "point x_1," and the second electron at "point x_2", is of practical significance. Function (11-15) is connected with the so-called pair correlation function

$$P(\mathbf{r}_1, \mathbf{r}_2) = \int \Gamma(x_1, x_2)\, d\sigma_1\, d\sigma_2 \tag{11-16}$$

which can be used for the study of correlation effects in electronic systems.[4]

For illustration, functions $\gamma(x_1)$ and $\varrho(\mathbf{r}_1)$ can be calculated for the two-electron system of the H_2 molecule, described by the VB wave function [cf. Eq. (10-137)]

$$\Psi(x_1, x_2) = \frac{1}{\sqrt{2}} \left[\chi_\mu(\mathbf{r}_1)\chi_\nu(\mathbf{r}_2) + \chi_\nu(\mathbf{r}_1)\chi_\mu(\mathbf{r}_2) \right] \frac{1}{\sqrt{2}} \left[\alpha(\sigma_1)\beta(\sigma_2) - \beta(\sigma_1)\alpha(\sigma_2) \right] \tag{11-17}$$

where, for the sake of simplification, it is assumed that the atomic orbitals are orthonormal. Obviously it holds that

$$\gamma(x_1) = 2\,\frac{1}{4}\int |\chi_\mu(\mathbf{r}_1)\chi_\nu(\mathbf{r}_2) + \chi_\nu(\mathbf{r}_1)\chi_\mu(\mathbf{r}_2)|^2 \times$$

$$\times |\alpha(\sigma_1)\beta(\sigma_2) - \beta(\sigma_1)\alpha(\sigma_2)|^2\, d\mathbf{r}_2\, d\sigma_2 =$$

$$= \frac{1}{2} \left[|\chi_\mu(\mathbf{r}_1)|^2 + |\chi_\nu(\mathbf{r}_1)|^2 \right] \left[|\alpha(\sigma_1)|^2 + |\beta(\sigma_1)|^2 \right] \tag{11-18}$$

and after integration over the spin variable σ_1,

$$\varrho(\mathbf{r}_1) = |\chi_\mu(\mathbf{r}_1)|^2 + |\chi_\nu(\mathbf{r}_1)|^2, \tag{11-19}$$

leading to the trivial result that $\varrho(\mathbf{r}_1)$ is equal to superposition of partial densities produced by the atomic orbitals localized on centres μ and ν. Eqs. (11-18) and (11-19) permit expression of the probability $\varrho^\alpha(\mathbf{r}_1)$ of finding the electron in spin state α (or β), whence

$$\varrho^\alpha(\mathbf{r}_1) = \varrho^\beta(\mathbf{r}_1) = \frac{1}{2}\varrho(\mathbf{r}_1) \tag{11-20}$$

Expressions defined by Eqs. (11-14) and (11-15) [or by (11-12) and (11-16)] can be understood as diagonal elements of matrices with continuous indices (matrices with continuous indices have already been described in Section 4.5):

$$\gamma(1; 1') = n \int \Psi(1, 2, \ldots, n)\, \Psi^*(1', 2, \ldots, n)\, d\tau_2 \ldots d\tau_n \tag{11-21}$$

$$\Gamma(1, 2; 1', 2') = \frac{n(n-1)}{2} \int \Psi(1, 2, 3, \ldots, n)\, \Psi^*(1', 2', 3, \ldots, n)\, d\tau_3 \ldots d\tau_n, \tag{11-22}$$

where it is intentionally not specified whether the wave functions depend on the space-spin or only on the space coordinates of the electrons, in order to include both possibilities. The given expressions define the density matrices of the first [Eq. (11-21)] and the second order [Eq. (11-22)].

The establishment of density matrices has the following practical importance. It has already been mentioned that theoretical expressions for physical quantities depend both on the wave function Ψ and on its complex conjugate Ψ^*; moreover, these relations can be expressed by matrix elements in which integration is performed only over the co-ordinates of a limited number of particles corresponding to the type of particle interaction. In applications, only two-particle interactions of coulombic character can occur and thus the number of coordinates $(n - 2)$ is not of direct importance for the calculation of physical quantities. Establishment of density matrices removes these disadvantages and therefore simplifies descriptions of electron systems.

Let us start from Eq. (5-18) for the Hamiltonian of a molecular electron system treated within the Born-Oppenheimer approximation. The energy expectation value of a molecule in a state defined by wave function Ψ can be expressed as

$$\langle \Psi \mid \mathcal{H} \mid \Psi \rangle = \int \Psi^*(1, 2, ..., n) \left[\sum_{i=1}^{n} \hbar(i) \right] \Psi(1, 2, ..., n) \, d\tau +$$
$$+ \int \Psi^*(1, 2, ..., n) \left[\sum_{i<j} g(i, j) \right] \Psi(1, 2, ..., n) \, d\tau, \qquad (11\text{-}23)$$

where $d\tau = d\tau_1 \, d\tau_2 ... d\tau_n$.

First the total energy contribution originating from matrix elements of the one-electron operator will be examined. Because the integration variable notation can be changed and because the product $\Psi\Psi^*$ is invariant under an arbitrary permutation of the coordinates, then any of the addends can be modified so that n identical integrals are obtained, i.e.

$$\int \Psi^*(1, 2, ..., n) \left[\sum_{i=1}^{n} \hbar(i) \right] \Psi(1, 2, ..., n) \, d\tau =$$
$$= n \int \Psi^*(1, 2, ..., n) \, \hbar(1) \, \Psi(1, 2, ..., n) \, d\tau =$$
$$= \int \left[\hbar(1) \gamma(1; 1') \right]_{1' \to 1} d\tau_1, \qquad (11\text{-}24)$$

where definition (11-21) was used. Primed variables were introduced to indicate that the operator acts only on the non-conjugate component of the density matrix. After operating the primes are dropped (symbolically denoted $1' \to 1$) and then it is necessary to carry out the integration. A similar modification can be made with the two-electron contributions of Eq. (11-23), to give

$$\langle \Psi \mid \mathcal{H} \mid \Psi \rangle = \int \left[\hbar(1) \gamma(1; 1') \right]_{1' \to 1} d\tau_1 +$$
$$+ \int g(1, 2) \, \Gamma(1, 2; 1, 2) \, d\tau_1 \, d\tau_2, \qquad (11\text{-}25)$$

where $g(1, 2)$ in the second term on the right-hand side of Eq. (11-25) is only a multiplication operator and thus primed coordinates need not be introduced (in contrast to operator $h(1)$, which, like other operators, also contains the Laplace operator involving differentiation).

Density matrices assume particularly simple expressions when wave function Ψ is represented by a single Slater determinant (cf. Section 5.5):

$$\Psi(1, 2, ..., n) = \Delta_0(1, 2, ..., n) = \frac{1}{\sqrt{n!}} \det \| \lambda_1(1), \lambda_2(2), ..., \lambda_n(n) \| \tag{11-26}$$

This happens in the Hückel and extended Hückel methods and in all SCF procedures. It will be assumed that the spin orbitals, λ_i, in Eq. (11-26) are orthonormal. Density matrices could be obtained by direct calculation. For the derivation, however, a method comparing known expressions will be employed. Using Table 5-2, the energy expectation value for Hamiltonian (5-18) and wave function (11-26) can be expressed by the relationship

$$\langle \Delta_0 | \mathcal{H} | \Delta_0 \rangle = \sum_{i=1}^{n} \langle \lambda_i(1) | h(1) | \lambda_i(1) \rangle +$$

$$+ \frac{1}{2} \sum_{i,j=1}^{n} [\langle \lambda_i(1) \lambda_j(2) | g(1, 2) | \lambda_i(1) \lambda_j(2) \rangle -$$

$$- \langle \lambda_i(1) \lambda_j(2) | g(1, 2) | \lambda_j(1) \lambda_i(2) \rangle] \tag{11-27}$$

Comparison of Eq. (11-27) with Eq. (11-25) leads to the expressions

$$\gamma(x_1; x_1') = \sum_{i=1}^{n} \lambda_i^*(x_1') \lambda_i(x_1) \tag{11-28}$$

$$\Gamma(x_1, x_2; x_1', x_2') = \frac{1}{2} \sum_{i,j}^{n} [\lambda_i^*(x_1') \lambda_j^*(x_2') \lambda_i(x_1) \lambda_j(x_2) -$$

$$- \lambda_i^*(x_1') \lambda_j^*(x_2') \lambda_j(x_1) \lambda_i(x_2)] =$$

$$= \frac{1}{2} [\gamma(x_1; x_1') \gamma(x_2; x_2') - \gamma(x_1; x_2') \gamma(x_2; x_1')] \tag{11-29}$$

for the density matrices (containing spin variables), so that, provided the first-order density matrix is known, the second-order density matrix can be constructed according to Eq. (11-29) (this conclusion is general enough to hold even for density matrices of higher orders and is one of the typical properties of the one-particle approximation). Therefore, the expression $\gamma(x_1; x_1')$ is denoted as *the Fock-Dirac density matrix* and is referred to as the "*fundamental invariant of the SCF solution*"[3].

If a system with closed shells is under study [cf. Eq. (5-43)], it is advantageous to employ the "spinless" density matrix $\tilde{\gamma}$, which can be obtained from Eq. (11-28) by integration over the spin variable:

$$\gamma(\mathbf{r}_1, \mathbf{r}'_1) = 2 \sum_{i=1}^{n/2} \varphi_i^*(\mathbf{r}'_1)\, \varphi_i(\mathbf{r}_1), \tag{11-30}$$

where φ_i are orbitals depending on space variables alone.

For further considerations, knowledge of some properties of the first-order density matrix, related to the one-determinant wave function, will be important. For derivation of the corresponding expressions, it will be useful to employ the matrix notation established in Section 4.5. In this notation, a set of spin orbitals can be written in matrix form

$$\lambda = \| \lambda_1(x), \lambda_2(x), \ldots, \lambda_n(x), \lambda_{n+1}(x), \ldots \|, \tag{11-31}$$

where the column index specifies spin orbitals arranged according to increasing relevant orbital energy E_i (or ε_i), and the row index (continuous) specifies the spin-space coordinates of the electron. It has already been mentioned that the property of orthonormality of a set of spin orbitals can be written [cf. Eq. (4-123)] as

$$\lambda^H \lambda = 1, \tag{11-32}$$

where 1 is a unit matrix of the same dimension as the spin-orbital space (given by the number of spin orbitals). Then density matrix (11-28) can be written in the form

$$\gamma = \lambda 1_n \lambda^H = \lambda_0 \lambda_0^H, \tag{11-33}$$

where 1_n is a square matrix in the spin-orbital space with the first n diagonal elements equal to 1 and all the other elements equal to zero; index 0 signifies that matrix λ_0 contains only occupied spin orbitals of number n. At the same time, number n gives the number of columns in λ_0.

Density matrix (11-33) has the following properties:

a) γ is *invariant to unitary transformation* of occupied spin orbitals. If a new set of spin orbitals, λ'_0, is established,

$$\lambda'_0 = \lambda_0 U, \tag{11-34}$$

where U is a unitary matrix, then

$$\gamma' = \lambda'_0 (\lambda'_0)^H = \lambda_0 U U^H \lambda_0^H = \lambda_0 \lambda_0^H = \gamma \tag{11-35}$$

Owing to this property, expectation values of physical quantities remain unchanged on transition from occupied molecular orbitals that are solutions of standard SCF equations to other orthogonal orbitals bound to the original occupied orbitals by a unitary transformation.

b) γ has the property of *a projection operator*, projecting an arbitrary function (defined in the spin-orbital space λ) into the space of the occupied orbitals. If the function f is of the type

$$f = \lambda c, \tag{11-36}$$

where c is a column matrix of the form

$$c = \left\| \begin{matrix} c_0 \\ c_v \end{matrix} \right\|, \tag{11-37}$$

c_0 (of dimension n) contains coefficients multiplying occupied spin orbitals and c_v denotes the contributions of the virtual functions, then it follows that, using Eqs. (11-32), (11-33), and (11-36), the equation

$$\gamma f = \lambda_0 c_0 \tag{11-38}$$

is valid. Therefore, after action of the density matrix on f, that part of the function lying in the space of occupied spin orbitals is obtained.

In closed shell systems, a simplification occurs in that the dimension of the density matrix is effectively reduced from n to $n/2$ if the "spinless" density matrix (11-30) is employed.

11.2.3 Localized orbitals

The methods of wave function analysis to obtain localized functions describing individual groups of electrons have been elaborated almost solely for the one-electron model[6,7]. Firstly, the properties of the Hartree-Fock manifold, *a)* and *b)* in Section 11.2.2, provided a natural basis for these methods. Secondly, in accordance with the Lewis interpretation of the chemical bond, pairs of electrons can be considered to be elementary localized groups. The simplest description of these pairs can be effected using a single function of space coordinates, which can be combined with spin functions α and β; the concept of a one-electron function is connected with the one-electron model.

To form localized one-electron functions describing bonds and lone pairs of molecules, properties *a)* and *b)* of the first-order density matrix can be employed. Therefore, methods applied in the analysis of wave functions will be divided into two groups. Both begin, of course, from the solved one-electron model of the problem and have molecular orbitals as input data for the calculation.

In the localization methods based on the invariance of the density matrix [property *a)*], use is made of the fact that Eq. (11-34) combines two equivalent sets of one-electron functions. If transformation matrix U

is of order m, then it contains $\frac{1}{2}m(m-1)$ independent parameters[8]. This number of degrees of freedom can be utilized for the introduction of additional conditions depending on a suitably defined criterion, thus fixing the elements of the respective transformation matrix. Since in this connection closed shell systems are of particular interest and only the spatial part of the one-electron function [cf. Eq. (11-9)] participates in the transformation, it holds that $m = n/2$ for this type of system and it is sufficient to investigate the relationship between the molecular orbitals, φ_i, and the localized functions, φ_i'.

Molecular orbitals are usually expressed as a linear combination of the atomic orbitals χ_μ, and it will be advantageous here to write the molecular orbitals as follows [cf. Eq. (5-63)]:

$$\varphi_i = \sum_I \sum_{\alpha \in (I)} c_{\alpha i}^I \chi_\alpha^I, \tag{11-39}$$

where the sum is carried out over all the atomic orbitals (denoted by subscript α) localized on atom I and over all atoms I forming the molecule. The molecular orbitals (the solution of the standard one-electron problem) has non-vanishing coefficients $c_{\alpha i}^I$ over the entire molecule; from Theorem 6-1 of Section 6.4 it follows that the coefficients assume values such that φ_i is a component of the basis corresponding to one of the irreducible representations of the symmetry group of the molecule. On the other hand, function φ_i' must be localized in a certain part of the molecule. It is, for example, optimal for the orbital describing the bond between atoms A and B that only coefficients $c_{\alpha i}^A$, $\alpha \in (A)$ and $c_{\beta i}^B$, $\beta \in (B)$ differ substantially from zero, while the contributions from the remaining atoms are negligibly small. Similarly, when φ_i' describes the inner electron shell or the lone electron pair on atom A, considerable contributions to orbital φ_i' originate from atomic orbitals situated on the single atom, A. Coefficients $c_{\alpha i}^A$, $\alpha \in (A)$, in both cases give information on the character of the hybrid orbital participating in the formation of the localized one-electron function.

The criteria used for the determination of transformation matrix **U** remain to be mentioned. With symmetrical molecules such as methane, for instance, the fact that the molecule contains some equivalent atoms or bonds can be used. If the properties of the $C-H$ bond (say, its dipole moment) in the CH_4 molecule were of interest, the original set of molecular orbitals would have to be transformed to give four physically equivalent orbitals describing the $C-H$ bonds in the methane molecule. It appears that, if the calculation is carried out by the MO-LCAO method with the minimum basis set of atomic orbitals (cf. the discussion of the

C_2H_4 molecule in Section 6.6), the symmetry of the problem suffices for determination of all the parameters of the transformation matrix.

Generally, of course, the symmetry of the problem is insufficient for determination of the transformation matrix; then other criteria based on particular physical concepts must be employed. For details, see Refs. 6−9.

The localized functions calculated on the basis of the invariance property of the density matrix are in a certain sense equivalent to the molecular orbitals. The thus-defined localized orbitals are, of course, not localized only in certain parts of the molecule, but have a certain non-zero electron density over practically the entire molecule; therefore the electron pairs cannot be isolated so that each occupies a region defined exclusively either by one or by two centres, although this would be an ideal property for orbitals transferable, for example, for a certain bond from molecule to molecule. Experiments on Compton X-ray scattering[10] yielded persuasive proof of the possibility of localization and transferability of bonds in some molecules, and demonstrated that theoretical analysis of molecular wave functions from this point of view is of practical importance.

It has been found[11] that the projection property of the Fock-Dirac matrix, i.e. property b) in Section 11.2.2, can be expediently utilized for the construction of functions localized only on a certain number of centres. For the sake of simplicity only closed shell electron systems will be considered. In view of Eqs. (11-38) and (11-30) the relationship

$$\eta^H \tilde{\gamma} \eta = k, \qquad 0 \leq k \leq 2, \tag{11-40}$$

holds for any normalized function η of type (11-36) (cf. notation in Section 4.5). It applies especially that, if η is identical with an occupied molecular orbital $\eta \equiv \varphi_i$, $1 \leq i \leq n/2$, then $k = 2$; if η is a virtual orbital, then $k = 0$. η can be considered to be a function of several undetermined parameters and to fulfil the localization condition in a certain region of the molecule (see below). These conditions can easily be realized if the one-electron functions are approximated in the LCAO form. With respect to the possible intepretation of quantity k as an occupation number, it is both physically and mathematically justified to require that the equation

$$k = \max \left[\eta^H \tilde{\gamma} \eta \right], \tag{11-41}$$

expressing maximization of the corresponding functional, be valid for the optimum localized functions.

If, for example, the bond between atoms A and B of the studied molecule is to be described, we can proceed in two ways:

a) We can assume that the bond orbital η_{AB} is of the form

$$\eta_{AB} = q(\eta_A + b\eta_B), \qquad (11\text{-}42a)$$

where $\eta_A = \sum_{\mu \in (A)} c_\mu \chi_\mu$ is a fixed hybrid orbital on atom A (i.e. its expansion coefficients c_μ are known), q is the normalization constant and b — a parameter (characterizing the polarity of the bond) which is to be optimized.

b) If the fixed hybrid assumption is abandoned, a more general problem can be solved, i.e. a function of the form

$$\eta_{AB} = \chi \mathbf{D}_{AB} \qquad (11\text{-}42b)$$

can be sought, where $\mathbf{D}_{AB}$ is a column matrix composed of linear coefficients multiplying the atomic orbitals localized on atoms A and B.

When criterion (11-41) is employed, both cases can be solved exactly and version *b*) can be used for the determination of "optimal" hybrid orbitals in the sense of the best approximation to the SCF solution of the respective problem. The assumption of the number of centres appearing in Eq. (11-42b) can, of course, be varied according to the type of problem. If an optimal hybrid describing a lone pair is sought, η is expanded in terms of the atomic orbitals localized on the corresponding atom.

This method therefore characterizes the electron pair by two quantities: by the localized function η and the occupation number k, which can be considered to be a quantitative measure of the localization. Its value, for bond orbitals for example, usually lies in the range 1.98 to 2.00 for both semiempirical and "ab initio" wave functions. In comparison with other methods of analyzing wave functions from the viewpoint of their localizability, its advantage lies mainly in its simplicity and small demands on computer time and, furthermore, in the fact that it permits study of a specific part of the molecule without explicit consideration of the remainder.

Let us summarize the results obtained by the analysis of wave functions from the point of view of orbital localizability. It has been shown that well-defined localized orbitals describe inner shell electrons, lone pairs and two-centre bonds. Localized functions exhibit a considerable degree of transferability between different (in a certain sense similar) molecules. The nature of hybridization of atomic orbitals which contribute to localized functions is also correlated with the position of the atom in the periodic table. Application to the wave functions of electron-deficient molecules confirmed the existence of three-centre two-electron bonds, BHB and BBB, in borohydrides. The application of the localization method to the π-electron systems of both butadiene isomers[12] provided interesting results. For the two-centre C1−C2 π bond the occupation number $k = 1.96$ results, and although it is smaller than the occupation numbers for σ orbitals, the difference is not sufficient to justify suggesting that the degree of localization is a crucial factor differentiating π and σ systems. The occupation numbers for orbitals of the C2−C3 and C1−C4 bonds are much smaller and the occupation number for the C1−C3 bond is the lowest, in agreement with Rumer's theorem, according to which valence schemes which exhibit bond crossing should be excluded from valence diagrams.

So far, the analysis of wave functions has been discussed from the point of view of the possibility of obtaining localized functions directly related to the concepts of the classical theory of the chemical bond. The positive results of these calculations confirmed or initiated the formulation of approximate methods and models, in which the localization of electron groups was assumed. Among the simplest methods of this type are semi-empirical one-electron methods describing the σ electron system by means of strictly localized orbitals and intended mostly for the calculation of the ground state physical properties of molecules, such as heats of atomization[13] and dipole moments[14] (cf. methods described in Section 10.2.2). The same idea, although in a more precise version, has been utilized in models constructing the wave function from molecular fragments. The approximation of separated electron groups[4], the theory of "atoms[15] (or molecules[16]) in molecules" and the method of molecular fragments[17] are of this type. Even within the framework of many-electron theories, the application of localized orbitals is useful for simplification of the calculation and for improvement of the convergence of both perturbation approaches and different versions of the configuration interaction method[18].

11.2.4 Electron distribution in molecules

Analysis of the molecular wave function from the viewpoint of the charge distribution in the individual parts (atoms) of the molecule allows quantitative expression of changes in the electron distribution when forming molecules from fragments (atoms). It is then possible to establish a qualitative interpretation of the quantum chemical calculation, based on concepts such as "ionicity", polarity and covalency of the bonds. Hereafter all charge values will be given as multiples of the elementary electron charge.

Electron redistribution during molecular formation can be studied using function $\delta(r)$, which was introduced by Roux and co-workers[19].

This function is defined by the relationship

$$\delta(\mathbf{r}) = \varrho(\mathbf{r}) - \sum_I \varrho_I(\mathbf{r}), \tag{11-43}$$

where the electron density function $\varrho(\mathbf{r})$ is given by Eqs. (11-14) and (11-12) and $\varrho_I(\mathbf{r})$ is the hypothetical free atom electronic density function unchanged by bond formation and localized on atom I of the molecule. The difference between the electron density in the molecule and the sum of the electron densities in the system of free atoms (or ions) must be calculated for each point in space. Most convenient is graphical representation of function $\delta(\mathbf{r})$ in the form of curves corresponding to the same density values in characteristic planes intersecting the molecule. $\delta(\mathbf{r})$ as a difference function describes the change in the electron densities which occur on formation of the bond better than electron density contour maps (cf. e.g. Fig. 9-2).

For some purposes the described representation of the electron distribution is unnecessarily detailed. Thus it is naturally sometimes necessary to condense the relevant information and to describe the charges localized on the atoms or to give data on the density of electrons in the individual bonds. The most common method for the calculation of atomic charges from a wave function of the MO-LCAO type is undoubtedly Mulliken's *"population analysis"*[20], which can be best illustrated on a two-centre one-electron (or two-electron) system. For a normalized molecular orbital φ we have

$$\varphi = c_\mu \chi_\mu + c_\nu \chi_\nu, \tag{11-44}$$

where subscripts μ and ν denote the centres (the nuclei of atoms I and J), and the number of electrons, k, contained in this orbital is given by

$$k = k(|c_\mu|^2 + 2c_\mu^* c_\nu \langle \chi_\mu | \chi_\nu \rangle + |c_\nu|^2), \tag{11-45}$$

where $\langle \chi_\mu | \chi_\nu \rangle$ is the overlap integral between the two orbitals. The term containing the overlap, called the *"overlap population"*, can be interpreted as a measure of the accumulation of the electron charge between the atomic partners and is therefore related to the strength of the corresponding bond. Assuming that the participating atoms influence the magnitude of this term to the same degree, the electron charge Q_I on atom I (whose nucleus is identical with centre μ) can be defined according to Mulliken[20] by the relationship

$$Q_I = k(|c_\mu|^2 + c_\mu^* c_\nu \langle \chi_\mu | \chi_\nu \rangle) \tag{11-46}$$

Derivation of the general expression for the atomic charge can begin with Eq. (11-28) or (11-30) for the density matrix. After substituting

284

Eq. (11-39) into Eq. (11-30) and after integration over the space coordinates of the electron, the relationship

$$n = 2 \sum_I \sum_J \sum_{\alpha \in (I)} \sum_{\beta \in (J)} \sum_{i=1}^{n/2} (c_{\alpha i}^I)^* (c_{\beta i}^J) \langle \chi_\alpha^I | \chi_\beta^J \rangle \qquad (11\text{-}47)$$

is obtained for molecular orbitals in the form of (11-39) and for a closed shell system. Assuming that the atomic orbitals localized on a single centre are orthonormal ($\langle \chi_\alpha^I | \chi_\beta^I \rangle = \delta_{\alpha\beta}$), the quantities Q_I can be introduced:

$$Q_I = \sum_{\alpha \in (I)} Q_{\alpha I}, \qquad (11\text{-}48)$$

where

$$Q_{\alpha I} = 2 \left[\sum_{i=1}^{n/2} (c_{\alpha i}^I)^* (c_{\alpha i}^I) + \sum_{J(\neq I)} \sum_{\beta \in (J)} \langle \chi_\alpha^I | \chi_\beta^J \rangle \sum_{i=1}^{n/2} (c_{\alpha i}^I)^* c_{\beta i}^J \right], \quad (11\text{-}49)$$

which, in the sense of the introductory example, can be interpreted as electron charges localized in atomic orbital $\chi_\alpha^I (Q_{\alpha I})$ and on atom $I(Q_I)$. The condition

$$n = \sum_I Q_I \qquad (11\text{-}50)$$

is, of course, valid. If Q_I is known, the total charge on atom I can be calculated as the difference $(Z_I - Q_I)$, where Z_I is the (effective) nuclear charge of the atom.

It is obvious that, for zero atomic orbital overlap, the term involving the overlap between the atomic orbitals of different atoms in expressions (11-46) or (11-49) can be disregarded and the value $k|c_\mu|^2$ is then a measure of the electron density on atom I, leading to the expression defined by Coulson[21] within the Hückel theory as the atomic charge for the π electrons. The electron density in the region of the bond, i.e. *the bond order*, is estimated from the value of the product $kc_\mu^* c_\nu$ (the Coulson bond order). In molecules with several occupied molecular orbitals, the (total) electron densities and bond orders are given by the sum of the contributions from the individual molecular orbitals.

Although the Mulliken definition of the atomic charge in a molecule is among the most frequently used, it has a number of shortcomings. First, the "overlap population" is equally distributed between atoms I and J, and is, in general, fulfilled only when the atoms are of the same kind. Further, the definitions given [cf. Eqs. (11-48) and (11-49)] do not exclude the possibility of negative electron densities if the contribution from the non-diagonal terms is negative and sufficiently large, and, on the contrary, it has even happened that the calculated charge was greater than 2. The results of population analysis are not invariant under transformation of atomic orbitals and, in addition, the calculated charges depend markedly on the choice of the AO basis set.

These drawbacks can best be illustrated on an example of calculation using expansion of the one-electron wave function (of the molecular orbitals) in the form of a linear combination of atomic orbitals, which are all localized on the nucleus of a single selected atom; one-centre methods seemed to afford good results in the calculation of physical quantities[22] for symmetrical molecules, such as the CH_4 molecule. Population analysis would nonetheless attribute the total electron charge to a single atom, the nucleus of which is chosen as the origin for expansion of the wave function. However, population analysis provides physically reliable results as long as the basis set of the atomic orbitals is chosen consistently with the electronic structure of the atoms forming the molecule.

The described shortcomings can be removed by the definition of the atomic charge proposed by Politzer et al[23], which sets out directly from the physical interpretation of the first-order density matrix. This method is based on the partition of the molecular space into regions corresponding to the individual atoms. The electron charge of atom I is given by the integral

$$Q_I = \int_I \varrho(\mathbf{r})\,d\mathbf{r}, \tag{11-51}$$

where $\varrho(\mathbf{r})$ is the electron probability density [cf. Eqs. (11-12) and (11-14)] and the integration is carried out over the region corresponding to atom I. It is evident that the method allows a certain amount of freedom in defining the spatial regions corresponding to the individual atoms forming the molecule. Nevertheless, the numerical application of this method has provided results identical with experimental data even when other methods yielded worse results, for example, with fluorinated hydrocarbons.

11.2.5 Dipole moment

Although it would be ideal to gain experimental information on the electron distribution in the regions corresponding to the individual atoms and in regions between neighbouring atoms, nonetheless less detailed information on the charge distribution, namely, the dipole moment, is also very useful. In *electroneutral systems* (molecules) there are, of course, many regions with a local excess or deficiency of electrons. From the point of view of the molecule as a whole, this distribution is equivalent to a distribution in which all the positive charges are concentrated in a single point charge (centre of charge); with negative charges the situation is analogous. The absolute magnitudes of these charges are, of course, identical (the electroneutrality condition). The dipole moment $\boldsymbol{\mu}$ is defined as follows (Fig. 11-8):

$$\boldsymbol{\mu} = \delta e \boldsymbol{r} \tag{11-52}$$

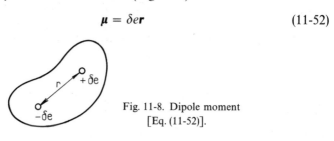

Fig. 11-8. Dipole moment [Eq. (11-52)].

In classical physics the dipole moment of a system of point charges $q_1, q_2, \ldots, q_n$ (where $\sum\limits_{i=1}^{n} q_i = 0$) equals

$$\mu = \sum_i q_i \boldsymbol{r}_i, \tag{11-53}$$

where $\boldsymbol{r}_i$ is the position vector of a given point charge from an arbitrary origin. For continuous charge distribution integration must be carried out:

$$\boldsymbol{\mu} = \int \varrho(\boldsymbol{r}) \, \boldsymbol{r} \, d\boldsymbol{r}, \tag{11-54}$$

where $\varrho(\boldsymbol{r})$ is the charge density ($\boldsymbol{r}$ expresses the dependence on the space coordinates) and $d\boldsymbol{r}$ is the volume element. In the study of molecules, both these expressions[24] are used: for purposes of calculation, the molecule is split into a set of positively charged nuclei [which have quite definite positions, Eq. (11-53)] and a continuously distributed electron density [Eq. (11-54)]. If the nuclear charge is equal to $Z_I e$ and if the electron charge density is expressed in terms of function $\varrho(\boldsymbol{r})$ [cf. Eqs. (11-12), (11-14) and (11-24)] multiplied by the electron charge, it follows that

$$\boldsymbol{\mu} = e \sum_I Z_I \boldsymbol{R}_I - e \int \varrho(\boldsymbol{r}) \, \boldsymbol{r} \, d\boldsymbol{r}, \tag{11-55}$$

where $\boldsymbol{R}_I$ denotes the position vector of nucleus I.

The expression for the calculation of the dipole moment can, of course, also be used when the calculation is confined to only a certain portion of the electrons. The most important example is the π-electron approximation applied to conjugated systems. In expression (11-55) Z_I then denotes the core charge of the I-th atom and the spinless density matrix $\varrho(\boldsymbol{r})$ describes the π-electron distribution. In this way the π-electron component of the total dipole moment is obtained.

The introduction of molecular orbitals as a linear combination of atomic orbitals into $\varrho(\mathbf{r})$, corresponding to a closed shell system [cf. Eqs. (11-12), (11-14) and (11-30)], can be written as follows (using the one-electron approximation):

$$\varrho(\mathbf{r}) = 2 \sum_{\mu,\nu} \chi_\mu^*(\mathbf{r}) \chi_\nu(\mathbf{r}) \sum_{i=1}^{n/2} c_{\mu i}^* c_{\nu i} \qquad (11\text{-}56)$$

If Eq. (11-56) is substituted into expression (11-54) the electronic part of the dipole moment (11-55) is obtained:

$$\boldsymbol{\mu}_{\mathrm{el}} = -e \int \varrho(\mathbf{r}) \mathbf{r} \, d\mathbf{r} = -e \sum_\mu \sum_\nu \mathbf{r}_{\mu\nu} P_{\mu\nu}, \qquad (11\text{-}57)$$

where $\mathbf{P}$ is the charge- and bond-order matrix defined by Eq. (10-5) and the matrix element $\mathbf{r}_{\mu\nu}$ is given by the equation

$$\mathbf{r}_{\mu\nu} = \int \chi_\mu^*(\mathbf{r}) \, \imath \chi_\nu(\mathbf{r}) \, d\mathbf{r} = \langle \chi_\mu | \imath | \chi_\nu \rangle, \qquad (11\text{-}58)$$

where $\mathbf{r} \equiv \imath$ denotes the position vector of the selected electron. All the position vectors are, of course, related to a particular origin of the coordinate system (in classical electrostatics it has been shown that, for an electroneutral system, the value of the dipole moment is invariant to the choice of the origin of the coordinate system).

The values of matrix elements (11-58) will now be evaluated. The expression for their calculation is considerably simplified if the zero differential overlap approximation [cf. Eq. (10-9)] is used:

$$\chi_\mu^* \chi_\nu = 0, \qquad \mu \neq \nu,$$

this being a typical feature of the π-electron approximation. Then all elements $\mathbf{r}_{\mu\nu}$, $\mu \neq \nu$, are also equal to zero. For diagonal elements $\mathbf{r}_{\mu\mu}$ it is expedient to express position vector $\mathbf{r}$ as the sum of two vectors,

$$\mathbf{r} = \mathbf{R}_I + \mathbf{r}', \qquad (11\text{-}59)$$

where $\mathbf{R}_I$ is a constant vector which gives the position of nucleus I on which χ_μ is centered with respect to the chosen origin and $\mathbf{r}'$ is a new variable vector related to the nucleus of atom I. It therefore follows that

$$\mathbf{r}_{\mu\mu} = \mathbf{R}_I \langle \chi_\mu | \chi_\mu \rangle + \langle \chi_\mu | \mathbf{r}' | \chi_\mu \rangle, \qquad (11\text{-}60)$$

which is a vector equation representing three equations for the individual components. For example, for the x-component

$$x_{\mu\mu} = X_I \langle \chi_\mu | \chi_\mu \rangle + \langle \chi_\mu | x' | \chi_\mu \rangle \qquad (11\text{-}61)$$

The coefficient of X_I is the norm of atomic orbital and thus equals 1; from symmetry considerations it follows that the second integral equals zero [cf. Eq. (6-70)]. Thus

$$r_{\mu\mu} = R_I \tag{11-62}$$

and finally

$$\mu_{el} = -e \sum_\mu R_\mu P_{\mu\mu}, \tag{11-63}$$

because the indices of the atomic nuclei and orbitals can be identified, because in the π-electron approximation each atom contributes a single atomic orbital to the total basis set of atomic orbitals. Since in this approximation $P_{\mu\mu}$ denotes the Coulson π-electron density Q_I on atom I, the final expression for the dipole moment can be written within the framework of the one-electron π electron approximation:

$$\mu = e \sum_I R_I(Z_I - Q_I) \tag{11-64}$$

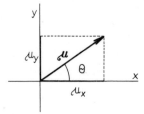

Fig. 11-9. Dipole moment vector and its μ_x and μ_y components.

Assuming that the studied system is planar (and lies in the x, y-plane), the absolute value of the dipole moment vector (cf. Fig. 11-9) is given by

$$\mu = \sqrt{(\mu_x^2 + \mu_y^2)}, \tag{11-65}$$

where μ_x and μ_y are the respective components of the vector, for example

$$\mu_x = e \sum_I X_I(Z_I - Q_I), \tag{11-66}$$

where X_I is the x-coordinate of nucleus I with respect to the fixed origin.

The dipole moment forms angle Θ with the positive part of the x-axis, for which it holds that

$$\tan\Theta = \frac{\mu_y}{\mu_x} \tag{11-67}$$

The total dipole moment of planar conjugated compounds can be expressed as the sum of the σ and π components:

$$\mu = \mu_\sigma + \mu_\pi \tag{11-68}$$

With some π-electron systems, the contribution of the σ component is practically negligible in comparison with the contribution of the π

component. The dipole moment of the σ component can be expressed by superimposing the dipole moment of the individual σ bonds, which are tabulated in the literature. The contributions originating from non-bonding atomic orbitals – lone pairs (for example on nitrogen in pyridine or oxygen in carbonyl compounds), which are more significant than the dipole moments of σ bonds – must also be included. If the π-electron density obtained by the HMO or SCF method is substituted into the expression for the dipole moment, the calculated values are too high. More correct values of μ are obtained if methods are employed for the calculation of the electron densities in which the α (or $α^c$) values are corrected for the charge densities in the respective positions. This is done in the ω-technique (the modified HMO method) and in the VESCF method (the modified SCF method). There are also very useful methods in which the electrons of the σ bonds are considered explicitly. In these cases, however, the simple relation (11-64) for the calculation of the dipole moment is not valid and it is necessary to employ a more complicated expression. Good agreement of theory and experiment has been achieved in a number of instances using the CNDO/2 method: the deviation of the calculated and experimental values is usually about 10%.

11.2.6 Nodal planes of molecular orbitals: the Woodward-Hoffmann rules

In the mid-sixties Woodward and Hoffmann published a method[25], which enabled prediction of the details of the stereochemical course of some cyclization reactions by means of the shape and nodal planes of the frontier molecular orbitals. It soon appeared, however, that the whole consideration must be given a more reliable physical basis, represented by correlation diagrams. Knowledge of the shape of molecular orbitals (the location of nodal planes) can, nevertheless, be used empirically for interpretation and prediction of the course of some reactions.

Electrocyclic reactions, i.e. reactions in which a σ bond is formed or broken between the ends of a linear conjugated system, will be treated here in greater detail. Examples are the formation of cyclobutene from 1,3-butadiene and the formation of 1,3-butadiene from cyclobutene:

$$\begin{array}{ccc}
\begin{array}{c}
\text{HC} \overset{\displaystyle CH_2}{\diagup} \\
\text{HC} \diagdown \\
 CH_2
\end{array}
& \rightleftharpoons &
\begin{array}{c}
\text{HC—CH}_2 \\
| | \\
\text{HC—CH}_2
\end{array}
\end{array} \qquad (11\text{-}69)$$

Woodward and Hoffmann represent this process schematically for a polyene with $n\pi$ electrons as follows:

$$(11\text{-}70)$$

Rotation of the terminal atoms of the open system out of the plane can occur in two ways, as is clear if we assume that the ends of the chain consist of different atoms A, B, C and D. The two possible types of rotation are called *disrotation* and *conrotation*:

Disrotation $(11\text{-}71)$

Conrotation $(11\text{-}72)$

It has been experimentally found for butadiene that disrotation occurs in the first excited state, whereas conrotation is decisive for the process in the ground state. For interpretation of this experimental data it is necessary to analyze the molecular orbitals of butadiene and cyclobutene which participate in the electrocyclic reaction. For butadiene these are four π-MO's, with cyclobutene two π-MO's and the two σ-MO's of the bond formed (Fig. 11-10). The cyclization can be followed using a correlation diagram which enables identification of orbitals with corresponding symmetry in the initial and final states. In addition to a two-fold symmetry axis, these states also have a symmetry plane (Fig. 11-11). The molecular orbitals of the initial and final states can be classified using symbols S (symmetric) and A (antisymmetric) (Fig. 11-11), depending on their behaviour on application of the particular symmetry operations. The symmetry of a molecular orbital is designated by two letters, where the first refers to reflection in the symmetry plane (σ) and the second refers to rotation about the symmetry axis (C_2) (Fig. 11-10).

It should be noted that the transition states which occur in disrotation and conrotation are of lower symmetry: to the transition states in disrotation corresponds only a symmetry plane (σ) and to the transition states in conrotation only a symmetry axis (C_2). The terminal atomic orbitals in the π-MO (φ_1) of butadiene permit to occur these processes

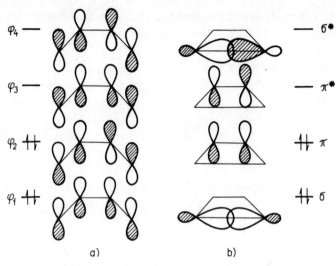

Fig. 11-10. π- and σ-molecular orbitals of butadiene (*a*) and cyclobutene (*b*).

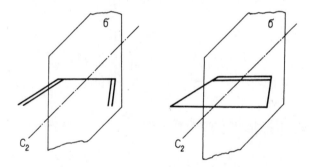

Fig. 11-11. Plane of symmetry (σ) and two-fold symmetry axis (C_2) in butadiene and cyclobutene.

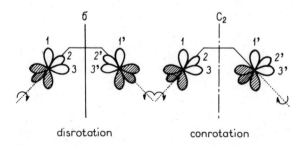

disrotation   conrotation

Fig. 11-12. Behaviour of the φ_1 molecular orbital of butadiene under disrotation and conrotation: only the p_z atomic orbitals in positions 1 and 4 are indicated. The initial position (1, 1′), position during rotation (2, 2′), position after rotation (3, 3′).

292

(Fig. 11-12). If states of equal symmetry are connected (Fig. 11-13), taking into account the principles usual in the construction of correlation diagrams[25,26] (conservation of symmetry of the molecular orbitals in the initial, transition and final states, the non-crossing rule, cf. Section 9.4), it follows that for the electrocyclic reaction of butadiene (thermal reaction) conrotation is preferable, because the bonding MO's of the reactant pass into bonding orbitals of the product (Fig. 11-13). A further occurrence is the initiation of an electrocyclic reaction by a photon. Whereas in the former case the reaction occurred in the electronic ground state, here the reaction occurs in an electronically excited state. The reaction course can be interpreted in terms of disrotation (cf. Fig. 11-13). Furthermore, it is necessary to investigate the effect of the chain length on the

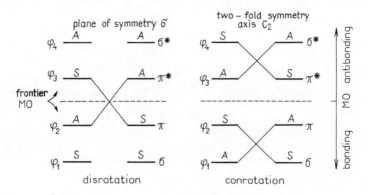

Fig. 11-13. Correlation diagrams of conrotation and disrotation in butadiene.

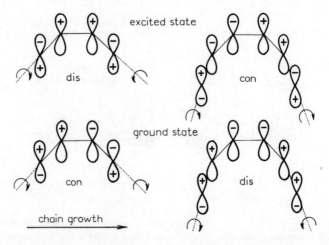

Fig. 11-14. The shape of frontier molecular orbitals in butadiene and hexatriene.

cyclization mechanism. First, however, it should be noted that the frontier MO's play a decisive role in these processes. This is usually the highest bonding orbital in the reaction in the ground state and the lowest antibonding orbital in photochemical reactions. It appears that, in order to determine the reaction course, it is sufficient to investigate whether, in the given type of rotation, the overlapping parts of the AO's have the same sign or whether the positive part of one terminal AO would overlap the negative part of the other. The situation can be clarified using the example of butadiene and hexatriene in the ground and excited states (Fig. 11-14). It is evident from the figure that knowledge of the nodal planes in the frontier MO is sufficient for formulation of the "selection rule". Thermally initiated electrocyclic reactions in systems with $4m$ π electrons in polyene ($m = 1, 2, 3, ...$) proceed through conrotation, whereas in systems with $4m + 2$ electrons disrotation takes place. In similar photochemical processes the opposite is true.

REFERENCES

1. Concerning more detailed information, the reader should refer to various textbooks on quantum chemistry, e.g. Daudel R., Lefebvre R., Moser C.: *Quantum Chemistry*. Interscience, New York 1959.
2. Turner D. W., Baker C., Baker A. D., Brundle C. R.: *Molecular Photoelectron Spectroscopy*. Wiley-Interscience, London 1970.
3. Löwdin P.-O.: *Phys. Rev.* **97**, 1474 (1955).
4. McWeeny R.: *Rev. Mod. Phys.* **32**, 335 (1960).
5. McWeeny R., Sutcliffe B. T.: *Methods of Molecular Quantum Mechanics*. Academic Press, London and New York 1969.
6. Weinstein H., Pauncz R., Cohen M.: *Advan. Atom. Mol. Phys.* **7**, 97 (1971).
7. Polák R.: *Chem. Listy*, **68**, 337 (1974).
8. Ruedenberg K.: in *Modern Quantum Chemistry* (ed. Sinanoğlu O.), part I. Academic Press, New York 1965, p. 85.
9. Gilbert T. L.: in *Molecular Orbitals in Chemistry, Physics and Biology* (ed. Löwdin P.-O. and Pullman B.). Academic Press, New York 1964, p. 405.
10. Eisenberger P., Marra W. C.: *Phys. Rev. Letters* **27**, 1413 (1971); Epstein I. R.: *J. Chem. Phys.* **53**, 4425 (1970).
11. Polák R.: *Theoret. Chim. Acta* **14**, 163 (1969); *Int. J. Quant. Chem.* **4**, 271 (1970); **6**, 1077 (1972).
12. Polák R.: *Collect. Czech. Chem. Commun.* **38**, 1450 (1973).
13. Brown R. D.: *J. Chem. Soc.* **1953**, 2615; Dewar M. J. S., Pettit R.: *J. Chem. Soc.* **1954**, 1625.
14. Del Re G.: *J. Chem. Soc.* **1958**, 4031.
15. Moffitt W.: *Proc. Roy. Soc.* (London) **A210**, 245 (1951).
16. von Niessen W.: *J. Chem. Phys.* **55**, 1948 (1971).
17. Christoffersen R. E., Maggiora G. M.: *Chem. Phys. Letters* **3**, 419 (1969).

294

18. Sinanoğlu O.: *Advan. Chem. Phys.* **6,** 315 (1964); Nesbet R. K.: *Advan. Chem. Phys.* **9,** 321 (1965); Diner S., Malrieu J. P., Claverie P.: *Theoret. Chim. Acta* **13,** 1 (1969).
19. Roux M., Besnainou S., Daudel R.: *J. Chim. Phys.* **53,** 218 (1956).
20. Mulliken R. S.: *J. Chem. Phys.* **23,** 1833 (1955).
21. Coulson C. A.: *Proc. Roy. Soc.* (London) **A169,** 413 (1939).
22. Saturno A. F., Parr R. G.: *J. Chem. Phys.* **33,** 22 (1960).
23. Politzer P., Harris R. R.: *J. Am. Chem. Soc.* **92,** 6451 (1970); Politzer P., Stout E. W., Jr.: *Chem. Phys. Letters* **8,** 519 (1971); Politzer P.: *Theoret. Chim. Acta* **23,** 203 (1971).
24. Dewar M. J. S.: *The Molecular Orbital Theory of Organic Chemistry.* McGraw-Hill, New York 1969.
25. Woodward R. B., Hoffmann R.: *Die Erhaltung der Orbitalsymmetrie.* Verlag Chemie, Weinheim 1970.
26. Čársky P.: *Chem. Listy* **66,** 1255 (1972).

12. EXAMPLES OF THE STUDY OF POLYATOMIC MOLECULES

12.1 Introductory comments

As far as quantum chemical studies of the electronic structure of molecules are concerned, modern techniques focus attention on semi-empirical and nonempirical methods, in which all the valence electrons (CNDO-type methods) or all the electrons in general ("ab initio" methods) are explicitly considered. The semiempirical methods are rather easily applicable, but they sometimes fail quantitatively or even qualitatively, their chief disadvantage being the limited region of application of the individual versions of the SCF method. It is evident, however, that there is a great number of problems in chemistry that can be successfully studied using simple empirical methods considering all the valence electrons or only the π electrons.

12.2 Inorganic compounds

More attention should be paid to the solid phase[1,2], whose importance is not confined to heterogeneous catalysis alone. Much effort is being devoted to theoretical studies of the electronic structure of the solid phase, both in non-metals and metals. In non-metals, molecular crystals (for example solid pentane, bromine, numerous organic compounds), covalent crystals (e.g. diamond, germanium) and ionic crystals (e.g. NaCl, $CuSO_4$) can be distinguished.

In molecular crystals, the individual molecules are held together by van der Waals forces (see Chapter 17). They are, roughly speaking, one to two orders of magnitude smaller than the forces responsible for chemical bonds. The heats of sublimation and melting roughly equal tens of kJ/mol. That the individuality of molecules is retained in the solid, liquid and gaseous phases is supported by the fact that vibrational spectra differ only insignificantly in these phases.

Covalent crystals represent giant molecules. It is expedient to describe them in terms of localized bonds, as the bonds in the crystal bulk greatly resemble bonds in normal covalent compounds. However, the bonding conditions in the surface layers often differ considerably from the conditions inside the crystal. It is rather interesting that both important types of hybridization known for carbon atoms in organic compounds, sp^3 and sp^2, are represented in covalent crystals, in diamond and graphite, respectively. Graphite can be considered to be a system of two-dimensional infinite benzenoid hydrocarbons. The distance between the individual layers is 0.335 nm, indicating the possibility that only van der Waals forces are operative. In graphite, the electrons originating from the p_z orbitals occupy MO's which extend over infinite areas: the high electrical conductivity of graphite is immediately evident from this description. It is a typical feature of all classifications of substances that, in addition to well-defined types, there are numerous transient types. This concept explains, in principle, the gradual transition to ionic crystals. Although zinc sulphide has a diamond lattice, it is certainly not a covalent crystal; the interaction between Zn^{2+} and S^{2-} participates in the bond formation to a considerable degree. There is even an almost ionic analogue of graphite, namely boron nitride, BN.

Classical representatives of ionic crystals are the halogenides of the alkaline metals. Cations M^+ and anions X^- are quite regularly arranged in cubic lattices: each anion is surrounded at distance r by six cations and vice versa. A further layer at a distance of $r\sqrt{2}$ contains twelve anions and finally, at a distance of $r\sqrt{3}$ there are eight cations. As a first approximation, the participation of the covalent bond can be neglected entirely and only the Coulomb interaction between M^+ and X^- need be considered. Whereas the potential energy of this interaction in a hypothetical diatomic molecule M^+X^- amounts to $e^2/4\pi\varepsilon_0 r$ (where r is the distance $M^+ \dots X^-$), the potential energy V of the interaction of ion M^+ with the two closest layers of anions (6 and 8 ions) and with

Table 12-1

Contributions to the Total Lattice Energy of Sodium Chloride, in eV [Ref. 1]

Electrostatic energy	8.92
Polarizability	0.13
Repulsion energy	−1.03
Zero-point vibrational energy	−0.08
Total lattice energy	7.94

the closest layer of cations (12 ions) amounts to

$$V = - \frac{e^2}{4\pi\varepsilon_0} \left(\frac{6}{r} - \frac{12}{r\sqrt{2}} + \frac{8}{r\sqrt{3}} \right) \tag{12-1}$$

Similar expressions apply to all further cations M^+ and anions X^-. If all these contributions are summed and the result is divided by two (otherwise each interaction would be counted twice), the expression for the total electrostatic energy of this type of crystal is obtained; this is usually stated in the following simplified form:

$$V = \frac{-Ae^2}{4\pi\varepsilon_0 r}, \tag{12-2}$$

where A is the Madelung constant, which has values roughly between 1.7 and 5. In reality, however, the covalent bond always participates in the bonding in ionic crystals. If the participation of the covalent bond is estimated, the "ionicity" of the bond is also determined and thus the factor by which the electrostatic energy must be reduced is obtained [Eq. (12-2)]. In more precise calculations, it is also necessary to consider the polarizability of the ions (particularly of easily deformable anions), the repulsion effects between clouds of electrons and the vibrational energy at absolute zero. Table 12-1 gives the magnitude of the individual contributions in sodium chloride. The experimental lattice energy amounts to 7.86 eV; the theory agrees well with the experimental value. The scope of this book does not permit a more detailed description or even brief comments on metals.

Among special illustrations, the results of MO studies in boranes[3] have an important place and have drawn the attention of both theoretical and experimental chemists for years. Several dozen boranes have been studied by different extended MO methods. Boranes are described as compounds with an electron deficit. This is not surprising, considering that formation of a normal σ bond requires two electrons. In the simplest borane B_2H_6, for example, there are eight atoms that are, in the extreme case, connected by seven bonds, which require 14 electrons. In diborane, however, only 12 valence electrons are available. However, if electron-deficient systems are considered to have unoccupied bonding MO's (as, for example, in many dications), then boranes cannot be included in this category since they have no such orbital. Many attempts have been made to interpret the nature of the bonding in boranes. For diborane (B_2H_6) the theory must take into account the experimentally determined structure (Fig. 12-1) which corresponds to the previous finding that only 4 of the 6 hydrogen atoms are equivalent

298

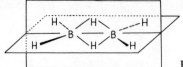

Fig. 12-1. Arrangement of atoms in diborane.

and can be exchanged. Only two theoretical concepts that seem to be most justified will be treated here. The first is the original idea of Pitzer, according to which diborane is some sort of diprotonated "ethylene" (I). The similarity

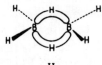

I

of the UV spectra of ethylene and diborane supports this hypothesis; it is contradicted, however, by the greater length of the $B-B$ bond (0.18 nm), as well as by the fact that the diborane hydrogens exhibit no acidity. The other description originated with Longuet-Higgins[4], who suggested tricentre bonds either of the $B-H-B$ type or of the $B-B-B$ type (in other types of boranes). Employing symmetry considerations it can be shown that the central part of the molecule contains molecular orbitals formed by the overlap of the hybrid orbitals of boron with the 1s orbital of hydrogen. The following hybrid orbitals are ascribed to the first boron atom:

II

$$\sigma_1 = \left(\sqrt{\frac{1}{6}}\right) s + \left(\sqrt{\frac{5}{6}}\right)\left[\left(\sqrt{\frac{2}{5}}\right) p_x + \left(\sqrt{\frac{3}{5}}\right) p_z\right] \quad (12\text{-}3)$$

$$\sigma_2 = \left(\sqrt{\frac{1}{6}}\right) s + \left(\sqrt{\frac{5}{6}}\right)\left[\left(\sqrt{\frac{2}{5}}\right) p_x - \left(\sqrt{\frac{3}{5}}\right) p_z\right] \quad (12\text{-}4)$$

Hybrid orbitals σ_3 and σ_4 are formed on the second boron atom. Symmetry orbitals (cf. Section 6.6) in the LCAO form are formed from these four orbitals as well as from the pair of 1s hydrogen orbitals (h_1, h_2) and the corresponding symbols of irreducible representations (Table 12-2) are assigned to them. The corresponding bonding and antibonding MO's will be formed by combination of AO's of the same symmetry (using

Table 12-2

Group Orbitals in Diborane (II) and Their Symmetry

Orbital	Symmetry	Orbital	Symmetry
$\left(\sqrt{\dfrac{1}{2}}\right)(h_1 + h_2)$	A_g	$\dfrac{1}{2}(\sigma_1 + \sigma_2 - \sigma_3 - \sigma_4)$	B_{3u}
$\left(\sqrt{\dfrac{1}{2}}\right)(h_1 - h_2)$	B_{1u}	$\dfrac{1}{2}(\sigma_1 - \sigma_2 - \sigma_3 + \sigma_4)$	B_{1u}
$\dfrac{1}{2}(\sigma_1 + \sigma_2 + \sigma_3 + \sigma_4)$	A_g	$\dfrac{1}{2}(\sigma_1 - \sigma_2 + \sigma_3 - \sigma_4)$	B_{2g}

orbitals of symmetry A_g and B_{1u}); the remaining B_{3u} and B_{2g} orbitals are nonbonding. For characterization of bonds in diborane both bonding MO's are of interest:

$$A_g: \quad \varphi_1 = c_1(h_1 + h_2) + c_2(\sigma_1 + \sigma_2 + \sigma_3 + \sigma_4) \qquad (12\text{-}5)$$

$$B_{1u}: \quad \varphi_2 = c_3(h_1 - h_2) + c_4(\sigma_1 - \sigma_2 - \sigma_3 + \sigma_4) \qquad (12\text{-}6)$$

It should be noted that the explanation of the bonding properties of boranes is, on the one hand, a problem of considerable importance in the theory of the chemical bond, and, on the other hand, it is important for interpretation of the physical properties and the reactivity of numerous compounds not only of boron, but also of beryllium and aluminium.

Great attention has been paid in recent years to experimental and also theoretical studies of formally conjugated inorganic compounds[5]. In addition to the already classical borazine (III), numerous cyclic compounds containing phosphorus and nitrogen or sulphur and nitrogen, often stabilized by fluorine bonded to phosphorus or sulphur, have been studied.

III   IV   V

In systems IV and V (examples of very extensive series of compounds) – considering possible conjugation – an atom with a p_z orbital (N) alternates with an atom with d orbitals (P, S). However, only d_{xz} and d_{yz} orbitals have suitable symmetry for overlap with a p_z orbital; d_{xy}, $d_{x^2-y^2}$ and d_{z^2} orbitals are symmetrical with respect to the xy plane,

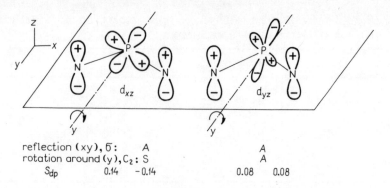

reflection (xy), $\bar{\sigma}$:		A		A	
rotation around (y), C_2: S		A		A	
S_{dp}	0.14	-0.14		0.08	0.08

Fig. 12-2. Symmetry of d_{xz} and d_{yz} orbitals from the viewpoint of possible overlap with p_z orbitals.

so that their overlap with p_z orbitals is ineffective. Although d_{yz} orbitals correspond to p_z orbitals (Fig. 12-2) even in symmetry with respect to rotation about the y axis, greater attention has been paid to models in which only the d_{xz} orbital is assigned to phosphorus atoms. According to Craig[5], such a system is called heteromorphous; a system of three $p_z(N)$ orbitals and three d_{yz} (P) orbitals, on the other hand, is described as homomorphous, because all the atomic orbitals have the same local symmetry. It is obvious that in a heteromorphous system (Fig. 12-3) there are two subsets of the same symmetry: three p_z orbitals and three d_{xz} orbitals. Within the framework of these sets MO's can be formed on the basis of symmetry consideration alone. The expression for the orbital energies of such a cycle can be given in the closed form ($\alpha_p = \alpha_d = \alpha$)

$$E_j = \alpha \pm 2 \sin\left(\frac{j\pi}{m}\right)\beta, \tag{12-7}$$

where j has the values 0, ± 1, ... $\pm m/2$ or $\pm(m-1)/2$ for m even or odd and m is the number of AO's in the respective subset (here, 3). If the Coulomb integrals of the p_z and d_{xz} orbitals differ by $\Delta\alpha$, then the diagram of orbital energies depicted in Fig. 12-4 corresponds to the system represented in Fig. 12-3. Thus the sequence of degenerate and non-degenerate levels is the opposite to that in benzene (cf. Fig. 6-6). If the Coulomb integrals of the p and d orbitals are identical, the highest occupied level would be degenerate and only incompletely occupied. It is not expedient to develop such considerations further, because experimental experience (spectral, geometrical and also thermochemical data) has shown that these systems do not have much in common with classical conjugated hydrocarbons. It should, however, be added that the theory leads to the result that, in general, for systems in which d orbitals are available on some atoms, on the one hand, the Hückel "condition of aromaticity" from the theory of planar monocyclic hydrocarbons (the number of electrons in the conjugation equals $4m + 2$, where $m = 0, 1, 2, \ldots$) is not valid and, on the other hand, even non-planar systems can, in principle, be conjugated.

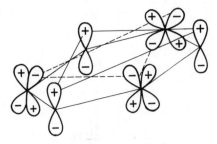

Fig. 12-3. Conjugated subsystems
of the heteromorphous cycle.

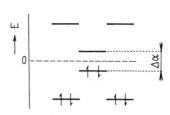

Fig. 12-4. Orbital energies
of a six-member p - d cycle ($\alpha_p \neq \alpha_d$).

Electronic spectra suggest that conjugation participates more significantly in non-substituted cycles composed of nitrogen and sulphur atoms. Several series of such compounds can be predicted theoretically and syntheses performed in recent years confirm that such considerations are justified. Firstly, cycles with an even number of atoms have been found; two subgroups can be distinguished here: that with an even number of nitrogen atoms and that with an odd number. Taking into account that nitrogen has five valence electrons (odd number), it follows that the neutral forms of systems of the second subgroup are radicals. Systems with an odd number of atoms have also been shown to exist. Experimental and also theoretical evidence tends to demonstrate that regularities common for conjugated hydrocarbons cannot be expected for these systems. On the contrary, often even formally very similar substances have different geometry and properties. In the S_4N_4 cycle, in agreement with X-ray analysis, EHT type calculations support the saddle form evident from Fig. 12-5. The system S_4N_3 was prepared in the cationic form and is obviously planar; its electronic spectrum was interpreted by means of the π-electron approximation, using the configuration interaction method (PPP method). The cation $S_5N_5^+$, which was prepared as the chloroaluminate, should also be mentioned. It results from the X-ray analysis that this is a nearly planar system, with the first absorption band in the region between the UV and visible regions (Fig. 12-6). The interpretation of both the physical properties and the reactivity of the system using the EHT and PPP theories proved successful. The heart-

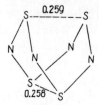

Fig. 12-5. Structure of S_4N_4.
The average length of the N−S bond
is 0.162 nm.

Fig. 12-6. Absorption curve of
$S_5N_5^+AlCl_4^-$ in concentrated H_2SO_4
(upper part) and LCI-SCF result of
calculations performed by
the standard LCI-SCF method (I)
and by the method allowing for
polarization of the σ-skeleton (II).
f denotes the oscillator strength;
forbidden transitions are indicated by
arrows.

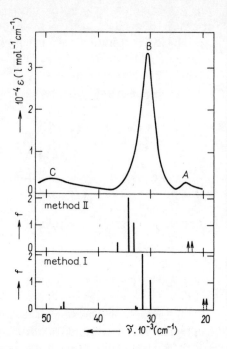

shaped form of this cation is remarkable and can be seen in the EHT molecular diagram (Fig. 12-7). That it is a conjugated system tends to be proved by its electronic spectrum, by the fact that it can be qualitatively interpreted using the π-electron approximation and by the generally uniform overlap population between the individual N and S atoms.

Another pair of experimentally and theoretically studied systems (EHT, PPP, CNDO/2) are cycles VI and VII. To a certain extent these systems can also be considered as conjugated.

VI

VII

According to the results of X-ray analysis, the structure which does not have the highest possible symmetry exists in some complex compounds. Complexes of divalent copper (Cu^{2+}, d^9 complex), for example, are not exactly octahedral but are somewhat prolonged in the direction of the z-axis. Jahn and Teller interpreted this observation. The unpaired copper electron seems to be located in one of the two degenerate E_g orbitals.

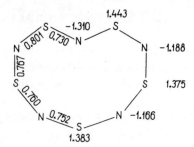

Fig. 12-7. The EHT molecular
diagram.

According to the Jahn-Teller theorem such a situation does not occur in real systems and the system can eliminate the degeneracy by a small change in the geometry: this is called a Jahn-Teller distortion. This distortion is manifested by removal of the degeneracy of incompletely occupied orbitals and it is accompanied by a decrease in the total energy compared with the original system. The prolongation (shortening) of the z-axis results in the $d_{x^2-y^2}$ (d_{z^2}) orbital having the highest energy and it is thus only singly occupied. Although this phenomenon does not apply to inorganic complexes alone, it plays a very important role in them.

Similarly, the final example can also be studied by the ligand field method; it concerns rhenium, which belongs in the third series of transition elements. The dianion $Re_2Cl_8^{2-}$ is characterized by a very short $Re-Re$ bond, only about 0.22 nm, around which no rotation occurs: the chlorine atoms are held in an energetically disadvantageous "eclipsed" position (Fig. 12-8)[6]. The figure demonstrates why the $Re-Re$ bond is so short: σ, π and δ bonds exist simultaneously between the rhenium atoms, so that this bond is of very high order. The participation of δ overlap (leading to the δ bond) requires an "eclipsed" position of the chlorines; the energy yield connected with the formation of this bond is obviously greater than the energy loss caused by repulsion of the chlorine atoms.

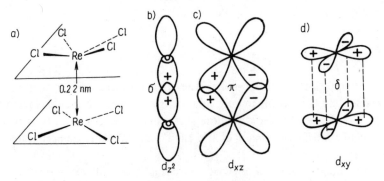

Fig. 12-8. Formation of σ (b), π (c), and δ (d) bonds in the dianion (a).

12.3 Organic compounds

For decades (1930 – 1960), bar some exceptions, the calculations of models of organic compounds were confined to π electron approximation in conjugated compounds. The breakthrough into the field of aliphatic compounds was made by the introduction of the extended Hückel theory (EHT) by Hoffmann in 1963. The basic features of this method were mentioned in Chapter 10. Despite all of its known shortcomings, this method can also be applied for the estimation of molecular geometry, although the data obtained are usually of only qualitative significance.

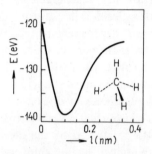

Fig. 12-9. Dependence of total energy
on the bond length l
(according to Hoffmann[7]).

Fig. 12-9 gives the dependence of the total energy of methane on the length of the $C-H$ bond. Use of the EHT method is, of course, not confined to aliphatic systems alone. An example[8] is the investigation of the effect of the orientation of the $C-H$ bond in vinylmethylene (VIII). The shift of the "methylene" hydrogen is described by the α and β angles, angle α being a measure of the deviation of the $C-H$ bond from the y-axis (in the arrangement in formula VIII, angle α equals 0).

$$
\begin{array}{c}
H \\
\diagdown \\
\quad C-C-H \cdots\cdots y \\
\diagup \diagup \\
C \\
\diagup \diagdown \qquad x \\
H \qquad H
\end{array}
$$

VIII

Positive values of α correspond to the *cis* arrangement (or rather to the "approach" to this arrangement); angle β is a measure of the displacement of the $C-H$ bond in the direction of the z-axis (which is vertical to the horizontal plane). The result of the EHT calculation is demonstrated in Fig. 12-10.

Extended methods (methods in which all valence electrons are considered) allow further physical properties to be estimated in aliphatic

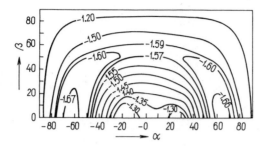

Fig. 12-10. EHT-energy contours in vinylmethylene: effect of orientation of the methylene bond (according to Ref. 8).

systems such as, for example, the ionization potential, the rotation barrier and the wave number of deformation vibrations.

Formaldehyde, for example, is a molecule in which both σ and π molecular orbitals play an important role, so that description of this molecule by the π-electron approximation is insufficient, although it is rather tempting due to its simplicity. Considered from a wider aspect, the situation would not even be improved by considering the electron repulsion within the framework of the PPP method. The difficulty lies in the fact that description of the formaldehyde molecule using a model considering only two electrons (out of 16) is too rough.

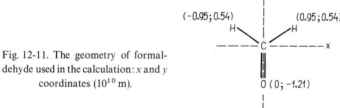

Fig. 12-11. The geometry of formaldehyde used in the calculation: x and y coordinates (10^{10} m).

With a comparatively small molecule such as formaldehyde, the application of extended methods is particularly desirable. Fig. 12-11 shows the coordinates of the formaldehyde atoms which were used in calculation by the EHT, CNDO/2 and "ab initio" methods. The orbital energies obtained are plotted in Fig. 12-12. In another figure (Fig. 12-13) are given the calculated charges and the dipole moments. There is no doubt that the EHT method exaggerates the charge distribution characteristics.

A great number of examples has convincingly shown that the different versions of the MO methods are suitable for interpreting the properties of conjugated systems. We should like to quote one example:

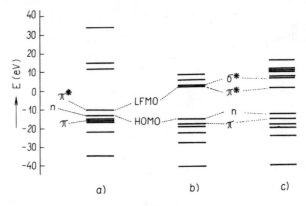

Fig. 12-12. EHT (*a*), CNDO/2 (*b*), and "ab initio" (DZ basis set) (*c*) orbital energies of formaldehyde.

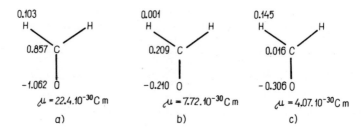

Fig. 12-13. EHT (*a*), CNDO/2 (*b*), and "ab initio" (DZ basis set) (*c*) charges and dipole moment of formaldehyde. The experimental dipole moment is 7.8×10^{-30} C m.

on the basis of knowledge of the HMO expansion coefficients of the frontier orbitals (HOMO, LFMO) and using first-order perturbation calculations, it is possible to draw qualitatively correct conclusions on the positions into which a substituent (of a chosen nature) should be introduced in order to increase (or decrease) the ionization potential or the electron affinity, or to induce a hypsochromic or bathochromic shift* of the first (longest wavelength) band in the electronic spectrum (cf. Section 13.3.1). For rapid orientation it is expedient to indicate the values of the expansion coefficients in the individual positions in the structural formulae (Fig. 12-14). The calculation is then easily performed using the expressions from Table 10-13 and Eq. (12-104) and Table 13-7. In the papers by E. Heilbronner and co-workers, it is possible to find a number of cases of skillful utilization of the perturbation treatment.

---

* Shift to shorter or longer wavelengths, respectively.

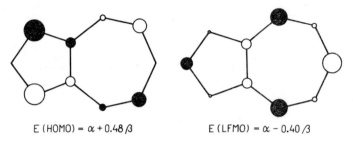

$$E \text{ (HOMO)} = \alpha + 0.48\,\beta \qquad E \text{ (LFMO)} = \alpha - 0.40\,\beta$$

Fig. 12-14. The circles correspond to the squares of the HMO expansion coefficients in HOMO and LFMO. Positive (○) and negative (●) coefficients.

The fact that MO calculations for dozens of systems and correct predictions of their stability were made *prior to experimental proof* is very encouraging. Amongst these systems are, for example, the molecules and ions designated IX − XIII. In recent years, quantum chemical methods have also begun to be used for structure elucidation in organic compounds.

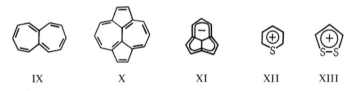

IX X XI XII XIII

The theoretical explanation of the conditions in numerous non-classical systems has often considerably influenced further experimental studies. This was true, for example, with paracyclophanes (XIV). Between the two benzene rings transannular interaction exists, which, on the whole, has no influence on the bonding conditions, but significantly affects the electronic spectrum. Mainly systems in which $m = n$ were studied theoretically; the influence of the CH_2 bridges is especially pronounced when $m = n = 2$. The splitting of the original benzene energy levels is illustrated in Fig. 12-15.

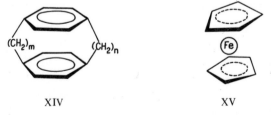

XIV XV

The last example is ferrocene, which is one of the longest known representatives of a very extensive group of organometallic compounds. Metals have been found to act as donors or acceptors of electrons in various organometallic compounds (Fig. 12-16). Using symmetry considerations it is possible to directly construct a qualitative scheme of the orbital energy levels. Still, it must be borne in mind that the π molecular orbitals of cyclopentadienyl correspond to the symmetry species A_1, E_1 and E_2 and the AO's of iron to $A_{1g}(4s, 3d_{z^2})$, $A_{2u}(4p_z)$, $E_{1g}(3d_{xz}, 3d_{yz})$, $E_{1u}(4p_x, 4p_y)$ and $E_{2g}(3d_{xy}, 3d_{x^2-y^2})$; of course, only orbitals of equal symmetry may be combined.

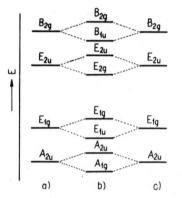

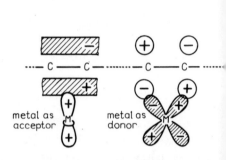

Fig. 12-15. Hartree-Fock orbital energies of benzene (*a*, *c*) and (*n*, *n*) paracyclophane (*b*). The levels are designated by the symbols of irreducible representations of the D_{6h} group (according to Ref. 9).

Fig. 12-16. Interaction of the d orbitals of a transition metal M with bonding and antibonding molecular orbitals (the occupied orbital is hatched).

Although calculations performed by various methods led to somewhat contradictory results for the electron distribution, all of them explain the kinetic and thermodynamic stability of ferrocene quite well. The situation is similar for numerous organometallic compounds.

12.4 Examples of systems studied in biochemistry

It is surprising how often one encounters conjugated systems when investigating compounds which are interesting or even occupy key positions in biochemistry more closely. This is all the more remarkable since the chief components of living organisms, proteins, sugars and fats, are unconjugated systems.

The conjugated components of nucleic acids (pyrimidines and purines), various coenzymes, porphyrines and bile pigments, pteridines and proteins belong among the most frequently studied components of living matter.

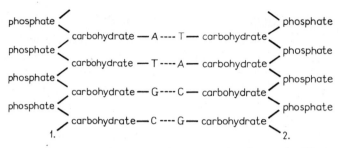

Fig. 12-17. Designation: A — adenine, T — thymine, G — guanine, C — cytosine. The 1st and 2nd phosphate-sugar chains are denoted by numbers (cf. Fig. 12-18).

From the classical point of view, proteins are, of course, unconjugated systems. If, however, we consider the possibility of hydrogen bond formation, the system becomes in a certain sense conjugated.

Two examples can be given for the sake of illustration. The first concerns calculation of the electronic structure of pairs of nucleic acid bases. Fig. 12-17 presents a scheme of the catenation of components of deoxyribonucleic acid (DNA): phosphate, sugar and bases (pyrimidines and purines). The *Watson-Crick model* of *DNA* (Fig. 12-18) shows the correct spatial arrangement (Fig. 12-18). The adenine-thymine and guanine-cytosine pairs are the most important; it is evident from their molecular diagrams (Fig. 12-19) and from the molecular diagrams of the free components that the interaction mediated by the hydrogen bonds is weak. This is true of the energy characteristics as well as of the electron distribution. In semiempirical calculations the presence of the hydrogen bond can be described in several ways: merely by considering mutual

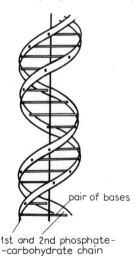

pair of bases

Fig. 12-18. The Watson-Crick DNA model.

1st and 2nd phosphate-
-carbohydrate chain

310

polarizations (without intermolecular electron transfer), by a suitable change in the values of the Coulomb integrals of the atoms lying in the vicinity of a particular atom, by attributing small values (about 0.2β) of the resonance integral of the hydrogen bond or, finally, by considering the hydrogen atom of a hydrogen bond explicitly in terms of its $2p_z$ orbital ($\alpha_H = \alpha - 1.8\beta$); the same value as previously (0.2β) is assigned to the $H-X$ (and $H \ldots X$) bonds. It is rather interesting to note that the differences between the SCF and HMO molecular diagrams are not significant.

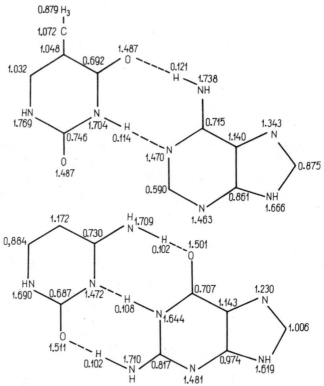

Fig. 12-19. SCF-electron densities of pairs of bases[11].

Hydrogen bonds play an important role in protein molecules. As early as in the forties Szent-Györgyi explained their semiconductivity qualitatively using a model which Evans and Gergely processed semi-quantitatively. It is assumed that the individual macromolecules are connected by hydrogen bonds in the peptide linkage region (Fig. 12-20).

For years attempts have been made to utilize quantum chemical characteristics in molecular pharmacology and toxicology. The attempts

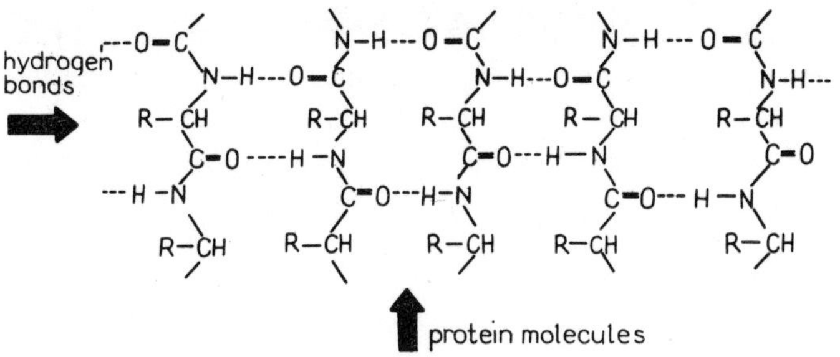

Fig. 12-20. Model of protein molecules.

of B. and A. Pullman to correlate the carcinogenic activities of benzenoid hydrocarbons with indices of the K and L spheres (9 – 10 bond in phenanthrene and the atom pair 9, 10 in anthracene) are amongst the oldest endeavours in this field. Theoretical characteristics have also been utilized for estimation of the carcinogenic activity of compounds of other structural types. The MO method has also been used for interpretation of the course of metabolism in vivo and in vitro. In recent years, the number of attempts to utilize theoretical characteristics in the search of biologically effective substances has been steadily increasing.

The number of studies published in the field of quantum bio-chemistry is increasing very rapidly. Great attention is being paid, e.g., to the conformation of biologically active substances and to the effects of the medium on the conformation. A large number of important works have been published in the Collections of Jerusalem Symposiums[12].

REFERENCES

1. Coulson C. A.: *Valence.* Clarendon Press, Oxford 1952.
2. Kittel C.: *Introduction to Solid State Physics,* Wiley, New York 1956.
3. Hoffmann R., Lipscomb W. N.: *J. Chem. Phys.* **37**, 2872 (1962).
4. Longuet-Higgins H. C.: *Quart. Rev.* **11**, 121 (1957).
5. Craig D. P.: *Theoretical Organic Chemistry.* Butterworths, London 1959, p. 20.
6. Pettit L. D.: *Quart. Rev.* **25**, 1 (1971).
7. Hoffmann R.: *J. Chem. Phys.* **39**, 1397 (1963).
8. Hoffmann R., Zeiss G. D., Van Dine G. W.: *J. Am. Chem. Soc.* **90**, 1485 (1968).
9. Koutecký J., Paldus J.: *Collect. Czech. Chem. Commun.* **27**, 599 (1962).
10. Pullman B., Pullman A.: *Quantum Biochemistry.* Interscience, New York 1963.
11. Pullman B., Weissbluth M. (ed.): *Molecular Biophysics.* Academic Press, New York 1965.
12. Pullman B. (ed.): *Environmental Effects on Molecular Structure and Properties.* D. Reidel, Dordrecht 1976. (8-th volume of Jerusalem Symposia on Quantum Chemistry and Bio-chemistry; cf. also other volumes.)

13. MOLECULAR SPECTROSCOPY

13.1 Phenomenological description

13.1.1 Introductory comments

In this chapter, the processes in which electromagnetic radiation plays a key role in addition to that played by the studied molecules will be discussed. We shall chiefly be interested in processes during which no structural changes occur in the molecule. Two types of processes will be considered:

a) the molecules accept energy either from the electric or from the magnetic component of the radiation; these are *absorption processes,*

b) molecules which are in an excited state (i.e., any state with energy higher than that of the ground state) return to the ground state with release of energy in the form of electromagnetic radiation; these are *emission processes.*

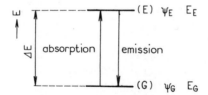

Fig. 13-1. Absorption and emission processes, G and E denote ground and excited states. Ψ and E denote the wave functions and energies of these states.

These processes are schematically illustrated in Fig. 13-1. In order that the molecule be capable of absorbing energy in the form of radiation, it must possess at least one further state with higher energy, an excited state, in addition to the ground state. This is usually some sort of rotational, vibrational or electronic state, which is "inherent" to the molecule and is attainable under normal conditions. However, there also exist certain degenerate states which are incompletely occupied and un-interesting from the point of view of absorption spectroscopy. It appears that this degeneracy can be removed by placing the studied molecule in a sufficiently strong external electric or magnetic field. In such cases, which are of decisive importance in nuclear magnetic and electron spin resonance, the given basic condition concerning the existence of several states with different energy is fulfilled in the presence of the external field.

13.1.2 Units and the spectral regions

In order that the system pass from the ground state G (described by the wave function Ψ_G) to the excited state E (described by the wave function Ψ_E), it must accept an amount of energy ΔE:

$$\Delta E = E_E - E_G \tag{13-1}$$

As long as radiation is considered to be of a corpuscular nature, the energy expression for the distance between the levels (G and E in Fig. 13-1) is appropriate; however, we are often obliged to characterize radiation as a propagated wave and it is then necessary to find a relation which would fit both the corpuscular and the wave characteristics; this is the common relationship

$$E = h\nu, \tag{13-2}$$

where E is the energy, ν is the frequency and h is Planck's constant.

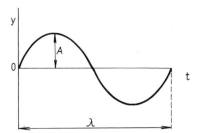

Fig. 13-2. Graphical representation of Eq. (13-3).

It is sometimes expedient to characterize the propagation of a wave by its wavelength; it should simultaneously be borne in mind that the electric and magnetic components of electromagnetic radiation can be described similarly to other periodic processes by the equation (Fig. 13-2)

$$y = A \sin \frac{2\pi}{\tau} t, \tag{13-3}$$

where A denotes the amplitude, τ is the time corresponding to one oscillation and t is time. According to the electromagnetic theory of light, the propagation of a light ray (in the direction of the t-axis in Fig. 13-3)

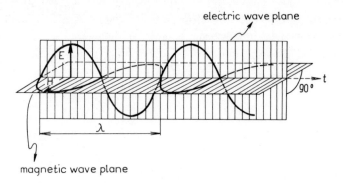

Fig. 13-3. Wave of plane polarized light. λ denotes the wavelength.

is connected with both the electric and the magnetic waves, where both are propagated in the same phase. The vectors of the electric (**E**) and magnetic (**H**) fields lie in mutually perpendicular planes; these vectors are simultaneously also perpendicular to the ray representing the direction of light propagation (cf. Fig. 13-3). We can now return to the original problem: for velocity v it holds by definition that $v = l/t$, where l denotes the path length and t time. The velocity of a periodical event can be expressed in terms of the frequency v and the wavelength λ. Here v denotes the velocity of electromagnetic radiation in a vacuum ($c \approx$ $\approx 3 \cdot 10^5$ km/s):

$$c = \lambda v \tag{13-4}$$

Equation (13-2) can then be rewritten in the form

$$E = hv = \frac{hc}{\lambda} \tag{13-5}$$

In place of the wavelength λ, its reciprocal value — the wave number (dimension length^{-1}) — is often employed. If the molecule accepts a photon of frequency v, its energy increases by hv. On emission of a photon from the excited molecule, its energy decreases by the same value. The energy of one mole of photons (i.e. 6.023×10^{23} photons) is called an einstein.

The kinetic energy, which is gained by an electron exposed to a potential of 1 V, is termed one electron volt (eV). If one mole of these elementary particles has this amount of energy, it corresponds to an energy of 96.49 kJ/mole.

It is one of the chief tasks of spectroscopy to determine experimentally and also theoretically the energy differences between the levels ·corresponding to the ground and the excited states. It follows from the above discussion that these differences can be expressed in the usual

Table 13-1

Units for Characterization of Radiation

Spectra	Energy	Frequency	Wave number	Wavelength
UV and visible	eV, J		cm^{-1}	$nm\,(\equiv 10^{-9}\,m)$
IR			cm^{-1}	$\mu m\,(\equiv 10^{-6}\,m)$
radiofrequency		$MHz\,(\equiv 10^{6}\,s^{-1})$		

energy units (joules, electron volts) or in the units of quantities that are proportional to the energy, i.e. in frequency units (Hz) or wave numbers (cm^{-1}), possibly in units of reciprocal wave numbers, i.e. in wavelength units (nm). Physicists prefer energy units, physical chemists often work with wave numbers and chemists employ wavelengths. Here the energy or quantities directly proportional to it will be employed, as one of our aims is the confrontation of theoretical and experimental data, where energies are obtained by solving the Schrödinger equation for the ground and excited states. In different wavelength regions of electromagnetic radiation, different quantities and different units are used for its characterization. Table 13-1 is only orientative.

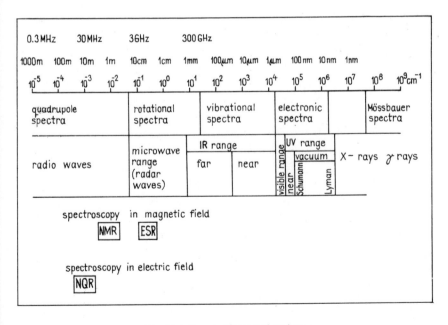

Fig. 13-4. Survey of spectral regions.

Table 13-2

Table of Equivalent Values of Energy, Wavenumber and Frequency

	eV	cm^{-1}	kJ/mol	MHz
1 eV	–	8065.5	96.487	2.4180×10^8
1 cm^{-1}	1.2399×10^{-4}	–	1.1963×10^{-2}	2.9979×10^4
1 kJ/mol	1.0364×10^{-2}	83.591	–	2.5060×10^6
1 MHz	4.1357×10^{-9}	3.3356×10^{-5}	3.9904×10^{-7}	–

It is necessary, of course, that equivalent quantities be quickly and reliably convertible; Table 13-2* serves this purpose. It would be unnecessary to learn the values of these conversion factors by heart, but it is very useful to have such a table available; the same applies to the survey of spectral regions given in Fig. 13-4.

13.1.3 Absorption and emission spectra, the population of excited states

For the absorption of radiation by a molecule passing from state Ψ_G to state Ψ_E (Fig. 13-1), radiation of frequency v which fulfils the condition

$$E_E - E_G = \Delta E = hv \tag{13-6}$$

must be employed. It might seem that the corresponding spectrum would consist of a single absorption line or of a number of lines, provided the system has a number of attainable excited states (Ψ_{E1}, Ψ_{E2}, etc.). In reality, the absorption spectra of molecules are, for various reasons, composed of absorption bands and not of absorption lines. With molecules, every electronic state corresponds to a number of vibrational states and each vibrational state again corresponds to a number of rotational states (Fig. 13-5). Thus, the bands corresponding to electronic excitations possess vibrational structure, the vibrational bands have rotational structure and the rotational bands quadrupole structure. According to relationship (13-6) the frequency of the radiation is clearly determined, corresponding to the energy emitted by the molecule in passing from state E to state G, provided the deactivation process is accompanied by emission. The excited state can also lose energy in another way; in the spectroscopy of solutions, collisions of excited molecules with molecules of the solvent are among the most important possibilities. Non-radiative deactivation then occurs.

---

* In infrared spectroscopy a quantity with the dimension of reciprocal length (i.e. the wave number) is sometimes called the frequency.

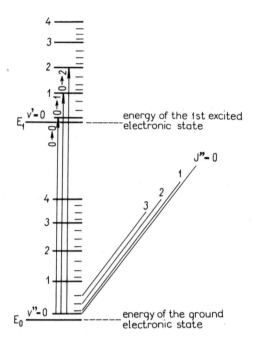

Fig. 13-5. Vibrational and rotational states corresponding to the E_0 and E_1 electronic states. The three indicated transitions explain the vibrational structure of the band. v and J are vibrational and rotational quantum numbers, respectively.

Fulfillment of the condition represented by Eq. (13-6) is, in itself, not sufficient for inducing absorption. First, the value of the quantity called the transition moment (see below) must be investigated. Only when this transition moment is non-vanishing will an absorption band exist in the spectrum. The transition moment is a quantity which characterizes a change in the electronic distribution (and thus a change in the dipole moment) during electronic excitation. It must be borne in mind that the electric dipole moment of the molecule in the state described by wave function Ψ_G is given by the expression [cf. Eq. (11-57)]

$$\mu_G = -e \int \Psi_G^* \mu \Psi_G \, d\tau, \tag{13-7}$$

where e is the proton charge and μ is the operator corresponding to the dipole moment of unit charge (which is the sum of the position vectors of the individual electrons, i.e. $\mu = \sum r_v$). For the electric dipole moment of the excited state, the relationship

$$\mu_E = -e \int \Psi_E^* \mu \Psi_E \, d\tau \tag{13-8}$$

similarly applies. Finally, the case in question (transition moment $Q_{G \to E}$) is described by the expression

$$Q_{G \to E} = -e \int \Psi_G^* \mu \Psi_E \, d\tau \tag{13-9}$$

318

In summary: only when integral (13-9) is non-vanishing can the studied molecule accept energy from the electromagnetic radiation which is incident upon it. The greater the value of integral (13-9), the more effectively the studied molecule absorbs energy. In the majority of cases discussed here, interaction of the molecule with the electric component of the electromagnetic radiation occurs. In other words, the interaction considered here can be characterized as interaction between the radiation and the oscillating electric dipole. There exist cases when the studied moment contains two such dipoles (a quadrupole) orientated so that their dipole moments are cancelled; nonetheless this pair causes an electrostatic field. Interaction of this field with radiation impinging upon the molecule occurs and is manifested by quadrupole absorption or emission. Interactions between the oscillating magnetic dipole and the impinging radiation also occur (magnetic dipole transition); both the above-mentioned types of transitions are less intense by several orders than electric dipole transitions and thus are encountered only rarely.

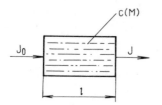

Fig. 13-6. Passage of electromagnetic radiation through a homogeneous solution of concentration c placed in a cuvette of length l.

It is now necessary to clarify more precisely the manner in which the value of the transition moment is connected with the magnitude of the absorption band. This can be most simply done by directly utilizing the Lambert-Beer law, according to which (Fig. 13-6) the relationship

$$A \equiv \log \frac{J_0}{J} = \varepsilon c l \qquad (13\text{-}10)$$

is valid, where J_0 is the intensity of the impinging radiation and J is the intensity of radiation emerging from the cuvette, c is the molarity of the investigated solution, l is the optical path length in cm, A is the absorptivity (optical density) and ε is the molar absorption coefficient (formerly called the molar extinction coefficient). According to Eq. (13-10), A (whose value can often be read directly on the spectrophotometer) is directly proportional to the molarity of the studied substance and the optical path length. Although the Lambert—Beer law is very often fulfilled, this cannot be automatically assumed (especially in more

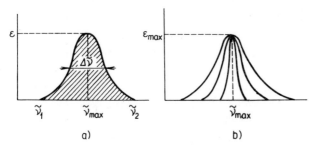

Fig. 13-7. Shapes of absorption bands in ε and $\tilde{\nu}$ coordinates.

concentrated solutions). In Fig. 13-7a the shape of the absorption band is given in $\varepsilon - \tilde{\nu}$ coordinates. The value of the molar absorption coefficient at the absorption maximum (ε_{max}) is frequently applied to characterize the absorptivity. The fact that this characteristic is not a very good measure of the band intensity is demonstrated in Fig. 13-7b, where three absorption bands with different shapes correspond to a maximum with the same ε_{max} value. Thus, a more correct measure of the transition intensity is the area enclosed by the absorption curve and the wave number axis; in Fig. 13-7a this area is shaded in. In fact, the classical theory of absorption affords an expression for the oscillator strength, f, in which the above-mentioned area is multiplied by a constant:

$$f = 2.303 \frac{4\varepsilon_0 mc^2}{N_A e^2} \int_{\tilde{\nu}_1}^{\tilde{\nu}_2} \varepsilon \, d\tilde{\nu} \qquad (13\text{-}11)$$

so that, after inserting the values of numerical constants, we obtain

$$f = 4.319 \times 10^{-9} \int_{\tilde{\nu}_1}^{\tilde{\nu}_2} \varepsilon \, d\tilde{\nu},$$

where the molar absorptivity ε is expressed in $l \, mol^{-1} cm^{-1}$ and $\tilde{\nu}$ in cm^{-1}. In Eq. (13-11) ε_0 denotes the permitivity of a vacuum, m is the electron mass, c is the velocity of light, N_A is the Avogadro constant, e is the proton charge and the factor 2.303 transforms natural to decadic logarithms. With symmetrical absorption bands, according to the mean value theorem, the integral in the expression for f (13-11) can be replaced by the product containing the band halfwidth $\Delta\tilde{\nu}$ (in cm^{-1}), i.e. the width of the band at $\varepsilon = \varepsilon_{max}/2$ (cf. Fig. 13-7a):

$$f = 4.319 \times 10^{-9} \varepsilon_{max} \Delta\tilde{\nu} \qquad (13\text{-}12)$$

The oscillator strength can also be calculated quantum mechanically and it then follows that it is proportional to the square of the transition moment. For transition from the ground state to the excited state it

320

holds that

$$f_{G \to E} = \frac{8\pi^2 mc}{3he^2} \tilde{v}_{G \to E} |\mathbf{Q}_{G \to E}|^2 = 1.085 \times 10^{-3} \tilde{v}_{G \to E} \frac{|\mathbf{Q}_{G \to E}|^2}{e^2} \qquad (13\text{-}13)$$

In this expression, $\tilde{v}_{G \to E}$ is the wave number of the absorption band maximum (cm^{-1}) and the ratio $|\mathbf{Q}_{G \to E}/e|$ is expressed in nm.

It is interesting that, even when using a very strong radiation source, the molecules cannot be quantitatively transferred from the ground state to the excited state. This is surprising since the transition from state G to state E appears to be an induced process (induced by radiation, which is absorbed by matter), whereas emission (transition $E \to G$) seems to be a simple spontaneous process. However, emission is also a process induced by radiation. Strictly speaking, spontaneous excitation $(G \to E)$ should also be considered, but for the majority of the spectral transitions discussed here, the probability of this process is negligible.

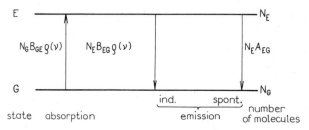

Fig. 13-8. Induced absorption and induced and spontaneous emission.

The probability of the occurrence of a spontaneous or induced transition is proportional to the number of molecules in the state from which the transition starts (N_G, N_E) and also to the Einstein probability coefficient (A in spontaneous and B in induced phenomena); with induced processes, it is also proportional to the radiation density $\varrho(v)$. The corresponding expressions are given in Fig. 13-8. At equilibrium the probability of excitation and emission is equal and therefore the relationship

$$N_G B_{GE} \varrho(v) = N_E [B_{EG} \varrho(v) + A_{EG}] \qquad (13\text{-}14)$$

is valid. Using the Boltzmann distribution law, the population of molecules N_i in the individual states i can be calculated. The equation

$$N_i = \text{const.} \, g_i \exp(-E_i/kT) \qquad (13\text{-}15)$$

is then valid, where g_i denotes the degeneracy of the i-th state ($g_i = 1$ if the state is non-degenerate), E_i is its energy, k is the Boltzmann constant and T is the absolute temperature.

13.2 Excitation within a single electronic level

13.2.1 Introductory comments on radiofrequency spectroscopy

The intrinsic angular momentum (spin), the magnetic moment and the distribution of the charge in atomic nuclei are of principal importance for all types of spectroscopy in the radiofrequency region.

The intrinsic angular momentum is expressed in units of $h/2\pi$, usually by means of the maximum value of its component in a certain direction – in the direction along some specified axis given, for instance, by an external field. In the expression $Ih/2\pi\,(\equiv I\hbar)$, I denotes the nuclear spin quantum number. For about 140 of the 280 stable isotopes, its value is zero for the others it lies between $\frac{1}{2}$ and $\frac{9}{2}$, assuming only half-integral or integral values.

Similar to electron rotation, the rotation of the nucleus (rotation of a charged particle) is also connected with its magnetic moment, for the magnitude of which the relationship

$$\mu = \frac{\gamma I h}{2\pi} \tag{13-16}$$

holds, where γ is the gyromagnetic ratio; μ is usually expressed in multiples of the nuclear magneton, β_N:

$$\beta_N = \frac{eh}{4\pi m_p} = 5.050 \times 10^{-27} \text{ A m}^2, \tag{13-17}$$

where e is the proton charge and m_p is the proton mass.

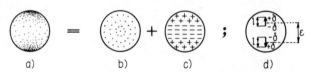

a)   b)   c)   d)

Fig. 13-9. Charge distribution in a nucleus with ellipsoidal symmetry.

The distribution of the charge is spherically symmetrical only in nuclei with zero spin or with spin equal to $\frac{1}{2}\hbar$. Otherwise, the charge is generally distributed so that in some regions there is an excess of positive and in others of negative charge. The situation in nuclei in which the distribution of the charge is not spherically symmetrical is demonstrated in Fig. 13-9. The density of the dots indicates (Fig. 13-9a) nonuniformity of the charge distribution. This situation can be imagined as superposition of a uniform distribution of positive charges (Fig. 13-9b) and of a distribution in which an excess of positive charge lies at the poles and of negative charge in the region of the equator (Fig. 13-9c). The centroids of the positive and negative charges are depicted in Fig. 13-9d. It can be seen that the system consists of two antiparallel dipoles, which form a quadrupole. Electric quadrupole moment eQ expresses the deviation of the real charge distribution from spherical symmetry:

$$eQ = 4l\varepsilon\delta e, \qquad (13\text{-}18)$$

where ε is the dipole length, δ is the absolute value of the charge and l is the distance between the centres of dipoles. In Eq. (13-18) it is assumed that the positive charge is concentrated around the poles; otherwise the quadrupole moment has a negative sign.

In Table 13-3 the magnetic characteristics of some chemically important atomic nuclei are given. It is apparent that two of the most important nuclei, C^{12} and O^{16}, possess zero magnetic moment and are, consequently, inactive as far as nuclear resonance is concerned. The H^1, C^{13}, F^{19} and P^{31} isotopes, with vanishing quadrupole moments,

Table 13-3

Characteristics of Atomic Nuclei

Isotope	Magnetic moment[a]	Spin number	$\dfrac{Q \cdot 10^{28}}{m^2}$ [b]
H^1	2.79268	$\frac{1}{2}$	0
H^2	0.85738	1	2.77×10^{-3}
C^{12}	0	0	0
C^{13}	0.70220	$\frac{1}{2}$	0
N^{14}	0.40358	1	7.1×10^{-2}
N^{15}	-0.28304	$\frac{1}{2}$	0
O^{16}	0	0	0
O^{17}	-1.8930	$\frac{5}{2}$	-4×10^{-3}
F^{19}	2.6273	$\frac{1}{2}$	0
P^{31}	1.1305	$\frac{1}{2}$	0
S^{33}	0.64274	$\frac{3}{2}$	-5.3×10^{-2}
Cl^{35}	0.82091	$\frac{3}{2}$	-7.9×10^{-2}
Br^{79}	2.0991	$\frac{3}{2}$	0.34
Br^{81}	2.2626	$\frac{3}{2}$	0.28
I^{127}	2.7937	$\frac{5}{2}$	-0.75

[a] In multiples of nuclear magneton, Eq. (13-17).
[b] The value Q characterizes the quadrupole moment, as this moment can be expressed as the product Qe (e denoting the proton charge).

belong among the most important active nuclei for nuclear magnetic resonance; they are characterized by especially sharp signals. For nuclear quadrupole resonance, the most important are the naturally occurring isotopes Cl^{35}, Br^{79}, Br^{81} and I^{127}, which have large quadrupole moments.

13.2.2 Nuclear quadrupole resonance (NQR)

The ellipsoidal distribution of the charge of some nuclei can alternatively be described by means of a monopole and a negative or positive quadrupole. With an prolate (*a*) and oblate (*b*) ellipsoid, a possible resolution is indicated in Fig. 13-10. The quadrupole moment is a measure of the ellipsoidal deformation of the nucleus. The energy of a quadrupole situated in an asymmetrical electric field depends on its orientation.

Fig. 13-10. Alternative description of conditions in a nucleus with ellipsoidal charge distribution.

Particularly important is the orientation of the nuclear quadrupole in the electric field of the electron in the p-type atomic orbital. The two extreme orientations are demonstrated in Fig. 13-11. If the studied nucleus has a spin of $I\hbar$, it follows that the nucleus can assume an orientation in relation to the p orbital only such that the quantum number of the spin component in the chosen direction acquires the value M, where $M = I$, $I - 1$, ... $-(I - 1)$, $-I$. Thus, a system consisting of a p electron and a nuclear quadrupole exists only in $(2I - 1)$ states of different energy. The chosen direction is represented by the direction

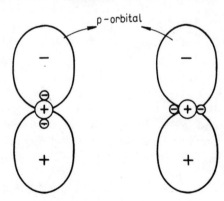

Fig. 13-11. Two important orientations of p orbitals and of the nuclear quadrupole.

324

of the asymmetrical electric field. For this reason, the number of orientations of the quadrupole in the electric field is limited; the individual "configurations" differ in energy. A transition between them is connected with absorption or emission of energy: these are the basic facts necessary for understanding the origin of NQR spectra. The interaction of the quadrupole with the other types of orbitals (s, d, f) is not significant, either due to their unsuitable symmetry or due to their diffuse character.

p-Type orbitals become ineffective as soon as they are completely occupied by electrons (altogether 6 electrons in np_x, np_y and np_z orbitals) as such a formation is then spherically symmetrical. Therefore the Br atom, for example, is active in NQR spectra, whereas the Br^- ion is ineffective. If it is considered that p orbitals in conjugated compounds are incompletely occupied, it seems the NQR offers unique possibilities, as far as the experimental determination of the electron distribution in various compounds, especially in conjugated compounds, is concerned. For a number of reasons, the situation is, unfortunately, far less favourable. In the first place, there is only a small number of suitable (active) atoms. Most of the studies carried out so far concerned compounds containing Cl and Br atoms and recently also N atoms. Furthermore, the signal connected with the transitions between the individual levels is so weak that work with solid samples is necessary, so that specific problems of work with a solid phase are added to the already quite complicated task.

13.2.3 The elementary theory of magnetic resonance

The nuclear angular momentum (spin, I) possesses similar properties as the electronic spin (S) except that the nuclear magnetic moment is usually parallel with the spin vector. With electrons, the opposite is true. To nuclei with a spin quantum number $\frac{1}{2}$ correspond two eigenfunctions, α and β, which possess the following properties:

$$\mathscr{I}^2\alpha = \frac{\hbar^2}{2}\left(\frac{1}{2}+1\right)\alpha \qquad \mathscr{I}_z\alpha = \frac{\hbar}{2}\alpha \qquad (13\text{-}19a, b)$$

$$\mathscr{I}^2\beta = \frac{\hbar^2}{2}\left(\frac{1}{2}+1\right)\beta \qquad \mathscr{I}_z\beta = -\frac{\hbar}{2}\beta \qquad (13\text{-}20a, b)$$

Here, $\mathscr{I}^2$ is the operator of the square of the nuclear spin momentum and $\mathscr{I}_z$ is the operator of the z-component of this momentum. Quite similarly as with electrons [cf. Eq. (13-41)], the operator of the magnetic moment of nuclei is given by the relationship

$$\mu = g_N\beta_N\mathscr{I}, \qquad (13\text{-}21)$$

where g_N is the nuclear gyromagnetic ratio and β_N is the nuclear magneton $(5.050 \times 10^{-27} \text{ A m}^2)$.

By placing a particle with a spin $\frac{1}{2}\hbar$ in the magnetic field, the degeneracy of the α and β states is removed. The form of the Hamiltonian for the interaction of a magnetic dipole of magnitude μ with a static magnetic field of induction B then becomes important. The classical expression for the energy corresponding to this interaction assumes the form

$$E = -\mu B \tag{13-22}$$

The direction of vector $\mathbf{B}$ is given by the direction of the z-axis, thus allowing us to treat this quantity as a scalar quantity; for enumeration of the scalar product, the projections of vectors $\mathbf{S}$ and $\mathbf{I}$ in the direction of vector $\mathbf{B}$ are then necessary. Operators $\mathscr{S}_z$ and $\mathscr{I}_z$ can be used to construct the "magnetic" Hamiltonian for the electron, $\mathscr{H}_e$, and for the nucleus, $\mathscr{H}_N$:

$$\mathscr{H}_e = g\beta_e B\mathscr{S}_z \tag{13-23}$$

$$\mathscr{H}_N = -g_N\beta_N B\mathscr{I}_z, \tag{13-24}$$

where β_e and g are constants defined in Eq. (13-41). The form of the Hamiltonian is now known and the energy corresponding to the interaction of an electron and the nucleus with the magnetic field can then be calculated:

$$\mathscr{H}_N\alpha = -g_N\beta_N B\mathscr{I}_z\alpha = -g_N\beta_N B\frac{1}{2}\alpha \tag{13-25}$$

$$\mathscr{H}_N\beta = g_N\beta_N B\frac{1}{2}\beta \tag{13-26}$$

For electrons and nuclei (with spin quantum number $\frac{1}{2}$) it can then be written that

$$E_\alpha = \pm\frac{1}{2}g_i\beta_i B \tag{13-27}$$

$$E_\beta = \mp\frac{1}{2}g_i\beta_i B \tag{13-28}$$

The upper signs apply to electrons ($g_i = g$, $\beta_i = \beta_e$), the lower signs to nuclei ($g_i = g_N$, $\beta_i = \beta_N$). The energy difference of the α and β states is proportional to the induction of the magnetic field. It is evident (cf. Fig. 13-12) that the energy which must be supplied to the electron (nucleus) during transition from the ground state to an excited

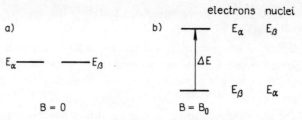

Fig. 13-12. Degenerate levels corresponding to spin functions (*a*). Removal of degeneracy due to external magnetic field (*b*).

state is given by the relationship

$$\Delta E = E_\alpha - E_\beta = h\nu = g_i\beta_iB_0 \tag{13-29}$$

Equation (13-29) represents the condition which must be fulfilled for resonance to occur. The theoretical basis of both magnetic resonance methods, electron spin resonance (ESR) and nuclear magnetic resonance (NMR), is the same.

13.2.4 Nuclear magnetic resonance (NMR)

Only atoms having nuclei with spin quantum number $\frac{1}{2}$ will be discussed. These are mainly the H^1, C^{13} and F^{19} isotopes; however, only proton magnetic resonance (PMR) will be treated in greater detail. The vector of the magnetic moment of a rotating nucleus can be orientated in an external magnetic field in two ways: either parallel or antiparallel to the vector of the induction of the magnetic field. The first possibility corresponds to a state with lower energy, the second to a state with higher energy. The energy difference between these two levels results from the resonance conditions (Fig. 13-13):

$$h\nu = g_N\beta_NB \tag{13-30}$$

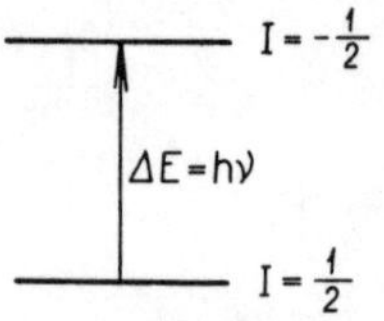

```
──────── I = -½
  ▲
  │ ΔE = hν
  │
──────── I = ½
```

Fig. 13-13. NMR resonance condition: Eq. (13-30).

The difference is so small ($\Delta E \ll kT$) that further discussion can be carried out within the framework of classical mechanics. In the classical description the expected precession motion of the studied system is characterized by the angular frequency ω (Larmor precession frequency), which is simply related to frequency v [Eq. (13-30)], and the induction of the magnetic field:

$$\omega = 2\pi v = \gamma B \tag{13-31}$$

Combining Eqs. (13-30) and (13-31) leads to the relationship

$$\gamma = \frac{g_N \beta_N}{\hbar} \tag{13-32}$$

Furthermore, it is evident that the magnitude of the energy gap (cf. Fig. 13-13) can be expressed by the equation

$$\Delta E = hv = g_N \beta_N B = \hbar \gamma B \tag{13-33}$$

The actual magnetic field operating in the neighbourhood of the nucleus is, of course, given by the vector sum of the external and internal magnetic fields. The magnitude of the internal field depends, on one hand, on the diamagnetic electron shielding (leading to chemical shifts) and, on the other hand, on the influence of the magnetic moments of the neighbouring nuclei (via spin–spin coupling). The reason for the appearance of chemical shifts is indicated schematically in Fig. 13-14.

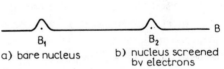

Fig. 13-14. Qualitative indication of the origin of chemical shift. In (a) and (b), conditions $hv_1 = (h/2\pi)\,B_1$ and $hv = (h/2\pi)\,B_2$, respectively, are satisfied.

a) bare nucleus

b) nucleus screened by electrons

The chemical shift. The internal field caused by the electronic motion is orientated in the opposite direction to the external field (B_0) and is proportional to this field. Thus the following relationship holds for the magnitude of the magnetic field at the nucleus B_n,

$$B_n = B_0(1 - \sigma_n), \tag{13-34}$$

where σ_n is the shielding constant of the nucleus. For the transition frequency, the relationship

$$v = \bar{\gamma} B_0(1 - \sigma_n) \tag{13-35}$$

is valid, where $\bar{\gamma} = \gamma/2\pi$. Since v and B_0 cannot be determined independently with sufficient accuracy, absolute vales of σ_n cannot be obtained.

328

Therefore, only relative measurements can be carried out and for this purpose the chemical shift, δ_n, is defined as follows:

$$\delta_n = \frac{B_m - B_r}{B_r} \times 10^6, \qquad (13\text{-}36)$$

where B_m (B_r) is the induction of the magnetic field corresponding to the peak of the measured (reference) proton (Fig. 13-15). The differences in the inductions of magnetic fields B_m and B_r are very small (with an order of magnitude of 10^{-6}) and therefore the definition of δ_n includes the factor 10^6 (δ_n then has values of the order of 10^c).

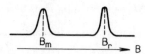

Fig. 13-15. NMR peak corresponding to the studied (m) and reference (r) protons. B denotes the induction of the magnetic field.

The induction of the magnetic field is expressed in tesla units or is recalculated into frequency in Hz by the relationship $v = \bar{\gamma}B$. (The numerical value of this frequency, of course, bears no relation to the frequency to which the sample is exposed during the entire measurement.) Tetramethylsilane is nearly always used as a reference substance; by convention, zero chemical shift is attributed to it. If the proton signal in the studied substance has lower B values, then δ is positive and thus the proton in this substance is more shielded than in $(CH_3)_4Si$. On this scale, most protons lie at lower values of induction B. With an alternate formerly used scale a value of 10.000 was conventionally attributed to protons in $(CH_3)_4Si$ and the chemical shift was given in units of a dimensionless quantity τ, defined as follows:

$$\tau = 10.000 - \delta \qquad (13\text{-}37)$$

Figure 13-16 gives the chemical shifts of protons of a number of solvents.

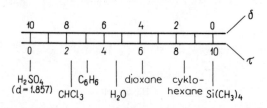

Fig. 13-16. Comparison of the δ scale and the τ scale; peak positions for several solvents are given.

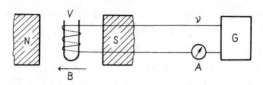

Fig. 13-17. Scheme of the experimental arrangement for NMR measurements; magnet poles (*N*, *S*), sample (*V*), ammeter (*A*), high-frequency generator (*G*) emitting radiation of frequency *v*.

Experimental arrangement. The resonance condition can be fulfilled either by exposing the specimen to a magnetic field of a given intensity and changing the radiation frequency or by irradiating the specimen at a constant frequency and changing the induction of the magnetic field until the resonance condition is fulfilled. The fulfillment of this condition is connected with significant absorption of energy from the high-frequency field and is manifested by an increasing current in the measuring instrument, *A* (Fig. 13-17). It should be noted that the specimen inside the induction coil is exposed to an alternating high-frequency field of constant frequency *v*. The induction of the magnetic field increases; only if the resonance condition is fulfilled a sharp (temporary) increase in the current passing through the coil is observed.

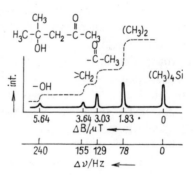

Fig. 13-18. NMR spectrum of an organic compound. Magnetic induction with respect to a standard (tetramethylsilane) is converted to frequency in Hz. The primary spectrum is denoted by a full line, the derivative recording is dashed.

Chemical applications of NMR spectroscopy. In Fig. 13-18 the NMR spectrum of 4-hydroxy-4-methylpentanone-2 is given. The assignment of peaks to the individual proton types is facilitated by the integral curve, which is, as a rule, recorded simultaneously. The ratios of the areas of the absorption bands are given by the ratios of the numbers of the individual types of protons. The observed chemical shifts, expressed in Hz, depend on the frequency of the applied electromagnetic field.

Another rather instructive case is [18] annulene, whose NMR spectrum has two groups of bands with an intensity ratio of 2:1.

The first corresponds to the external and the second to the internal protons (Fig. 13-19). The internal protons are strongly shielded, leading to the very considerable difference between the two groups. This strong shielding is attributed to the diamagnetic effect of the π electrons circulating under the influence of the external field in the carbon ring (ring current). The aromaticity of the annulene can be estimated from the position of the peripheral proton signal.

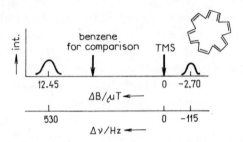

Fig. 13-19. NMR spectrum of [18] annulene.

Interaction constants (*spin − spin coupling*). The interaction between the magnetic moments of active nuclei (which are also important in determining the value of the "micromagnetic" field at the nuclei) complicates NMR spectra considerably, but simultaneously makes them far more interesting and valuable when studying small differences in the structures of substances. In analysis of spin−spin coupling peaks, it is necessary to begin from the fact that the splitting of the peak of the given protons depends on the number of equivalent protons in neighbouring positions; n protons bonded in the position next to the studied proton lead to splitting of the line of this proton into $(n + 1)$ lines.

The NMR spectrum of acetaldehyde can serve as an example. Fig. 13-20 illustrates recording of the spectrum at smaller and larger

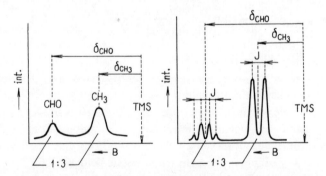

Fig. 13-20. NMR spectrum of acetaldehyde recorded with low (left) and high (right) resolving power. Standard: tetramethylsilane (TMS). Numerical data refer to area ratios.

resolving powers. It is evident from the figure that hydrogen atoms neighbouring the H atom in the CHO group (i.e. the three hydrogens of the methyl group) lead to splitting of the original peak of the H atom in the CHO group into a quadruplet; the H atom of the aldehyde group, in turn, causes splitting of the CH_3 band into a doublet. The chemical shift of the multiplet is given by the position of its centre; the distance of the multiplet lines in Hz equals the interaction constant J. The value of this constant does not depend on the intensity of the external field. The lines corresponding to different multiplets can be distinguished by recording the spectrum at two different field strengths.

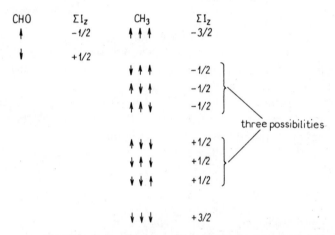

Fig. 13-21. Spin orientation of protons in acetaldehyde.

A group of equivalent protons forms a magnetic field proportional to $\sum_i I_{z,i}$, where $I_{z,i}$ is the z-th component of the spin momentum of the i-th nucleus. The proton spin orientation depicted in Fig. 13-21 for the proton in the CHO group and for the protons of the CH_3 group, is valid. It is evident that there are two fields by which the proton of the CHO group can influence the protons in the CH_3 group; these two fields correspond to two kinds of possible orientation of the nuclear spin. In the specimen of the studied substance these two forms are present in a ratio of $1:1$. This leads to the splitting of the methyl peak into a doublet; both peaks of the doublet, of course, have the same intensity. The formation of the quadruplet and the ratio of the intensities of its peaks, $1:3:3:1$, can be explained in a similar manner.

Remarks on the analysis of NMR spectra. Spin – spin coupling makes NMR spectra very interesting, but very difficult or even impossible to interpret under certain circumstances. This difficulty is decided by the ratio of the chemical shift δ and the interaction constants J (both in Hz). If the chemical shift is comparatively large, compared to the interaction constant, then the spectrum can be interpreted (a first-order spectrum). This is true when the condition

$$\frac{\delta}{J} \geqq 10 \qquad (13\text{-}38)$$

is satisfied. The situation is much less favourable with higher-order spectra, when

$$\frac{\delta}{J} < 10 \qquad (13\text{-}39)$$

Certain rules are valid for the interpretation of spectra of the first order. These rules will be demonstrated on the system $A_m X_n$, which has $m + n$ interacting nuclei. All m nuclei of atom A are magnetically equivalent as are the n nuclei of atom X.

It is important, first of all, that interaction among magnetically equivalent nuclei is not reflected in the NMR spectra. The multiplicity of the peaks of the nuclei of the A atoms depends both on the number of nuclei of atoms X, of which there are n, and on the nuclear spin quantum number, I_A; this multiplicity is given by the expression

$$2nI_A + 1, \qquad (13\text{-}40)$$

so that if the nuclei H^1, C^{13}, F^{19} and P^{31} $(I = \frac{1}{2})$ are responsible for the splitting, the number of lines in the multiplet equals $n + 1$. If the nuclei of atoms A interact with another group of nuclei Y_p (which has p nuclei), then the multiplicity of the peak of nuclei A is given by the product $(2nI_X + 1)(2pI_Y + 1)$.

The multiplet corresponding to the A atoms is symmetrical and is formed by a series of equidistant peaks, the intensity of which is given by the coefficients of the binomial series (provided that $I_X = \frac{1}{2}$):

n^* ratio of the intensities of peaks in the multiplet

1	1 : 1
2	1 : 2 : 1
3	1 : 3 : 3 : 1
4	1 : 4 : 6 : 4 : 1
5	1 : 5 : 10 : 10 : 5 : 1

---

* Number of equivalent protons (or nuclei with spin $\frac{1}{2}$) causing the splitting.

This simple situation becomes complicated when both multiplets lie relatively close together, i.e. when the first-order spectrum changes into a higher-order spectrum.

13.2.5 Electron spin resonance (ESR)

The magnetic moment of an unpaired electron can be oriented in two ways with respect to the external magnetic field, parallel or antiparallel. The corresponding states have different energies; the transitions between them is manifested by the formation of peaks in electron spin (para-magnetic) resonance (ESR or EPR) spectra. As a rule, these spectra have a fine structure, which is usually called a hyperfine structure. This is due to the interaction of an unpaired electron with other particles having a magnetic moment, for example, with protons.

The magnetic moment of the electron is given by the relationship

$$\mu_s = -g\beta_e S, \qquad (13\text{-}41)$$

where g is a constant analogous to the spectroscopic "splitting" factor ($g = 2.0023$ for the free electron), β_e is the Bohr magneton* and S is the spin with value of $\pm\frac{1}{2}\hbar$.

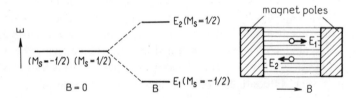

Fig. 13-22. Schematic representation of ESR spectroscopy.

In an external magnetic field of induction B the magnetic moment of the electron is orientated either parallel or antiparallel to this field; to the first state corresponds a lower energy level (E_1), to the second, a higher energy level (E_2). Before application of the magnetic field, the same energy corresponded to both spin states of the electron, which were thus degenerate. The application of the magnetic field results in splitting of the levels (Fig. 13-22).

The energy of an electron is given by the product of the magnetic moment and the induction of the magnetic field:

---

* $\beta_e = eh/4\pi m_e = 9.273 \times 10^{-24}$ A m^2 (e is the elementary charge and m_e the rest mass of the electron).

$$E_1 = -g\beta_e BS_1 \qquad (13\text{-}42a)$$

$$E_2 = -g\beta_e BS_2, \qquad (13\text{-}42b)$$

where S_1 and S_2 correspond to the two possible orientations in the magnetic field. For the excitation energy necessary for the transition of an electron from state E_1 to state E_2 in a magnetic field of induction B, the expression [cf. Eq. (13-29)]

$$E_2 - E_1 = \Delta E = h\nu = g\beta_e B$$

is valid. If, for example, $B = 1\,T$, then resonance occurs in the presence of radiation with a wavelength of about 1 cm. Fig. 13-23 illustrates the way in which the magnitude of the excitation energy ΔE depends on the induction of the external magnetic field. If electromagnetic radiation of a frequency which satisfies the resonance condition is applied to a radical placed in a sufficiently strong magnetic field (for example 0.3 T), a considerable increase in the current passing through the coil in which the specimen is placed results. In practice an arrangement is employed which works with a fixed frequency ν and the specimen is placed in a magnetic field whose induction gradually increases; if the resonance condition is satisfied, a maximum (Fig. 13-24a) appears on the absorption curve. It is customary to record derivative curves; the usual shape of an ESR spectrum without hyperfine structure is depicted in Fig. 13-24b. For organic radicals the value of constant g is approximately the same as for free electrons. With ions of the transition metal elements (where there is an unpaired electron in the d orbital), however, the value of this constant is very different. If the ESR spectra of free radicals or organic

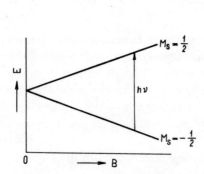

Fig. 13-23. Magnitude of splitting of the energy levels in ESR spectroscopy as a function of the magnetic induction.

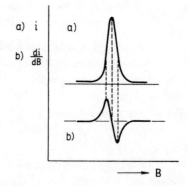

Fig. 13-24. Primary (a) and derivative (b) recording to the ESR spectrum.

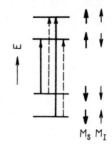

Fig. 13-25. Allowed (————→) and forbidden (– – – →) transitions in the ESR spectrum. M_S and M_I denote the spin quantum numbers of an electron and a proton, respectively.

compounds were not influenced by any other factors, it would be very simple and similar for all the substances and thus uninteresting. In reality, interaction between the magnetic moment of the unpaired electron and that of the nuclei of the atoms, with non-vanishing nuclear spin, occurs and this interaction is manifested in the formation of hyperfine structure in the ESR spectrum. For example, in the methine radical, $C-H$, the nuclear spin of C^{12} is zero and the proton spin quantum number equals $\pm\frac{1}{2}$. The magnetic moment of the proton can also be orientated parallel or antiparallel to the external magnetic field. As a result of the interaction between the magnetic moments of the electron and of the nucleus, the magnetic field in the vicinity of the electron is somewhat changed — it is a little larger or smaller than the magnetic field of the free electron. Both possible magnetic moments of the electron and proton may combine and this leads to four states of different energy (Fig. 13-25). But only some of the transitions between these states are allowed (Fig. 13-25). The following selection rules are valid:

$$\Delta M_S = \pm 1$$

$$\Delta M_I = 0,$$

where M_S and M_I denote the quantum numbers characterizing the z-components of the electron and the nuclear spin, respectively. In Fig. 13-25 the allowed transitions are depicted by full arrows and the forbidden transitions by dashed arrows. Due to this interaction, the ESR spectrum of CH consists of two bands.

The interaction between the unpaired electron and the nucleus may be considered to be the interaction of two small magnets. The expression for the energy of this interaction has two terms. The first is the classical expression for interaction between the dipoles of the electronic and nuclear moments and is non-vanishing only in crystals. The second expression is non-classical and its value is proportional to the square of the electronic wave function at the atomic nucleus.

336

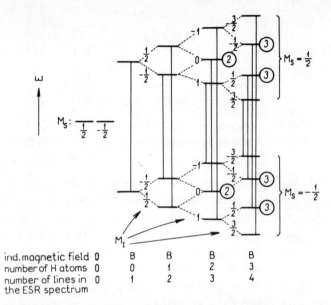

Fig. 13-26. Energy spectrum produced by interaction of an unpaired electron with an increasing number of equivalent protons. Numbers beside the levels (left) denote the nuclear spin quantum number M_I and (right in circles) the degeneracy of the level.

The combination of these two contributions leads to the complexity of the ESR spectra of solid substances.

Let us return to the interaction of an unpaired electron with a certain number of equivalent protons. The interaction of an electron with one proton has already been discussed. The addition of each further proton leads to the splitting of all the existing levels; as the only allowed levels are those which have M_I values equal to a multiple of $\pm\frac{1}{2}$, some levels are degenerate. When the selection rules are taken into account, the selection of the allowed transitions in the energy spectrum becomes a simple task (Fig. 13-26). It is evident that the number of lines is larger by 1 than the number of equivalent protons. The ratio of the intensities is given by the ratio of the number of degenerate levels between which the transitions occur. In this way, the reasons why the ESR spectrum of the CH_2 radical consists of three lines in an intensity ratio of $1:2:1$ and why the spectrum of the (planar) CH_3 radical consists of four lines with intensities of $1:3:3:1$ become apparent. During interaction of an unpaired electron with n equivalent protons, the number of lines generally equals $n + 1$ and the ratio of their intensities is given by the coefficients of the binomial series of order n. Fig. 13-27 schematically depicts the ESR spectrum of the benzene radical anion. The distance between the lines is constant and is termed the hyperfine splitting constant (a_H). This constant is sometimes denoted hfs (*hyperfine splitting*) and is given in teslas.

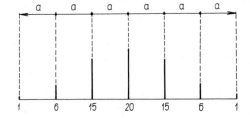

Fig. 13-27. ESR spectrum of benzene radical anion. *a* is the constant of hyperfine splitting (hfs). Numbers below the picture indicate the intensity ratio.

The ESR spectrum becomes much more complicated if protons bonded in non-equivalent positions are present in the studied radical. The naphthalene radical anion with 4 protons in the α positions and 4 protons in the β positions is a good example. In interaction of the unpaired electron with four α protons, five lines (with intensity ratios of $1:4:6:4:1$) are obtained (Fig. 13-28a); each of these five lines is split by the interaction with the four β protons into a series of five lines (the ratio of their intensities is again $1:4:6:4:1$; see Fig. 13-28b); thus from general considerations it can be expected that the ESR spectrum will consist of 25 lines. The experimentally measured spectrum (Fig. 13-29a) confirms this expectation except that it is not as simple as might be expected, as the hfs constants are equal to 0.49 and 0.183 mT, so that overlapping of both the quintets occurs. As the spectrum is most complicated in the centre, its analysis is usually begun at the short or long wavelength parts. In Fig. 13-29b the reconstructed spectrum is represented; the values of the hfs constants given above were utilized here.

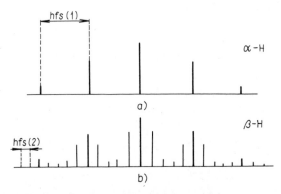

Fig. 13-28 Expected ESR spectrum of the naphthalene radical anion: (*a*) splitting due to α-hydrogens, (*b*) splitting due to β-hydrogens.

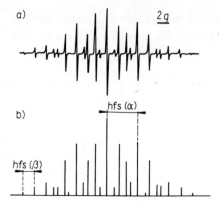

a)

2q

b)

hfs (α)

hfs (β)

Fig. 13-29. Measured ESR spectrum of the naphthalene radical anion (*a*); theoretical ESR spectrum constructed under the assumption that hfs constants are 0.490 mT and 0.183 mT for α and β positions, respectively (*b*).

The hyperfine splitting constant is proportional to the spin density on the hydrogen causing the splitting. The extended methods (EHT, CNDO, and in particular INDO) allow calculation of these densities and there appears to be good correlation between the theoretical and experimental values. As the very numerous radicals and radical ions which have been studied are conjugated systems, the possibility of using the π-electron approximation is worth mentioning. The spin density on the hydrogen appears to be proportional to the electron density of the unpaired electron (i.e. the electron affecting the radical character of the studied system) on the carbon or some other atom to which this hydrogen is bonded. Using the HMO method and Pople's SCF method, calculation of this electron density is very simple: it equals the square of the expansion coefficient in the respective position in the MO occupied by an unpaired electron. In Table 13-4 are given data for a series of conjugated radical anions. The graphical representation of the data is based on McConnell's relationship,

$$a_H = Q\varrho,$$ (13-43)

where a_H is the experimental hyperfine splitting constant (hfs), ϱ is the spin density, approximated successfully by the quantity $c^2_{\mu m}$ ($c_{\mu m}$ is the HMO expansion coefficient, m denotes the singly occupied MO in the radical and μ specifies the atom) and Q is a constant which is approximately equal to 2.5 mT (Fig. 13-30). Several thousand ESR spectra have been interpreted using HMO electron densities; this is considered to be one of the great successes of the simple MO theory, the usefulness of the HMO method is particularly remarkable. The situation in this area should, however, be taken as a warning that the applicability of the one-electron approximation has certain limitations, since the analysis of experimental

Table 13-4

Comparison of Theoretical and Experimental ESR Data

Radical-anion	Position (μ)	$a_\mu/(\text{mT})$	$c^2_{\mu m}$
Naphthalene	1	0.495	0.181
	2	0.183	0.069
Anthracene	1	0.274	0.097
	2	0.151	0.048
	9	0.534	0.193
Tetracene	1	0.155	0.056
	2	0.115	0.034
	5	0.425	0.147
Pyrene	1	0.475	0.136
	2	0.109	0
	4	0.208	0.087
Perylene	1	0.308	0.083
	2	0.046	0.013
	3	0.353	0.108
Coronene	1	0.147	0.056
Diphenyl	2	0.273	0.090
	3	0.043	0.020
	4	0.546	0.158

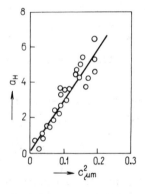

Fig. 13-30. Dependence of the experimental hyperfine splitting constants in ESR spectra on the HMO density of an unpaired π-electron on carbons linked to the corresponding proton.

ESR spectra sometimes tends to lead to negative spin densities. Since the smallest value of the square of the LCAO expansion coefficient is zero, it is obvious that the simple methods then fail qualitatively. If, however, spin densities are calculated using the configuration interaction method, correct results are obtained. McLachlan derived an expression which permits characterization even of negative spin densities using HMO quantities:

$$\varrho_\mu = c^2_{\mu m} + \lambda \sum_v c^2_{vm} \Pi_{\mu v}, \tag{13-44}$$

where $c_{\mu m}$ ($c_{\nu m}$) are the HMO expansion coefficients in the m-th molecular orbital occupied by an unpaired electron in position μ (ν), $\Pi_{\mu\nu}$ is the atom—atom polarizability and λ is a parameter resulting from this theory. This expression has also frequently been successfully applied.

13.2.6 Pure rotational spectra

This type of transition occurs in the far IR region and in the microwave region. Rotational spectra are not yet of great importance in structural determinations in large molecules; they are, however, very important for ascertaining the geometry of small molecules. The analysis of the rotational spectrum permits determination of the moment of inertia of a molecule, i.e., of a quantity defined by the mass and coordinates of the atoms forming the molecule. Only simplest rotators will be discussed here, i.e. diatomic molecules. If we assume that such a molecule behaves as a rigid rotator, the above-described wave functions and their corresponding energies describing the rotation (cf. Section 3.3.4) can be employed.

When a transition occurs from the state with quantum number J'' to the state with quantum number J' (the quantities corresponding to the states with lower and higher energy are denoted by a double prime and a single prime, respectively) the transition energy can be described by the expression

$$\Delta E = E' - E'' = \frac{h^2}{8\pi^2 I} \left[J'(J' + 1) - J''(J'' + 1) \right] \tag{13-45}$$

The value of the transition moment decides whether the transition is allowed (cf. Section 6.5):

$$P_{J'' \to J'} \approx \int Y^*_{J''m}(\Theta, \Phi) \, \mu Y_{J'm}(\Theta, \Phi) \, d\tau, \tag{13-46}$$

where Y_{Jm} are the spherical harmonics and μ is the dipole moment operator. Enumeration of integral (13-46) leads to the very simple selection rule that only those transitions are allowed during which the rotational quantum number changes by one, i.e.

$$J' - J'' = \pm 1 \tag{13-47}$$

J'' can thus be expressed in terms of J' so that

$$\Delta E = \frac{h^2}{8\pi^2 I} 2(J'' + 1) \tag{13-48}$$

On dividing Eq. (13-48) by hc, the wave number expression is obtained:

$$\tilde{v} = \frac{\Delta E}{hc} = \frac{h}{8\pi^2 cI} 2(J'' + 1) = 2B(J'' + 1) \qquad (13\text{-}49)$$

Quantity B is a characteristic constant of the molecule and is called *the rotational constant*. According to Eq. (13-49), the rotational spectrum of a diatomic molecule consists of a series of equidistant lines at a distance of $2B$. This value can be read from the spectrum and can be used for calculation of the moment of inertia,

$$I = \frac{h}{8\pi^2 cB}, \qquad (13\text{-}50)$$

from which the bond length of a diatomic molecule can be calculated:

$$a = \sqrt{\frac{I}{\mu}}, \qquad (13\text{-}51)$$

where μ is the reduced molecular mass.

It follows, however, from analysis of the rotational spectra that the lines are not exactly equidistant, as their distance decreases somewhat with increasing J values. A better description employing the non-rigid rotator approximation interprets the experimental data very well.

13.2.7 Vibrational spectroscopy

The discussion will again begin with the investigation of a diatomic molecule, assuming that its vibrational behaviour can be described by the harmonic oscillator approximation. The solution (the wave functions and the corresponding energies) obtained above for the harmonic oscillator can then be used directly (cf. Section 3.3.3).

First, the expression for the energy can again be written as [cf. Eq. (3-74)]

$$E_n = \frac{h}{2\pi} \left(\sqrt{\frac{k}{m}} \right) \left(n + \frac{1}{2} \right) = \left(n + \frac{1}{2} \right) hv$$

It is evident that the vibrating molecule has an energy of $\frac{1}{2}hv$, $\frac{3}{2}hv$, $\frac{5}{2}hv$ etc. The calculation of the transition moment will be discussed in detail later. Here, it is sufficient to state that, similar to rotational spectra, transitions are only allowed between states whose quantum numbers differ by one:

$$n_1 - n_2 = \Delta n = \pm 1 \qquad (13\text{-}52)$$

Obviously, the distance between the energy levels between which the transition can occur is constant and for this reason only a single line corresponds to the diatomic molecule in the vibrational spectrum, within the framework of the approximation.

The transition moment can be calculated by means of the generally valid expression. During vibration, the dipole moment of a heteronuclear diatomic molecule must, of course, change. If the expression for the dipole moment is expanded into a Taylor series and if we neglect the terms containing higher derivatives, then

$$\boldsymbol{\mu} = \boldsymbol{\mu}_0 + \left(\frac{\partial \boldsymbol{\mu}}{\partial q}\right)_0 q, \tag{13-53}$$

where $\boldsymbol{\mu}_0$ is the permanent dipole moment and q is the vibrational coordinate. If excitation from state i to state j occurs, then the expression

$$\mathbf{R}_{i \to j} = \int \Psi_i^* \left[\boldsymbol{\mu}_0 + \left(\frac{\partial \boldsymbol{\mu}}{\partial q}\right)_0 q \right] \Psi_j \, dq =$$

$$= \int \Psi_i^* \boldsymbol{\mu}_0 \Psi_j \, dq + \int \Psi_i^* \left(\frac{\partial \boldsymbol{\mu}}{\partial q}\right)_0 q \Psi_j \, dq \tag{13-54}$$

can be written. Since $\boldsymbol{\mu}_0$ is a constant and Ψ_i and Ψ_j are orthogonal functions, a non-zero change in the dipole moment during vibration is obviously a necessary condition for the transition to be allowed.

If the vibrations in real molecules were harmonic, a single band in the spectrum would correspond to each vibrational mode. In the spectra, however, further bands (one or two) of smaller intensity and higher wave number occur in addition to these fundamental bands. These are called *higher harmonic transitions*; their wave numbers correspond to transition to states with a higher vibrational quantum number (usually 2 or 3). The anharmonicity of vibrations in the expression for the potential energy in the Hamiltonian must be taken into account when solving the problem for an oscillator. Taylor's expansion of potential energy V can be utilized for this calculation:

$$V = V_0 + \sum_i \left(\frac{\partial V}{\partial q_i}\right)_0 q_i + \frac{1}{2} \sum_{ij} \left(\frac{\partial^2 V}{\partial q_i \, \partial q_j}\right)_0 q_i q_j +$$

$$+ \frac{1}{6} \sum_{ijk} \left(\frac{\partial^3 V}{\partial q_i \, \partial q_j \, \partial q_k}\right)_0 q_i q_j q_k + \text{higher terms} \tag{13-55}$$

The subscript zero appears in expressions referring to the equilibrium state of the molecule, which is, of course, characterized by an energy minimum; the q_i's represent the generalized coordinates and V_0 denotes the potential energy of the equilibrium state of the molecule and is the conventional zero of the energy scale. In the energy minimum the first derivative equals zero. The second derivatives represent the force ($i = j$) and the interaction ($i \neq j$) constants. If the partial derivatives (which are constants) are denoted as b_{ij}, b_{ijk}, which is common in the literature, the relationship

$$V = \frac{1}{2} \sum_{ij} b_{ij} q_i q_j + \frac{1}{6} \sum_{ijk} b_{ijk} q_i q_j q_k + \text{higher terms} \qquad (13\text{-}56)$$

is obtained. For diatomic molecules the expression simplifies to give

$$V = \frac{1}{2} k_1 q^2 + \frac{1}{6} k_2 q^3 + \text{higher terms} \qquad (13\text{-}57)$$

It is apparent that, in the extreme case, expression (13-57) is reduced to the expression which was obtained assuming the validity of Hook's law (cf. Section 3.2.2). Solution of the Schrödinger equation for an oscillator using expression (13-57) for the potential energy leads to eigenfunctions, which, after substitution in the expression for the transition moment, do not result in such strict selection rules. In other words, there are also some allowed transitions in which the change of the vibrational quantum number is not equal to ± 1.

The eigenvalues can be written in simple form:

$$E_n = \left(n + \frac{1}{2} \right) hc\omega_e - \left(n + \frac{1}{2} \right)^2 hc\omega_e X + \text{higher terms} \qquad (13\text{-}58)$$

In Eq. (13-58), X denotes the anharmonicity constant and ω_e is the equilibrium frequency of the diatomic molecule. This expression permits calculation of the wave numbers of the fundamental and higher harmonic vibrations of a diatomic molecule.

For a molecule containing N atoms, the relationships are rather more complicated. Here the procedure for calculating the fundamental frequencies will only be outlined. We are chiefly interested in calculation of the frequencies in connection with interpretation of infrared spectra and also for the purposes of statistical thermodynamics. In the expressions for the equilibrium and for the velocity constants, the vibrational partition function appears, among other factors. Of $3N$ degrees of freedom of the N-atomic molecule, 3 correspond to translation and 3 to rotation of the molecule (2 for a linear molecule). The remaining $3N - 6$ ($3N - 5$) degrees of freedom correspond to vibrations. If the vibrations of the atoms correspond to harmonic oscillations and if all the atoms vibrate with the same frequency and in phase (in other words, if the all atoms pass through the equilibrium position simultaneously), *normal modes of vibration* are involved. Their calculation requires knowledge of the interatomic forces (described by the force constants) and of the configuration of the molecule (valence angles and bond lengths). Wave number $\tilde{v}$ (called the fundamental frequency here) is calculated from quantities λ by means of the relationship

$$\lambda = 4\pi^2 c^2 \tilde{v}^2 \tag{13-59}$$

Quantities λ are obtained by solving the determinant equation

$$\begin{vmatrix} \Sigma G_{1i}F_{i1} - \lambda, & \Sigma G_{1j}F_{j2}, & \ldots \Sigma G_{1k}F_{kn} \\ \Sigma G_{2i}F_{i1}, & \Sigma G_{2j}F_{j2} - \lambda, \ldots \Sigma G_{2k}F_{kn} \\ \vdots & \vdots & \vdots \\ \Sigma G_{ni}F_{i1}, & \Sigma G_{nj}F_{j2}, & \ldots \Sigma G_{nk}F_{kn} - \lambda \end{vmatrix} = 0 \tag{13-60}$$

The indicated summations refer to the internal indices. F_{ij} and G_{ij} from Eq. (13-60) are elements of the potential and the kinetic energy matrices (**F** and **G**), defined as follows:

$$\mathbf{F} = \mathbf{U}\mathbf{f}\mathbf{U}^T, \tag{13-61}$$

$$\mathbf{G} = \mathbf{U}\mathbf{g}\mathbf{U}^T, \tag{13-62}$$

where **U** is the transformation matrix between the symmetry and internal coordinates, **f** is the force constant matrix and **g** is the kinetic energy matrix (including masses of the atoms and configuration of the molecule).

For enumeration of the symmetry coordinates, it is necessary to know the internal coordinates of the molecule, r_k (which are of four types and describe changes in the bond length, bond angle and out-of-plane deformations and torsion angles). The symmetry coordinates S_i are given by linear combinations of the internal coordinates:

$$S_i = \sum_k U_{ik}r_k \tag{13-63}$$

Transformation matrix **U** is unitary (cf. Section 4.5). In solving Eq. (13-60), enumeration of the elements of the **F** and **G** matrices does not lead to particular difficulties. The diagonal elements of the matrix **f**, namely f_{ii}, are the force constants of the individual bonds and angles which occur in the studied molecule, and the non-diagonal elements, f_{ij}, are the so-called interaction constants.

Spectroscopy in the infrared region is vibrational rotational spectroscopy. It is of extraordinary importance for experimental chemistry and, until NMR was discovered, it was the only almost universally applicable and unusually effective method for structure determinations. Although the measured vibrational states are, of course, the result of the conditions in the entire molecule, the individual functional groups have *characteristic frequencies* in regions which often are not too dependent on the structure of the remaining part of the molecule. The utilization of IR spectroscopy for structure determination is based on this observation. A great deal of attention has been paid to this subject in the literature.

13.2.8 Raman spectroscopy

For a certain vibration to be active in the Raman spectrum, a change in the induced dipole moment must occur as a result of excitation. The condition for this change is a change in the polarizability of the molecule.

The induced dipole moment μ is proportional to the electric field vector E:

$$\mu = \alpha E \tag{13-64}$$

Proportionality constant α is called the polarizability. Equation (13-64) is valid only for spherically symmetrical systems, such as atoms. Vector equation (13-64) can also be written for the individual components:

$$\mu_x = \alpha E_x$$
$$\mu_y = \alpha E_y \tag{13-65}$$
$$\mu_z = \alpha E_z$$

If a molecule in the vibrational ground state accepts an amount of energy which transfers it into a state which is unstable at laboratory temperature, the return to the ground state is often connected with Rayleigh scattering; the energy of the scattered photon is the same as the energy of the absorbed photon. The excited molecule can, however, also pass into one of the higher (excited) vibrational states. Fig. 13-31a indicates a transition into a vibrational state with quantum number $v = 1$ and also excitation of a molecule from this vibrational state into a metastable state and transfer back into the vibrational ground state. In the first case the energy of the scattered photon is obviously smaller and in the second case greater than the energy of the exciting radiation by a value exactly equal to the difference between the energies of the ground and the first excited (generally higher) vibrational states.

In Raman spectroscopy, energy corresponding to the mercury line with a wave number of 22,945 cm^{-1} can be used for excitation; in the scattered radiation, in Raman spectra, lines of smaller and larger wave number occur, which are called Stokes and anti-Stokes lines; their energy differences (which possess the same value for both types of lines) represent the vibrational spectrum of the studied molecule (Fig. 13-31b). Since the

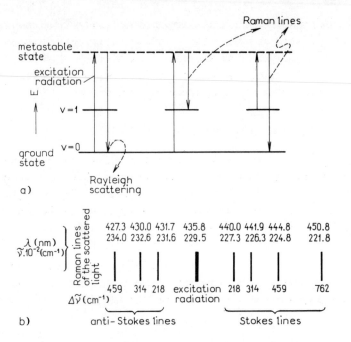

Fig. 13-31. (*a*) Origin of the Rayleigh scattering and of Raman lines. (*b*) Raman spectrum.

vibrational state with $v = 1$ is not very strongly populated at laboratory temperature, the intensity of the Stokes lines (see below) is usually greater than the intensity of the anti-Stokes lines. For spherically asymmetrical molecules the polarizability is a function of the direction. In the general case it holds for the components of the induced dipole moment that

$$\mu_x = \alpha_{xx}E_x + \alpha_{xy}E_y + \alpha_{xz}E_z$$
$$\mu_y = \alpha_{yx}E_x + \alpha_{yy}E_y + \alpha_{yz}E_z \qquad (13\text{-}66)$$
$$\mu_z = \alpha_{zx}E_x + \alpha_{zy}E_y + \alpha_{zz}E_z$$

The total polarizability is represented by the matrix

$$\boldsymbol{\alpha} \equiv \begin{Vmatrix} \alpha_{xx} & \alpha_{xy} & \alpha_{xz} \\ \alpha_{yx} & \alpha_{yy} & \alpha_{yz} \\ \alpha_{zx} & \alpha_{zy} & \alpha_{zz} \end{Vmatrix} \qquad (13\text{-}67)$$

This matrix is symmetrical ($\alpha_{zx} = \alpha_{xz}$ etc.). The matrix which mediates the linear relationship between two vectors is called *a tensor*.

The intensity of the Raman scattering and also the selection rules are given by matrix elements of the type

$$\int (\Psi_1^v)^* \, \boldsymbol{\alpha}_{ij} \Psi_2^v \, d\tau, \qquad (13\text{-}68)$$

where Ψ_n^v is the vibrational wave function of the n-th state. The intensity of the Raman transition, A, is proportional to the square of the matrix elements of the polarizability:

$$A \sim \left| \int (\Psi_1^v)^* \, \alpha_{ij} \Psi_2^v \, d\tau \right|^2 \tag{13-69}$$

In conclusion, it should be mentioned that, in addition to vibrational spectra, there are also Raman rotational spectra.

13.3 Excitation within the framework of several electronic levels

13.3.1 Absorption spectra in the ultraviolet, and visible regions

The energy gap between the ground state and the first electronically excited state is usually greater by one-half to two orders of magnitude than the gap between the vibrational levels. The great majority of molecules whose spectra will be of interest possess a singlet ground state. If the molecule accepts an amount of energy corresponding to the gap between the ground and the excited state, this process can be interpreted within the one-electron approximation as transition of the electron from one of the MO's occupied in the ground state into an unoccupied MO. The electronic excitation requiring the least energy is depicted in Fig. 13-32. In addition to the orbital one-electron representation of electronic transitions, this figure also depicts a better representation of the ground state

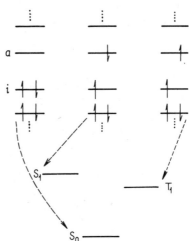

Fig. 13-32. Schematic representation of transitions with the lowest energy demands.

and the excited states by means of a term scheme. The levels in this scheme correspond to the energies of n-electron functions of the individual states. The connection between the orbital and term representation is indicated by dotted lines.

The approach based on the orbital scheme is justified in molecules with a variety of structural types. Therefore, expressions valid within the SCF theory for transitions of the system from the ground state to various excited states will be discussed first, followed by the physical chemical aspect of electronic spectroscopy and its application to compounds of different structural types.

For the excitation energy of transitions indicated in Fig. 13-32, the expression

$$^1\Delta E_{i \to a} = {}^1E_{i \to a} - E_0 \tag{13-70}$$

$$^3\Delta E_{i \to a} = {}^3E_{i \to a} - E_0 \tag{13-71}$$

can be written for *excitation from the i-th to the a-th MO*. When using the term scheme, the relationships

$$^1\Delta E = {}^1E_1 - {}^1E_0 \tag{13-72}$$

$$^3\Delta E = {}^3E_1 - {}^1E_0 \tag{13-73}$$

can be written. Within the orbital description we can enumerate Eqs. (13-70) and (13-71) either at the SCF level (Roothaan, Pople) or at the level of simple MO methods. If we choose the SCF approximation, we must calculate the energy of the corresponding Slater determinant of the ground state $^1E_0$ and of the determinant (or the linear combination of two determinants) of the excited state ($^1E_1$ or $^3E_1$) and find the difference between these two energies. The expression for the energy of a closed shell system has already been derived (in Section 5.5):

$$^1E_0 = 2\sum_i H_i + \sum_i \sum_j (2J_{ij} - K_{ij}),$$

where J_{ij} is the Coulomb integral and K_{ij} is the exchange integral.

For an excited state configuration, the possibility of transition of an electron with a spin function α or β must be considered (Fig. 13-33). The corresponding determinants have the following form:

$$\Delta_A = \left| \varphi_1, \bar{\varphi}_1, ..., \varphi_i, ..., \varphi_{n/2}, \bar{\varphi}_{n/2}, \bar{\varphi}_a \right| \tag{13-74}$$

$$\Delta_B = \left| \varphi_1, \bar{\varphi}_1, ..., \bar{\varphi}_i, ..., \varphi_{n/2}, \bar{\varphi}_{n/2}, \varphi_a \right| \tag{13-75}$$

$$\Delta_C = \left| \varphi_1, \bar{\varphi}_1, ..., \varphi_i, ..., \varphi_{n/2}, \bar{\varphi}_{n/2}, \varphi_a \right| \tag{13-76}$$

$$\Delta_D = \left| \varphi_1, \bar{\varphi}_1, ..., \bar{\varphi}_i, ..., \varphi_{n/2}, \bar{\varphi}_{n/2}, \bar{\varphi}_a \right| \tag{13-77}$$

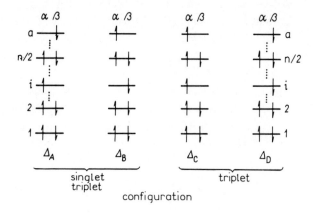

Fig. 13-33. Schematic representation of singlet and triplet configurations.

The absence of a bar indicates occupation of the orbital by an electron with spin α and a bar refers to occupation by an electron with spin β.

Investigation of determinants (13-74) to (13-77) shows that they correspond to the following eigenvalues of operator $\mathscr{S}_z$: 0, 0, 1, -1. From the pair of functions (13-74) and (13-75) (Δ_A, Δ_B), linear combinations (cf. Section 6.7) can be formed:

$$\Delta^- = \frac{1}{\sqrt{2}}(\Delta_A - \Delta_B) \tag{13-78}$$

$$\Delta^+ = \frac{1}{\sqrt{2}}(\Delta_A + \Delta_B) \tag{13-79}$$

($1/\sqrt{2}$ is the normalization factor).

Function Δ^- corresponds to the singlet state and functions Δ^+, Δ_C, Δ_D to the triplet state, as is evident from the eigenvalues of the operator $\mathscr{S}^2$:

$$\mathscr{S}^2 \Delta^- = 0 \qquad \mathscr{S}^2 \Delta_C = \frac{2h^2}{4\pi^2} \Delta_C$$

$$\mathscr{S}^2 \Delta^+ = \frac{2h^2}{4\pi^2} \Delta^+ \qquad \mathscr{S}^2 \Delta_D = \frac{2h^2}{4\pi^2} \Delta_D$$

The wave function of the excited singlet state formed by the $i \rightarrow a$ transition is thus described by two determinants, similarly as with one of the three wave functions of the triplet state; one determinant to each of the two remaining functions belongs:

$$^1\Psi_{i \rightarrow a} = (\Delta_A - \Delta_B)/\sqrt{2} \tag{13-80}$$

$$^3\Psi_{i \to a} = (\Delta_A + \Delta_B)/\sqrt{2} \qquad (13\text{-}81)$$

$$^3\Psi_{i \to a} = \Delta_C \qquad (13\text{-}82)$$

$$^3\Psi_{i \to a} = \Delta_D \qquad (13\text{-}83)$$

The calculation of the energy corresponding to the Slater determinant wave function has already been discussed in Section 5.4; the energies corresponding to the wave functions of triplet states are identical and generally lower than the energies of the singlet state. For $^1\Delta E_{i \to a}$ it is possible to write

$$^1\Delta E_{i \to a} = \tfrac{1}{2} \int (\Delta_A - \Delta_B)\, \mathscr{H}(\Delta_A - \Delta_B)\, d\tau - \int \Delta_0 \mathscr{H} \Delta_0 \, d\tau, \quad (13\text{-}84)$$

where Δ_0 is the Slater determinant of the ground state. Calculation of the energy corresponding to a linear combination of the determinants leads to the expression

$$^1E_{i \to a} = 2\sum_{j \neq i} H_j + \sum_j \sum_{k \neq i}(2J_{jk} - K_{jk}) + H_i + \sum_{j \neq i}(2J_{ij} - K_{ij}) +$$
$$+ H_a + \sum_{j \neq a}(2J_{ju} - K_{ja}) + J_{ia} + K_{ia} \qquad (13\text{-}85)$$

In agreement with Eq. (13-84), the expression for E_0 [cf. Eq. (5-60)] must be subtracted from this expression. To facilitate this process, those terms which allow completion of the summations are added to Eq. (13-85) (these terms must, of course, also be subtracted so that the value of the expression remains unchanged). Thus, the relationship

$$^1E_{i \to a} = 2\sum_j H_j + \sum_j \sum_k (2J_{jk} - K_{jk}) - H_i - \sum_j (2J_{ij} - K_{ij}) + H_a +$$
$$+ \sum_j (2J_{ja} - K_{ja}) - (J_{ia} - K_{ia}) + K_{ia} \qquad (13\text{-}86)$$

is obtained.

The energy of the excited triplet state can be calculated employing a similar procedure. For the SCF energy of the singlet-singlet (S−S) and singlet-triplet (S−T) transitions the relationships

$$^1\Delta E_{i \to a} = \varepsilon_a - \varepsilon_i - J_{ia} + 2K_{ia} \qquad (13\text{-}87)$$

$$^3\Delta E_{i \to a} = \varepsilon_a - \varepsilon_i - J_{ia} \qquad (13\text{-}88)$$

are obtained, where ε_a (ε_i) is the orbital energy of the a-th (i-th) MO and J_{ia} and K_{ia} denote the Coulomb and exchange integrals, respectively.

As K_{ia} is a positive quantity, in agreement with Hund's rule the S−S transition requires greater energy than the S−T transition; in the extreme case ($K_{ia} = 0$) the energy of both transitions would be the same. The magnitude of the S−T splitting is given by the difference between Eqs. (13-87) and (13-88):

$$^1\Delta E_{i\to a} - {}^3\Delta E_{i\to a} = 2K_{ia} \qquad (13\text{-}89)$$

Bearing in mind the definition of the K_{ia} integral,

$$K_{ia} = \iint \varphi_i^*(1)\,\varphi_a(1)\,g(1,2)\,\varphi_i^*(2)\,\varphi_a(2)\,d\tau_1\,d\tau_2,$$

the condition which must be fulfilled for the splitting to be large or small then follows. Obviously, the larger the space in which both orbitals φ_i and φ_a simultaneously expand significantly, the greater the value of K_{ia} and the greater the splitting. The opposite case is well illustrated by an example from molecular spectroscopy: in the $n\to\pi^*$ transitions (see below) the corresponding integral has a very small value, because the non-bonding atomic orbital and the antibonding π molecular orbital often occupy different regions in space (Fig. 13-34). It has also been ascertained experimentally that the S−T splitting is very small for $n\to\pi^*$ transitions.

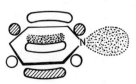

Fig. 13-34. An example of a system with a low value of exchange integral K (K_{nn^*} in pyridine).

The expression for the excitation energy within the empirical MO methods can easily be obtained from Eqs. (13-87) and (13-88) by dropping the terms for the electron repulsion, i.e. integrals J_{ia} and K_{ia}. We thus arrive at the expression given above; at the same time it is immediately evident that, if the repulsion is neglected, there is no difference between the singlet and triplet transitions:

$$^1\Delta E_{i\to a} = {}^3\Delta E_{i\to a} = E_a - E_i, \qquad (13\text{-}90)$$

where E_a and E_i are the HMO or EHT energies of the orbitals between which the transition occurs. For HMO data, taking into consideration the form of the expression for the orbital energy ($E_j = \alpha + k_j\beta$), it can then be written that

$$^1\Delta E_{i\to a} = {}^3\Delta E_{i\to a} = (k_a - k_i)\,\beta \qquad (13\text{-}91)$$

Equality of the energy values for S→S and S→T transitions is a qualitative shortcoming of theories which do not explicitly take the electron repulsion into account.

Classification of transitions. In the near UV region (200 to 400 nm) and the visible region (400 to 800 nm), two chief groups of substances manifest significant absorption. They are organic conjugated compounds and complex compounds which contain a transition element ion as the central atom. With compounds of the first type, the absorption of light is mainly caused by transitions between the bonding (π) and antibonding (π^*) MO's, as well as by transitions between nonbonding AO's (e.g. the lone pair on the nitrogen atom in pyridine) and π^* orbitals. These are termed $\pi \to \pi^*$ (or N$\to$V) and n$\to \pi^*$ (or N$\to$Q) transitions (Fig. 13-35a). These two types can be distinguished experimentally, for example, by studying the influence of the permittivity of the solvent on the position of the band. Whereas an increase in the relative permittivity (for example, on changing from hexane to water) causes a relatively large shift to shorter wavelengths (the hypsochromic shift, 20 to 40 nm) of the n$\to \pi^*$ band, in the $\pi \to \pi^*$ bands it induces a small bathochromic shift to longer wavelengths (3 to 10 nm). In addition, the intensity of the n$\to \pi^*$ bands is usually smaller ($\varepsilon \simeq 10$ to $1000 \ \mathrm{l \ mol^{-1} \ cm^{-1}}$) than the intensity of the $\pi \to \pi^*$ bands ($\varepsilon \simeq 500$ to $100\,000 \ \mathrm{l \ mol^{-1} \ cm^{-1}}$).

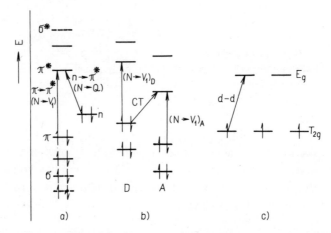

Fig. 13-35. Graphical representation of $\pi \to \pi^*$ and n$\to \pi^*$ transitions in conjugated compound (a). Intermolecular excitation: transition of an electron from the donor HOMO (D) to the acceptor LFMO (A), charge-transfer (CT) band (b). Transition from the T_{2g} orbital of the transition element atom to the E_g orbital. Splitting of the degenerate d levels is due to a ligand field of octahedral symmetry; d–d transition (c).

Interaction of a substance having a low ionization potential (for example, anthracene) with a substance of a high electron affinity (for example, chloranil) is frequently connected with intermolecular excitation, which is manifested by an absorption band at relatively long wavelengths which has no analogy in the spectra of the two components: this is termed a (intermolecular) "charge-transfer" band (Fig. 13-35b). These bands are usually of relatively low intensity ($\varepsilon \simeq 100 \ \mathrm{l \, mol^{-1} \, cm^{-1}}$), flat and without vibrational structure.

The absorption in the visible region by the complexes of transition metal ions (for example, Fe^{2+}, Co^{3+}) with different ligands (for example, H_2O, Cl^-) is dependent on transitions between the incompletely occupied d orbitals of the central ion, the degeneracy of which has been removed in part by the influence of interaction with ligands. These d$-$d transitions are usually of low intensity (Fig. 13-35c).

The value of the transition moment (cf. Section 6.5)

$$\mathbf{Q}_{i \to j} = \int \Psi_{i \to j} (\sum_{\mu} \imath_{\mu}) \, \Psi_0 \, d\tau \qquad (13\text{-}92)$$

determines whether the transition is allowed theoretically. In Eq. (13-92) $\Psi_{i \to j}$ and Ψ_0 denote the wave functions of the excited and ground states. $\sum \imath_{\mu}$ is obviously the operator of the dipole moment for a unit charge and $\mathbf{r}_{\mu}$ ($\mu = 1, 2, ..., n$) are the position vectors of the electrons. The direction of the transition moment $\mathbf{Q}_{i \to j}$ is called *the direction of polarization* of the electronic transition $i \to j$.

If the mentioned integral vanishes, the transition is termed forbidden; otherwise it is called allowed. In symmetrical molecules it is possible to decide whether the transition is allowed or forbidden simply from knowledge of the symmetry of the orbitals (or states) between which the transition occurs. The rules derived from this analysis are called selection rules. For comparison of the theoretical values with the experimental ones, a quantity called the oscillator strength f is introduced, which is proportional to the square of the transition moment:

$$f_{i \to j} = 1.085 \times 10^{-3} \tilde{v} \left(\frac{|\mathbf{Q}_{i \to j}|}{e} \right)^2, \qquad (13\text{-}93)$$

where $\tilde{v}$ is the wave number in $\mathrm{cm^{-1}}$ of the absorption maximum and Q/e is expressed in nm. The oscillator strength can also be calculated from experimental data using the expression [cf. Eq. (13-11)]

$$f^{(\mathrm{exp})} = 4.319 \times 10^{-9} \int_{\tilde{v}_1}^{\tilde{v}_2} \varepsilon \, d\tilde{v}, \qquad (13\text{-}94)$$

where the integral corresponds to the area of the absorption curve in ε ($\mathrm{l \, mol^{-1} \, cm^{-1}}$) and $\tilde{v}$ ($\mathrm{cm^{-1}}$) coordinates.

The situation is, in reality, far more complicated as several vibrational states (corresponding to functions $\chi_{i,v}$, where the first index denotes the electronic state and the second the vibrational state) are associated with each electronic state (corresponding to wave function Ψ_i). If excitation

354

occurs from the electronic (Ψ_0) and the vibrational (χ_{00}) ground states, then the expression

$$\mathbf{Q}_{00 \to 1k} = \mathbf{Q}_{0 \to 1} S_{00 \to 1k} \qquad (13\text{-}95)$$

is valid for the transition moment of the vibronic transition (a transition which is simultaneously electronic and vibrational), where

$$S_{00 \to 1k} = \int \chi_{00}^* \chi_{1k} \, d\tau \qquad (13\text{-}96)$$

is the overlap integral and χ_{1k} denotes the wave function of the k-th vibrational state in the first excited electronic state and the integration is performed over the electron coordinates. The square of the overlap integral (13-96) then appears in the expression for the oscillator strength. In expression (13-96), functions χ_{00} and χ_{1k} are not orthogonal because they do not pertain to the same electronic Hamiltonian.

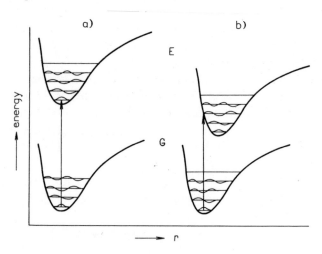

Fig. 13-36. Vibrational wave functions superimposed on potential energy curves. Equilibrium distance (r) is the same as in the ground state (a) or larger (b). In case (a) the $0-0$ vibrational transition will be most intense, in case (b) it will be the $0-2$ transition.

Equation (13-95) is a representation of the Franck-Condon principle. Analysis of the vibrational wave functions depicted in Fig. 13-36 aids in its clarification. Strictly speaking, this figure refers to a diatomic molecule; it appears, nevertheless, that the situation in polyatomic molecules is similar. The time required for excitation of an electron is short in comparison with the time required for changing the coordinates of the nuclei (the time of vibration) and thus immediately after the electronic excitation the molecule has the same geometry as before the excitation. If the minima of both potential curves (G, E) correspond to the same

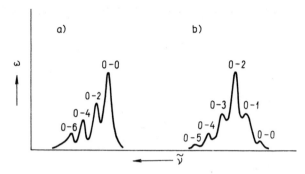

Fig. 13-37. Vibrational structure of the electronic band (for the corresponding transitions see Fig. 13-36). In case (*a*) the vertical transition is the $0-0$ transition, in case (*b*) it is the $0-2$ transition.

interatomic distance, then *the vertical transition* (or *the Franck-Condon transition*, the transition with unchanged bond length) is identical with the $0-0$ vibrational transition. It is clear from Fig. 13-36a that the overlap integral (13-96) has a maximum value for the $0-0$ transition and the corresponding absorption curve assumes the form demonstrated in Fig. 13-37a. In turn, when the minimum of the potential energy curve E lies at a greater interatomic distance, then the vibration band with the longest wavelength ($0-0$ transition) is not the most intense. The most intense band is that corresponding to the vertical transition for which the overlap integral (13-96) acquires maximum values (Fig. 13-37b).

In connection with the potential energy curves for the ground and excited states, a phenomenon called *predissociation*, which appears in the spectra of some polyatomic molecules, can be mentioned. It appears that, in a certain wavelength region, the rotational or vibrational band structure (the rotational structure of electronic bands is, of course, only detectable using an instrument with high resolving power) disappears in the absorption spectrum of a molecule in the gaseous phase and an absorption continuum results. This phenomenon is connected with the crossing of the potential curves of excited states, which occurs very frequently with large molecules. Fig. 13-38 indicates one of the possible situations which can occur during predissociation. The $G \rightarrow E_1$ transition is allowed and the $G \rightarrow E_2$ transition is forbidden. After the $G \rightarrow E_1$ transition the vibrating molecule can assume the configuration corresponding to point P, which is common to both the potential curves of the excited states. Because there is a finite probability of transition from state E_1 to state E_2 at this point, dissociation of the molecule can obviously occur, as the dissociation energy in the E_2 state is (presumed to be) considerably lower than in the E_1 state. This transition occurs in a shorter time than that required for one rotation or vibration, so that these motions cease to be quantized and a continuum appears on the absorption curve. In the case demonstrated in Fig. 13-38, the condition for formation of a continuum is that the energy used for the excitation should be equal to or greater than the energy represented in the figure by the line segment.

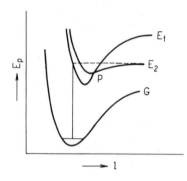

Fig. 13-38. Predissociation caused by crossing of the potential energy curves of the 1st and 2nd excited states (E_1, E_2).

S − T and T − T transitions. The absorption bands in electronic spectra generally correspond to singlet-singlet (S − S) transitions. S − T transitions (between the singlet ground state and the excited triplet state) are forbidden, because the expression $\int \alpha\beta\, d\sigma$, which has zero value (orthogonality of spin functions), appears in the spin part of the transition moment integral. The spin multiplicity rule is rigorous, so that S − T transitions are not observed under normal conditions. Only when the spin quantization is disturbed do these transitions become allowed. This can be achieved either by adding paramagnetic substances to the solution or by introducing a heavy atom into the vicinity of the substance whose spectrum is being measured. Among the available paramagnetic substances, radicals (for example NO) and also biradicals (for example O_2) are frequently used. Either a heavy atom can be introduced directly into the studied substance as a substituent (for example, by transforming naphthalene into 1-iodonaphthalene), or the spectrum of the studied substance can be measured in a solvent containing a heavy atom (for example, in ethyl iodide). This is termed the internal or external heavy atom effect.

When a heavy atom is present the spin quantization is disturbed by spin-orbit coupling. When describing this interaction, a term must appear in the Hamiltonian resulting from relativistic quantum mechanics and has the form [cf. Eq. (4-88)]

$$\mathscr{H}_{(SL)} = \zeta \mathscr{L} \mathscr{S},\qquad(13\text{-}97)$$

where $\mathscr{L}$ and $\mathscr{S}$ are the angular momentum operators of the orbital motion and spin, respectively, and quantity ξ is connected with the type of potential field in which the electron moves. It suffices here to investigate the matrix elements of the corresponding operator (13-97) using simple AO's:

$$\chi_{n,l,m,s} = R(n,\,l)\,\Phi(r'), \tag{13-98}$$

where R is the radial and Φ is the angular part of the AO and r' denotes angular momentum quantum numbers. The matrix elements can be factorized:

$$\langle \chi_{n,l,m,s} \,|\, \mathscr{H}_{(SL)} \,|\, \chi_{n,l,m,s} \rangle = \langle R \,|\, \xi \,|\, R \rangle \langle \Phi \,|\, \mathscr{L}\mathscr{S} \,|\, \Phi \rangle \tag{13-99}$$

The most important contribution appears to come from the first term and has the following form:

$$\xi_{n,l} \approx \frac{Z^4}{n^3 l(l+1)(l+1/2)}, \tag{13-100}$$

where n and l are the quantum numbers and Z is the atomic number. From this discussion, the significance of spin-orbit coupling with heavy atoms becomes apparent.

In contrast to $S-T$ transitions, $T-T$ transitions are spin allowed. These transitions can be measured as follows: very intense radiation is employed to establish a relatively high concentration of the first excited triplet state in the studied substance specimen (by means of the $S_1 \rightarrow T_1$ intersystem crossing) and the measurement of absorption spectrum is then carried out in the usual way. Since the excitation energy for many $T-T$ transitions is usually smaller than the energy of the first $S-S$ transitions, identification of the longwave $T-T$ transitions causes no difficulties. It ought to be added that development of flash photolysis in the pico-second region has recently enabled measurement of spectral transitions between the first excited singlet state and higher excited single states ($S_1 \rightarrow S_x$ transition).

Electronic spectra and their interpretation in compounds of different structural types will form the subject of the following section. Although we are, in principle, interested only in molecular spectroscopy, it will be useful in connection with the transition metal complexes to begin with atomic spectroscopy.

Selection rules for atoms are easily formulated. The following selection rules must be fulfilled (in addition to satisfying the condition $\Delta E = h\nu$) for the transition to occur:

$$\Delta J = 0 \text{ or } \pm 1; \qquad \Delta M_J = 0 \text{ or } \pm 1$$

(for definition of the symbols, see Chapter 8).

Provided that there is no perturbing influence of the external field and that the excitation can be described by transition of an electron from one AO to another AO—and this is frequently permitted—the following condition must be fulfilled:

$$\Delta l = \pm 1$$

For illustration, transitions in the sodium atom $(1s^2\,2s^2\,2p^6\,3s^1)$ will be discussed. Transitions of the 3s valence electron are most important for optical spectroscopy. The quantum number l can acquire values of 0, 1, 2, corresponding to the s, p and d orbitals. The states of the atom as a whole are therefore S, P and D, respectively. In order to arrive at symbols for the states, numbers L must be vectorially summed with spin numbers S:

L	0	1	2
S	$\frac{1}{2}$	$\frac{1}{2}$	$\frac{1}{2}$
J	$\frac{1}{2}$	$\frac{1}{2}, \frac{3}{2}$	$\frac{3}{2}, \frac{5}{2}$
$2S + 1$	2	2	2
Symbol of state	$^2S_{1/2}$	$^2P_{1/2}, {}^2P_{3/2}$	$^2D_{3/2}, {}^2D_{5/2}$

In Fig. 13-39 are depicted the electronic transitions in the sodium atom.

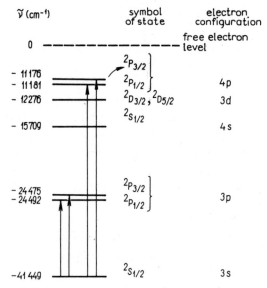

Fig. 13-39. The excitation energies of allowed and forbidden transitions are given by differences in the energies of the individual states and the energy of the $Na^+ + e$ system, set equal to zero. Allowed transitions are indicated by arrows.

Table 13-5

Colour of Hydrates of Ions $[Me(H_2O)_6]^{n+}$ in Aqueous Solutions of Sulphates and Perchlorates of Transition Elements

Number of d electrons	Number of unpaired electrons	Metallic ion	Colour of the solution
0	0	K^+, Ca^{2+}, Sc^{3+}	colourless
1	1	Ti^{3+}	pink-violet
2	2	V^{3+}	green
3	3	Cr^{3+}	violet
4	4	Cr^{2+}	blue
5	5	Mn^{2+}	pale pink
6	4	Fe^{2+}	green
7	3	Co^{2+}	pink
8	2	Ni^{2+}	green
9	1	Cu^{2+}	blue
10	0	Cu^+, Zn^{2+}, Ga^{3+}	colourless

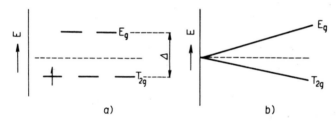

Fig. 13-40. Electronic transition in an octahedral complex of an ion with one d electron. (a) Energy of one-electron functions ($l = 2$, $s = \frac{1}{2}$), (b) the term scheme.

The interpretation of the electronic spectra of complexes provides an important task for quantum chemistry; in addition, the strength of the ligand field can be determined from spectral data. It is evident from Table 13-5 that the octahedral complexes of transition metal ions absorb in different parts of the visible region.

In systems where only one d electron on the central atom is exposed to the field of octahedral symmetry (Fig. 13-40), the relations are very simple. The $Ti(H_2O)_6^{3+}$ complex is an example of a d^1 complex; in d^9 complexes (for example Cu^{2+}) the relationships are very similar. The Cu^{2+} spectrum is somewhat more complicated than the very simple Ti^{3+} spectrum, as a significant splitting of the levels results due to the Jahn-Teller effect.

Fig. 13-41 demonstrates the absorption curve of the $Ti(H_2O)_6^{3+}$ complex with one band at $20\,000\ cm^{-1}$ in the visible region. The model

360

of this complex can be treated better using ligand field theory, which was established by combination of the crystal field theory with MO theory. In addition to the electrostatic ion-ligand interaction, it also takes the covalent components of the chemical bond into account (cf. Section 10.6.3).

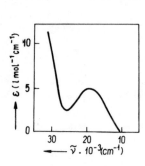

Fig. 13-41. Electronic absorption spectrum of $Ti(H_2O)_6^{3+}$.

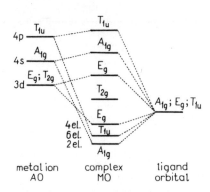

Fig. 13-42. Correlation diagram: formation of the molecular orbitals of a complex from the atomic orbitals of the metal ion and the orbitals of the ligand. 12 electrons are located in the molecular orbitals.

To a first approximation, our discussion will be confined to assuming one non-bonding orbital with two electrons in each of the 6 water molecules (ligands) and five 3d, one 4s and three 4p orbitals. Combining these 15 orbitals yields 15 MO's. The symmetries of the individual orbitals (cf. Section 6.6) are given in Fig. 13-42. The T orbitals are threefold degenerate and the E orbitals twofold. Twelve electrons forming six σ bonds occupy the following MO's: A_{1g} (2 el.), T_{1u} (6 el.) and E_g (4 el.), the thirteenth electron (d electron) is in the non-bonding T_{2g} MO; on excitation it passes into the anti-bonding E_g MO. The relationships in the d^9 system are equally simple. In the other cases the situation is more complicated and thus the discussion will be confined to a number of generalizations:

  a) The Δ value (energy difference between the T_{2g} and E_g levels) is similar in complexes of a given ligand with ions of elements of the same series in the periodic table and of the same valence.

  b) The Δ value increases rapidly with increasing valence of the metal ion [$\Delta(Me^{2+}) \approx 20\,000$ cm^{-1}, $\Delta(Me^{3+}) \approx 30\,000$ cm^{-1}].

  c) The Δ value increases (in the corresponding complexes) by about 30% on going from complexes of metals of the first transition series to complexes of the second series.

d) Ligands can be arranged in a series according to the relative magnitude of the Δ values which they produce. This sequence does not depend on the ion studied (spectrochemical series):

$$I^-, Br^-, Cl^-, F^-, C_2H_5OH, H_2O, NH_3, H_2NCH_2CH_2NH_2, NO_2^-, CN^-$$

(I^- corresponds to the minimum and CN^- to the maximum Δ value).

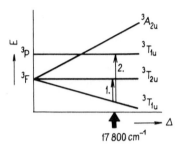

Fig. 13-43. Interpretation of the absorption spectrum of the $V(ox)_3^{3-}$ complex. The 1st and 2nd transitions (corresponding to the ligand field strength $\Delta = 17\,800\ cm^{-1}$) are indicated; ox denotes the dianion of oxalic acid.

For ions with two d electrons the situation is more complicated. The energy levels are not influenced by the ligand field alone, but also by the interaction of the d electrons. The V^{3+} ion is a good example. For the reasons given above, the ground state is a ^3F state which splits in the octahedral field into three states: A_{2u}, T_{1u}, T_{2u}. In the visible region two absorption bands with maxima at 17 000 and 24 000 cm^{-1} were observed experimentally in the $V(ox)_3^{3-}$ spectrum. If a value of 17 800 cm^{-1} is attributed to the ligand field "strength" (Fig. 13-43), the first two bands can be interpreted as $^3T_{1u} \rightarrow {}^3T_{2u}$ and $^3T_{1u} \rightarrow {}^3T_{1u}$ transitions. It is then evident that some transitions which are forbidden in the free atom are allowed in the complex. These are the following:

a) Transitions connected with a change of spin multiplicity $[2S + 1]$ are forbidden, i.e. the number of unpaired electrons must not change during excitation. Weak interaction between the spin and the orbital moments of the electrons results in transition, which would be strictly forbidden otherwise, being slightly allowed (see above). For example, with d^5 complexes in a weak field all the transitions are spin forbidden and yet occur (of course, the ε values are small, equalling about 1 l mol^{-1} cm^{-1} or less).

b) All transitions between d orbitals are forbidden, as $\Delta l = 0$ (the condition $\Delta l = \pm 1$ is the Laporte selection rule). This rule applies to atoms in the gaseous phase; with ions of transition elements in complexes, a certain mixing of d orbitals with p and f orbitals occurs; then d−d transitions occur in these complexes (ε is relatively small, $\varepsilon < 50$ l mol^{-1} cm^{-1}).

Table 13-6

Positions of the First Intense Bands and Colour of Conjugated Hydrocarbons

Substance	$\dfrac{\lambda^{\,b}}{nm}$	Colour[a]	Substance	$\dfrac{\lambda^{\,b}}{nm}$	Colour[a]
Ethylene	163	a	Benzene	207	a
1,3-Butadiene	217	a	Naphthalene	285	a
1,3,5-Hexatriene	251	a	Anthracene	375	a
1,3,5,7-Octatetraene	304	a	Tetracene	471	orange-yellow
1,3,5,7,9-Decapentaene	334	a	Pentacene	580	violet

[a] Letter a means colourless.
[b] Position of the first absorption band.

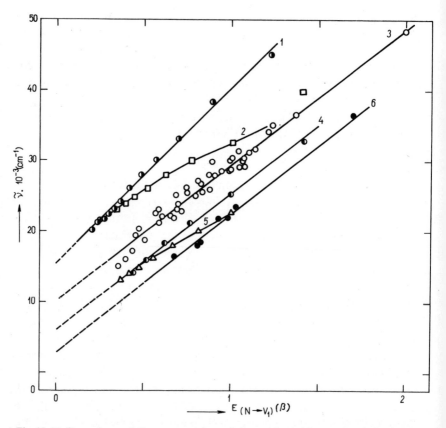

Fig. 13-44. Dependence of the wave numbers of the maxima of intense bands in electronic spectra of conjugated hydrocarbons and ions on the HMO energies of the $N \rightarrow V_1$ transitions. Designation: 1 — polyenes, $2 - \alpha$, ω-diphenylpolyenes, 3 — benzenoid hydrocarbons, 4 — odd polyenyl cations, 5 — odd α, ω-diphenylpolyenyl cations, 6 — tropylium and its benzoderivatives.

Some further examples refer to electronic spectroscopy of conjugated organic compounds. The position of the first absorption band, which often determines the colour of the compound, generally (but not always) shifts to the longer wavelength region as the size of the conjugated system increases (Table 13-6).

It has been found that, for correlation of the positions of the first intense (longest wavelength) absorption bands in structurally related systems HMO data (the energy of $N \rightarrow V_1$ transitions) can be utilized successfully. Fig. 13-44 demonstrates how carefully the structural types have to be considered. Extensive material on this subject is available in the literature. It is therefore sufficient to stress that deviation from the theoretically expected, simple, linear dependence between $\tilde{v}$ (cm^{-1}) and $E(N \rightarrow V_1)$ is caused by neglecting electron repulsion in the HMO theory. If, instead of the HMO excitation energies, the values resulting from SCF theory are employed [the expressions for the excitation energy are given by Eqs. (13-87) and (13-88)], a single linear dependence between the experimental and calculated excitation energies, with a slope equal to one and passing through the origin of the coordinates, is obtained.

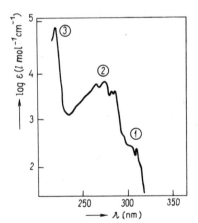

Fig. 13-45. Absorption curve of naphthalene in ethanol. The positions of the first three absorption bands are indicated.

It appears, however, that sometimes not only the HMO theory but even the SCF theory (in the Pople approximation) fails qualitatively. This occurs when more excited state configurations correspond to the same energy, i.e., if they are degenerate. This situation can also be illustrated by the HMO data, for example, for naphthalene. Fig. 13-45 shows its absorption curve in which three distinct regions (1, 2, 3) can be distinguished, which can be assigned intuitively to the transitions from the ground state to the 1st, 2nd, and 3rd excited singlet states. This assignment is, of course, only tentative and its correctness can be decided only after performing the approximate calculations.

In Fig. 13-46 the HMO orbital energies of naphthalene are given and the occupation of the individual levels is depicted for the ground state and for the four singly excited configurations of lowest energy; in addition the figure also depicts the relative energies of these configurations, where the energy of the ground state is set equal to zero by convention. It is obvious that the $\Psi_{2 \to -1}$ and $\Psi_{1 \to -2}$ configurations are degenerate. A diagram constructed using SCF data looks quite similar. It thus seems that the band at 310 nm corresponds to the $\Psi_0 \to \Psi_{1 \to -1}$ transition, the band at 280 nm to the two $\Psi_0 \to \Psi_{2 \to -1}$; $\Psi_0 \to \Psi_{1 \to -2}$ transitions, and finally the band at 220 nm to the transition $\Psi_0 \to \Psi_{2 \to -2}$. It is, however, readily seen from quantitative comparison of the SCF excitation energies, of the oscillator strengths and of polarization directions of the individual transitions with experimentally determined quantities that this assignment is incorrect.

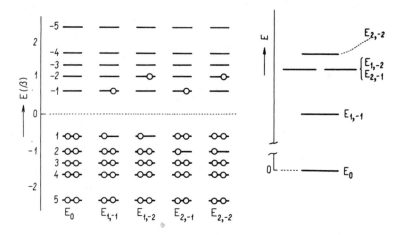

Fig. 13-46. Occupation of HMO levels in naphthalene in the ground state and in the first four singly excited configurations. In the right-hand part of figure the relative energies of these four singly excited configurations with respect to the ground state are given. E_0, $E_{1,-1}$, $E_{2,-1}$, $E_{1,-2}$, and $E_{2,-2}$ energies correspond to determinants Δ_0, Δ_1, Δ_2, Δ_3, and Δ_4.

This difficulty can easily be overcome if the original degenerate functions are substituted by linear combinations of functions. If denotation according to Fig. 13-46 is employed, we can write

$$\Psi^+ = \frac{1}{\sqrt{2}}(\Delta_2 + \Delta_3) \tag{13-101}$$

$$\Psi^- = \frac{1}{\sqrt{2}}(\varDelta_2 - \varDelta_3) \qquad (13\text{-}102)$$

Thus, interaction between the configurations of degenerate states is considered, termed *configuration interaction* (CI). CI between degenerate functions is termed first-order configuration interaction. Without considering this CI, interpretation of the electronic spectrum is, in general, difficult. Within the Pariser – Parr limited configuration interaction method, interactions between a certain number of singly excited configurations are assumed. If the SCF-MO expansion coefficients are used for evaluation of the matrix elements, it can then be shown (cf. Section 5.5) that the matrix element between the ground and singly excited state is zero, thus

$$\langle \Psi_0 \,|\, \mathscr{H} \,|\, \Psi_{i\to j}\rangle = 0 \qquad (13\text{-}103)$$

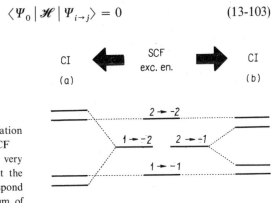

Fig. 13-47. Effect of configuration interaction on degenerate SCF excitation energies. (*a*) The very intense CI effect results in that the 1→1′ transition does not correspond to the 1st band in the spectrum of naphthalene. (*b*) Moderate CI effect in the spectrum of anthracene.

For naphthalene, the interaction of the configurations is manifested very dramatically (Fig. 13-47a). It appears that data thus obtained interpret the experimental spectrum perfectly. In other cases the influence of the first order CI is not conspicuously manifested. With anthracene, for example, the first intense band in the spectrum corresponds to the $1 \to -1$ transition and the result of the CI calculation can be seen in Fig. 13-47b.

Figure 13-48 compares the result of LCI-SCF calculations with experimental data for the non-alternant hydrocarbon acenaphthylene.

A few words should be added on the use of the simple perturbation treatment for investigation of influence of introducing a substituent into the conjugated system. It can easily be shown that, for the energy change of the $N \to V_1$ transition caused by a change $\delta\alpha_\mu$, of the Coulomb integral in the μ position it holds that

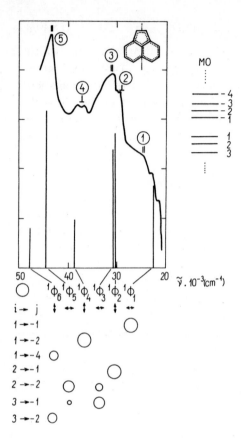

MO

Fig. 13-48. Comparison of LCI-SCF calculations with the absorption curve of acenaphthylene in hexane. Experimentally determined relative polarization directions ($\|$, $\perp$) of the transitions (measured by the fluorescence method) are noted on the individual absorption bands. Below the figure data on the theoretical polarization directions ($\leftrightarrow$, $\updownarrow$) related to the axis given in the formula and data on participation of the configurations in LCI wave functions are listed. The circle below number 50 indicates 100% contribution.

Table 13-7

Values of $\delta\alpha_\mu$-Constants Characterizing the Inductive Effect of Substituents (in cm^{-1})[a]

Substituent[b]	Non-alternant systems	Alternant systems	Substituent[b]	Non-alternant systems	Alternant systems
OH	–	13 780	CN	−3 330	−6 710
OMe	–	11 870	COOH	−4 750	−11 850
NH_2	13 750	13 880	CHO	−5 420	−17 280
Cl	3 540	5 370	NO_2	−8 710	–
Br	2 960	5 480	N[c]	−18 750	−17 500
CH_3	3 290	4 290			

[a] From J. M. Murrell: *The Theory of the Electronic Spectra of Organic Molecules*, Methuen, London 1963, p. 254.

[b] Inductive effect of substituent considered.

[c] This nitrogen atom replaced one $=CH-$ group in the conjugated system.

$$\delta E(N \rightarrow V_1) = (c_{\mu j}^2 - c_{\mu i}^2)\, \delta\alpha_\mu, \qquad (13\text{-}104)$$

where indices j and i denote the MO's into which and from which the transition of the electron occurs, respectively. Table 13-7 gives the numerical values of constants recommended by Murrell for estimation of the influence of the substituent on the position of the $N \rightarrow V_1$ transition; only the inductive effect of the substituent is assumed in this approximation.

13.3.2 Luminescence phenomena
(fluorescence, phosphorescence)

So far, processes during which the studied molecule accepts a quantum of energy and passes from a state of lower energy into various states of higher energy have beeen discussed. Now deactivation processes in the electronically excited molecule will be treated. These are two types: radiative (fluorescence, phosphorescence) and non-radiative (internal conversion). With fluorescence, the multiplicity of the state does not change during emission (for example, during the $S_1 \rightarrow S_0$ transition); with phosphorescence, the multiplicity of the state changes. The most important transition of this type is the $T_1 \rightarrow S_0$ transition.

For emission of radiation, by far the most important states are the first two excited states, S_1 and T_1. Transition from higher excited states (S_x, T_x) into these states is very rapid and occurs without radiation (except for some rare exceptions, for example, the $S_2 \rightarrow S_0$ fluorescence in azulene). With molecules in the singlet ground state, the T_1 state always lies lower than the S_1 state; thus the phosphorescence

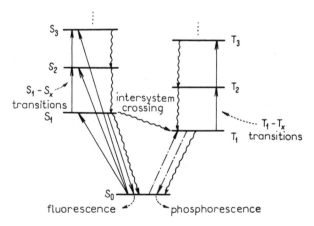

Fig. 13-49. The generalized Jablonski diagram indicating allowed (———) and forbidden ($-\cdot-\cdot-$) transitions and non-radiating ($\sim\!\sim\!\sim$) absorption and emission transitions.

band appears at longer wavelengths than the fluorescence band. Schemes of different types of absorption and deactivation processes are shown in Fig. 13-49. Fig. 13-50 depicts the S_0, S_1 and T_1 states. In the figure can be seen the reason for the frequently observed symmetry of the vibrational structure of the $S_0 \rightarrow S_1$ ($S_0 \rightarrow T_1$) absorption and the $S_1 \rightarrow S_0$ ($T_1 \rightarrow S_0$) fluorescence (phosphorescence). Measurement of the absorption and emission curves facilitates localization of the $0-0$ vibrational transition.

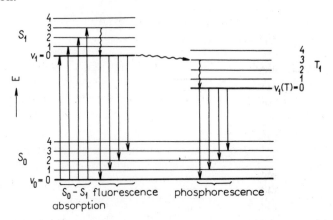

Fig. 13-50. Vibrational structure of absorption (S_0—S_1) and emission (fluorescence, phosphorescence) transitions.

The first excited singlet and triplet states have very different lifetimes—for S_1 this is usually 10^{-9} to 10^{-7} s and for T_1, 10^{-4} to 10 s. Fluorescence quenching proceeds according to the law of the first-order reactions. For the relation between the half-life of the fluorescence state (τ_F) and the rate constant of fluorescence quenching (k_F) it holds that

$$k_F = 1/\tau_F \tag{13-105}$$

The observed half-life of the $S_1(\tau)$ state is usually the result of different deactivation processes: fluorescence, the $S_1 \rightarrow S_0$ non-radiative transition and transition between the S_1 and T_1 states (intersystem crossing). The intensity of phosphorescence also decreases exponentially with time.

The great difference between the lifetimes of the S_1 and T_1 states can be understood by considering the relationship between the lifetime τ_0 and a quantity related to the oscillator strength, Einstein's coefficient of spontaneous emission $A_{2 \rightarrow 1}$:

$$A_{2 \rightarrow 1} = C \int \varepsilon \, d\tilde{\nu} = \frac{1}{\tau_0}, \tag{13-106}$$

where

$$C = \frac{8\pi c \tilde{v}^2 n^2}{N_A} \times 2.303$$

and c is the velocity of light, $\tilde{v}$ is the wave number of the emission maximum (cm^{-1}), n is the refractive index of the medium, ε is the molar absorption coefficient and N_A is the Avogadro constant. The small values of integral $\int \varepsilon \, d\tilde{v}$ in the $T_1 \rightarrow S_0$ transitions explain the relatively long lifetimes of the T_1 states.

The last quantity to be mentioned is the quantum yield (φ_F, φ_P), which is defined as follows:

$$\varphi_F = \frac{n_F(hv)}{n_A(hv)}, \qquad (13\text{-}107)$$

$$\varphi_P = \frac{n_P(hv)}{n_A(hv)}, \qquad (13\text{-}108)$$

where n_F (n_P) is the number of quanta emitted during fluorescence (phosphorescence) and n_A is the number of quanta absorbed during the $S_0 \rightarrow S_1$ transition.

Provided that the studied system does not decay photochemically and that the quantum yield of internal conversion (through vibrational deactivation) is small, the balance condition can be written as follows:

$$\varphi_F + \varphi_P \approx 1 \qquad (13\text{-}109)$$

Thus fluorescence and phosphorescence are obviously interrelated processes.

Measurement of the absolute values of quantum yields is very difficult. Far easier and also very useful is measurement of the relative quantum yields, for example, using rhodamine B solutions.

It is necessary to add that, while fluorescence is often observed even in solutions at laboratory temperature, phosphorescence appears under these conditions only exceptionally. Thus measurement is usually performed in solid "glasses" (solidified transparent mixtures of several solvents) at the temperature of liquid nitrogen.

13.3.3 Photochemistry

Excitation from the ground state into the S and T excited states is usually accompanied by extensive electronic redistribution. This is of utmost importance for photochemistry which follows the fates of excited molecules mainly in the first singlet and triplet states. The relationships

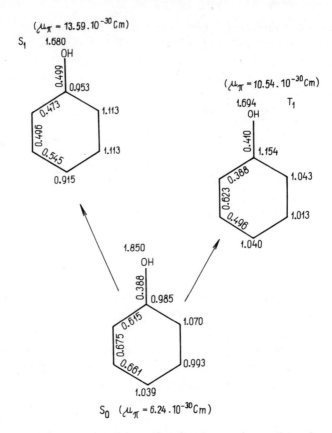

Fig. 13-51. LCI-SCF molecular diagrams for phenol: ground state (S_0) and excited states (S_1 and T_1).

in phenol are noteworthy: Fig. 13-51 depicts the molecular diagrams for the S_0, S_1 and T_1 states. The higher excited states are less topical since they change very quickly (during a time interval of the order of 10^{-14} s) via non-radiative transitions into the S_1 and T_1 states. It is evident from data on the electron densities, bond orders and free valences that the changes in the electron distribution are very deep-seated. In general, in spite of the short lifetimes of the excited states, they correspond to definite particles whose physical and chemical properties differ from the properties of the same molecule in the ground state. This difference is frequently so great that it is impossible to transfer empirical rules concerning ground states into the region of excited states. For the sake of illustration it can be added that the values of the dissociation constants change up to a millionfold, dipole moments change by orders of magnitude (excitation is sometimes also accompanied by a change in the orientation of the vector of the dipole moment) and centres of nucleophilic substitution become centres of electrophilic substitution and vice versa. The relatively short lifetime of excited particles, of course, does not usually permit application of classical methods to the determination of physical properties, but, fortunately, a number of indirect procedures have been developed.

More detailed data on lifetimes, deactivation processes and other characteristics are given in the chapter on emission phenomena. A few remarks should be added here on selected physical properties (geometry of the molecule, dipole moment, spectroscopy) and on the reactivity of molecules in the excited state (equilibrium and rate processes).

If we start from the general expression for the dipole moment, it is obvious that calculation for the excited state is just as simple as for the ground state. The experimental determination is usually more complicated. If the molecules possess a sufficiently large permanent dipole moment ($\mu > 3 . 10^{-30}$ C m) they become orientated in a strong electric field and this is manifested by dichroism. The quantity L_χ can be defined as follows:

$$L_\chi = - \frac{\Delta I}{I} \frac{1}{2 . 3\varepsilon_r E^2},$$
(13-110)

where χ is the angle between the vector of the electric component of the linearly polarized light (used for spectroscopy) and the direction of the external electric field, $\Delta I/I$ is the relative change in the luminous flux density under the influence of the field, ε_r is the relative permittivity of the medium and E is the intensity of the applied electric field. Quantity L_χ is a rather complex function of the change in the dipole moment (which accompanies the transition from the ground to any excited state) and thus, when the dipole moment in the ground state is known, it can be used for calculation of the value in the excited states. This technique also permits calculation of the polarizability of the studied molecule in the excited states.

With a fluorescing molecule, even modest experimental equipment suffices for determination of the dipole moment of the first excited singlet state. The difference between the absorption and emission maxima ($\Delta\tilde{\nu}$) in media with different relative permittivities depends on the change in the dipole moment:

$$\Delta\tilde{\nu} = \tilde{\nu}_a - \tilde{\nu}_f = \frac{2\Delta f}{hca^3} (\Delta\mu)^2 + \text{const},$$
(13-111)

where $\Delta\mu = \mu_E - \mu_G$ (E and G are the indices of the excited and the ground states, resp.), h is the Planck constant, c is the velocity of light,

a is the effective diameter of a sphere with the volume of the molecule and for Δf it holds that

$$\Delta f = \frac{\varepsilon_r - 1}{2\varepsilon_r + 1} - \frac{n^2 - 1}{2n^2 + 1}, \qquad (13\text{-}112)$$

where ε_r is the relative permittivity and n is the refractive index of the solvent. According to Eq. (13-111) the dependence of $\Delta \tilde{v}$ on Δf is linear and $\Delta \mu$ can be calculated from its slope:

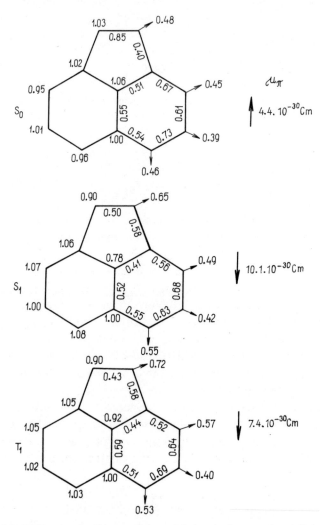

Fig. 13-52. LCI-SCF molecular diagram of acenaphthylene in the ground state (S_0) and excited states (S_1 and T_1). The π-electron contributions to the dipole moment and the orientation of its vector are also given.

$$k = \frac{2(\Delta\mu)^2}{hca^3} \tag{13-113}$$

or

$$\mu_E = \mu_G + (khca^3/2)^{1/2} \tag{13-114}$$

It is evident from the molecular diagrams of acenaphthylene (Fig. 13-52) that during excitation significant changes in the bond orders, and thus also changes in corresponding bond lengths, occur. The general shape of large molecules is retained whereas with small molecules excitation is accompanied by a marked change in their geometry. For example, numerous triatomic linear molecules are bent in the excited state, for example CO_2:

(S_0 and S_1 are symbols for the ground and first excited singlet states). With ethylene the CH_2 planes are rotated:

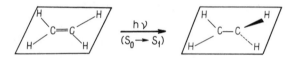

In spite of the short lifetime of the first excited triplet state (mostly 10^{-3} to 10^0 s), it is possible to measure the electronic absorption spectrum of these excited particles. It is essential, of course, that the concentration of excited particles (population of the triplet state) increases sufficiently in the studied solution. This increase in the concentration must be achieved in a very short time; flash photolysis introduced by Porter is very suitable for these purposes. The principle of the device is quite simple (Fig. 13-53). Spectrophotometry is, of course, begun only after termination of the flash. The demands on the device recording the spectrum are considerable.

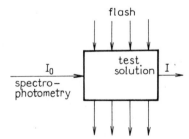

Fig. 13-53. Principle of a device for measurement of $T-T$ spectra.

374

Numerous bands in the T − T spectrum are intense, as they are spin-allowed transitions. In Fig. 13-54 the triplet − triplet spectrum of anthracene is given for illustration together with the results of the semiempirical calculation. It remains to be added that, using picosecond flash photolysis, the absorption spectra of molecules (benzenoid hydrocarbons) have been measured in the first excited singlet state, the lifetime of which equals about 10^{-8} s. The differences in the energies of the excited states calculated by the PPP method can be used both for analysis of ordinary electronic spectra ($S_0 \rightarrow S_x$ transitions) and for intepretation of $S_1 \rightarrow S_x$ transitions.

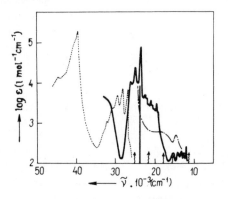

Fig. 13-54. Triplet-triplet absorption curve of anthracene (ethanol-methanol, 113 K) and the result of the LCI-SCF calculation (? denotes a forbidden transition). For comparison the $S_0 - S_x$ (......) and $S_0 - T$ (— · — · —) absorption curves are also given.

The lifetime of excited states is sufficiently long for the establishment of acid-base equilibria. Förster proposed a procedure which allows calculation of pK values in excited states (pK^*) using a simple thermochemical equation. Calculation of pK^* requires knowledge of pK (in the ground state) and of the excitation energy for transition into the first excited state (position of the first absorption band) of the acid and the conjugate base. Assuming that the change in the entropy of protonation is the same in the ground and excited states ($\Delta S = \Delta S^*$). the relationship for the dissociation constants can be written in the form

$$\log \frac{K^*}{K} = \frac{\Delta H - \Delta H^*}{RT} = \frac{\Delta E_{AH} - \Delta E_{A^-}}{RT}, \qquad (13\text{-}115)$$

where ΔE_{AH} (ΔE_{A^-}) is the excitation energy of the first transition of the acid (conjugate base); the other symbols have the usual meaning. Equilibrium constant K refers to the process

$$AH \ \rightleftarrows \ A^- + H^+$$

$$K = \frac{a_{A^-} a_{H^+}}{a_{AH}} \qquad (13\text{-}116)$$

The form of Eq. (13-115) becomes clear if we represent the process graphically (the Förster cycle) (Fig. 13-55). The determination of quantities ΔH, ΔE_{AH} and ΔE_{A^-} causes no difficulties; the required ΔH^* value follows from the condition

$$\Delta H^* = \Delta H + \Delta E_{A^-} - \Delta E_{AH} \qquad (13\text{-}117)$$

The pK changes are usually so large that, for example, very weak acids (comparable to phenol) become as strong as the mineral acids.

The easy oxidations and reductions in the excited states of many substances (photooxidation, photoreduction) can be explained simply.

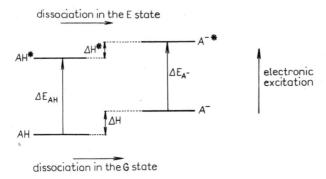

Fig. 13-55. The Förster cycle for calculation of pK in the excited state.

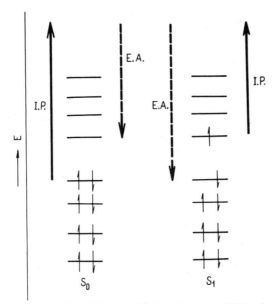

Fig. 13-56. Ionization potential and electron affinity in the ground (S_0) and excited (S_1) states.

376

It can be seen from the one-electron scheme in Fig. 13-56 that the ionization potential (I) usually decreases and the electron affinity (A) increases on excitation, thus facilitating both oxidation and reduction. Transition to electronically excited states is generally accompanied by a decrease in the bond orders (overlap populations) and thus also by an increase in the free valences and, consequently, in the chemical reactivity (cf. Section 16.5.1), which concerns not only conjugated substances. The dimerization of acenaphthylene in the excited state with formation of substance (I) can be understood by considering the significant increase in the free valences

(I)

of atoms 1 and 2. More interesting, however, is the fact that the region around the five-membered ring becomes a centre for nucleophilic reactions in the excited state, whereas in the ground state it is a centre for electrophilic reactions. This change of centres is an almost general phenomenon. It seems that in experimental chemistry — with a suitable experimental arrangement — remarkable possibilities can be expected.

Finally, it is worth mentioning that the $N \rightarrow V_1$ excitation in the HMO approximation is equivalent to the simultaneous formation of radical anions and radical cations. This fact explains some of the striking similarities of the chemistry of radical ions and the chemistry of particles in the excited state.

REFERENCES

1. West W. (ed.): *Chemical Application of Spectroscopy*. Interscience, New York 1956 and 1968.
   (The older edition concerns spectroscopy from radio — to electronic frequency regions: with the new edition, up to now the volume concerning electronic spectroscopy is only available.)
2. Murrell J. N.: *The Theory of the Electronic Spectra of Organic Molecules*. Methuen, London 1963.
3. Colthup N. B., Daly L. H., Wiberley S. E.: *Introduction to Infrared and Raman Spectroscopy*. Academic Press, New York 1964.
4. Sandorfy C.: *Die Elektronenspektren in der theoretischen Chemie*. Verlag Chemie, Weinheim 1961.
5. Parr R. G.: *The Quantum Theory of Molecular Electronic Structure*. Benjamin, New York 1964.
6. Gerson F.: *Hochauflösende ESR-Spektroskopie*. Verlag Chemie, Weinheim 1967.
7. For *spectroscopy of inorganic complexes* see references to Chapter 10, refs. 76, 77, 78.

14. MAGNETIC PROPERTIES
OF MOLECULES

In this chapter some magnetic properties of systems with closed electronic shells[1-4], namely, *the magnetic susceptibility* and *the anisotropy* of this susceptibility will be discussed very briefly. Proton behaviour under NMR conditions has already been mentioned (Section 13.2.4).

The external magnetic field induces electronic currents of various types: atomic or interatomic and diamagnetic or paramagnetic. The sensitivity of a system to magnetization is expressed by its susceptibility, most frequently by the specific value of the susceptibility, $\varkappa_{sp}$:

$$\varkappa_{sp} = \frac{\varkappa}{\varrho}, \tag{14-1}$$

where ϱ is the density of the measured substance and $\varkappa$ is the magnetic susceptibility. $\varkappa$ is defined as follows:

$$\varkappa = \frac{M}{H} \tag{14-2}$$

In Eq. (14-2), M denotes the magnetization and H, the intensity of the external magnetic field; for the sake of simplicity it is assumed that the direction of the external magnetic field is identical with the direction of one of the axes of the Cartesian coordinate system. Thus the problem can be formulated using scalar quantities without decreasing the generality of the conclusions. The product of the specific susceptibility, $\varkappa_{sp}$ [Eq. (14-1)] and the relative molecular weight is referred to as the molar susceptibility $\varkappa_m$. Depending on the $\varkappa$ value, we speak of diamagnetism (about 10^{-5}), paramagnetism (about 10^{-3}) and ferromagnetism (10^{-1} to 10^5).

In organic substances diamagnetism (ignoring in this chapter the behaviour of paramagnetic substances, radicals, under ESR conditions) and in inorganic substances (more accurately, in complexes of the transition elements) paramagnetism will be of greatest interest.

In the latter group of substances, the connection between the experimentally detectable molar susceptibility and the magnetic moment

$\boldsymbol{\mu}_{\mathrm{ef}}$, which can be calculated from theoretical characteristics, is important. For atoms and ions, whose electronic systems are in a state defined by quantum number J, it holds for $\varkappa_{\mathrm{m}}$ ($\mathrm{m}^3\,\mathrm{mol}^{-1}$) that

$$\varkappa_{\mathrm{m}} = \frac{N_A g^2 J(J + 1)\,\beta_e^2 \mu_0}{3kT}, \qquad (14\text{-}3)$$

where N_A, k and T have the usual significance, β_e is the Bohr magneton [cf. Eq. (13-41)],

$$\beta_e = \frac{eh}{4\pi m_e}, \qquad (14\text{-}4)$$

μ_0 is the permeability of a vacuum and g is the Landé factor.* If expression $g\sqrt{[J(J + 1)]}$, which is the effective magneton number, is denoted by $\mu_{\mathrm{ef}} = |\boldsymbol{\mu}_{\mathrm{ef}}|$, then Eq. (14-3) can be written in the form

$$\varkappa_{\mathrm{m}} = \frac{N_A \beta_e^2 \mu_0}{3kT}\,\mu_{\mathrm{ef}}^2 \qquad (14\text{-}5)$$

However, it appears that the measured values never correspond to expression (14-5), due to cancelling out of the orbital contributions under the influence of the crystal field; consequently, μ_{ef} is usually described by a simpler expression,

$$\mu_{\mathrm{ef}} \approx g\sqrt{[S(S + 1)]} = 2\sqrt{[S(S + 1)]}, \qquad (14\text{-}6)$$

which can easily be enumerated provided that the number of unpaired electrons in the atom is known. In the general case, however, quantum numbers L and S must be considered and thus the quantum mechanical rule for calculation of the magnetic moment of the electron in the free ion takes the form

$$\boldsymbol{\mu} = \int \chi_j^* \boldsymbol{\mu} \chi_j \, d\tau, \qquad (14\text{-}7)$$

where the AO, χ_j, describes the state of the electron in the ion and the one-electron operator $\boldsymbol{\mu}$ represents the vector quantity $\boldsymbol{\mu} = \mathbf{L} + 2\mathbf{S}$. From deviations between the experimental values of the magnetic moments, and those calculated from expression (14-6), conclusions on the geometry of the studied systems can be drawn.

Magnetic moment $\boldsymbol{\mu}$ is induced in a substance placed in a static magnetic field of intensity H; the proportionality constant is the specific

---

* This factor can be simply expressed for the free ion assuming the validity of Russell–Saunders coupling in terms of its quantum numbers (L, S, J):

$$g = 1 + \frac{J(J + 1) - L(L + 1) + S(S + 1)}{2J(J + 1)}$$

susceptibility (m being the mass of the specimen):

$$\mu = \varkappa_{sp} mH \tag{14-8}$$

In Eq. (14-8), the possibility of different orientation of the substance with respect to the external field is taken into account. The magnitude of this moment can be determined, using a Gouy balance (Fig. 14-1), from the difference in the weight of the specimen with and without the external magnetic field. The greater the susceptibility of the specimen, the greater is the strength holding the specimen in the magnetic field. The balance is calibrated using substances whose susceptibilities are known with sufficient accuracy, for example, cupric sulphate or water. The molar susceptibility of molecules can be measured in the direction of the principal axes of the molecule, i.e. quantities $\varkappa_{mx}$, $\varkappa_{my}$ and $\varkappa_{mz}$ if x, y, and z denote the principal axes of the molecule. For simplicity these quantities will be denoted $\varkappa_x$, $\varkappa_y$ and $\varkappa_z$.

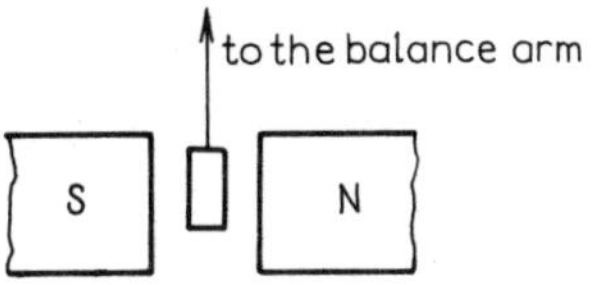

Fig. 14-1. Measurement of magnetic moment. Magnetic poles: N, S.

Table 14-1

The Pascal Constants $\varkappa_A$ of Selected Elements

Atom	$\dfrac{\varkappa_A \cdot 10^6 \cdot (4\pi)^{-1}}{cm^3\,mol^{-1}}$	Atom	$\dfrac{\varkappa_A \cdot 10^6 \cdot (4\pi)^{-1}}{cm^3\,mol^{-1}}$
H	-2.93	B	-7.0
C	-6.00	F	-11.5
N (aliphatic)	-5.57	Cl	-20.1
N (cyclic)	-4.61	Br	-30.6
N (amide)	-1.54	S	-15.0
N (imide)	-2.11	Se	-23.0
O (alcohol, ether)	-4.61	P	-10.0
O (aldehyde, ketone)	$+1.72$	Li	-4.2
O (carboxyl, ester)	-3.36	Na	-9.2

Pascal showed that the molar susceptibility of substances can be expressed as the sum of the atomic susceptibilities, bond susceptibilities and structural contributions (Tables 14-1 and 14-2):

$$\varkappa_m = \Sigma \varkappa_A + \Sigma \lambda, \tag{14-9}$$

where λ are the Pascal structural parameters (Table 14-2). This equation

Table 14-2

The Pascal Structural Parameters λ

Atom; bond	$\dfrac{\lambda \cdot 10^6 \cdot (4\pi)^{-1}}{\text{cm}^3 \text{ mol}^{-1}}$	Atom; bond	$\dfrac{\lambda \cdot 10^6 \cdot (4\pi)^{-1}}{\text{cm}^3 \text{ mol}^{-1}}$
$>$C$=$C$<$	$+5.5$	$=$C$-$Cl	$+3.1$
$-$C$\equiv$C$-$	$+0.8$	$=$C$-$Br	$+4.1$
$>$C$=$C$-$C$=$C$<$	$+10.6$	$C_3(\alpha, \gamma, \delta, \varepsilon)^a$	-1.3
$-$N$=$N$-$	$+1.8$	$C_4(\alpha, \gamma, \delta, \varepsilon)^a$	-1.55
$-$C$\equiv$N	$+0.8$	$C_3(\beta)$, $C_4(\beta)^a$	-0.5
$C_{\text{arom.}}$ (in one cycle)	-0.24	C (cyclic: 3)	$+4.1$
$C_{\text{arom.}}$ (in two cycles)	-3.1	C (cyclic: 4)^b	$+3.05$
$C_{\text{arom.}}$ (in three cycles)	-4.0	C (cyclic: 5)^b	-0.98
		C (cyclic: 6)^b	$+0.86$

^{a} $C_i(\alpha)$ denotes the carbon atom in the α position (with respect to oxygen) bonded to i other atoms (except for hydrogen, which is not considered).
^{b} In parentheses: number of carbon atoms in the non-aromatic cycle.

has been used for structure elucidations. It is applicable when several structures which correspond to sufficiently different values of $\varkappa_m$ [calculated using Eq. (14-9)] can be attributed to a certain substance (whose experimental $\varkappa_m$ value is known); that structure is considered most probable for which there is the best agreement of $\varkappa_{m,\text{exp}}$ with $\varkappa_m$ [Eq. (14-9)]. A considerable increase in the diamagnetic susceptibility and high anisotropy are typical for delocalized cyclic compounds. A peculiarity arises in that the absolute values of $\varkappa_m$ calculated for conjugated (aromatic) compounds for a single Kekulé structure are much smaller than the experimental values. In benzene, for example, the values $-503 \cdot 10^{-6} \text{ cm}^3 \text{ mol}^{-1}$ and $-691 \cdot 10^{-6} \text{ cm}^3 \text{ mol}^{-1}$, respectively, have been found. The anisotropy of the diamagnetic susceptibility $\Delta\varkappa_m$ of a planar molecule is defined as follows:

$$\Delta\varkappa_m = \varkappa_z - \frac{\varkappa_x + \varkappa_y}{2}, \qquad (14\text{-}10)$$

where $\varkappa_z$ denotes the molar susceptibility in the direction of the axis perpendicular to the plane of the planar molecule. With benzene the value of $\Delta\varkappa_m$ is approximately equal to $-754.10^{-6}\,cm^3\,mol^{-1}$; in polynuclear benzenoid hydrocarbons the anisotropy values increase with increasing size of the molecule. In general, there is a parallel between these quantities and the respective delocalization energies. Pauling interpreted large anisotropy values (as well as large susceptibility values in conjugated hydrocarbons) by suggesting electron flow between centres of the conjugated system under the influence of the external field. If an external magnetic field of intensity H is applied in the direction of the z-axis [Eq. (14-10)] directed upwards from the benzene ring, then the induced ring current flows clockwise (Fig. 14-2) around the z-axis. To this ring current corresponds an induced moment that is also oriented in the direction of the z-axis but in the opposite sense. The extraordinary high anisotropy values in planar conjugated molecules can be explained by noting that to this induced moment corresponds a susceptibility (which represents the contribution to the "original" value) oriented perpendicular to the ring plane.

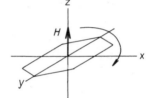

Fig. 14-2. Orientation of external magnetic field H and direction of induced circular current.

To study the quantitative behaviour of an electron under the influence of an external magnetic field, it is necessary to use a Hamiltonian in the form of Eq. (14-12) (where A is the vector potential) in place of the former Hamiltonian $\mathscr{H}$ [Eq. (14-11)]:

$$\mathscr{H} = \frac{1}{2m}\left(\frac{h}{2\pi i}\nabla\right)^2 + \mathscr{V} \tag{14-11}$$

$$\mathscr{H} = \frac{1}{2m}\left(\frac{h}{2\pi i}\nabla + \frac{e}{c}A\right)^2 + \mathscr{V} \tag{14-12}$$

The vector potential A is connected with the intensity of the magnetic field H by the relationship

$$H = \mathrm{rot}\,A \tag{14-13}$$

Using the thus-defined Hamiltonian, secular equations modified with respect to the external magnetic field can be obtained by a rather complicated procedure. The expression for the HMO orbital energies k'_j $(E_j = \alpha + k'_j\beta)$, applicable to cyclic polyenes (the dashed quantities correspond to a molecule exposed to the external field), can be given in the form

$$k'_j = 2\beta \cos\left[\frac{2\pi}{m}(j + f)\right],\qquad(14\text{-}14)$$

whereas, for cyclic polyenes in the absence of an external magnetic field, the usual expression is valid:

$$k_j = 2\beta \cos\frac{2\pi j}{m},\qquad(14\text{-}15)$$

where $j = 0, \pm 1, \pm 2, \ldots$
$$\begin{cases} \pm\dfrac{1}{2}(m-1) & \text{for odd } m \\[2mm] \pm\dfrac{1}{2}m & \text{for even } m \end{cases}$$

For f in Eq. (14-14) it holds that

$$f = \frac{eS}{h}B,\qquad(14\text{-}16)$$

where e is the elementary charge, h is the Planck constant, S is a quantity which has the dimension of area and B is the induction of the external magnetic field. By substituting numerical values into Eq. (14-16) a value of $f \approx 10^{-4}$ is obtained for a field of induction 1 T, which is obviously a very small value.

The magnetic moment in a system with spin S, which has n unpaired electrons, is given for compounds of the transition elements by the simple expression

$$\mu = 2\sqrt{[S(S+1)]}\,\beta_e = \sqrt{[n(n+2)]}\,\beta_e,\qquad(14\text{-}17)$$

in which β_e is the Bohr magneton [Eq. (14-4)] and n is the number of unpaired electrons. The simplicity of expression (14-17) for μ results from the fact that, in these compounds, the contribution originating from the orbital motion of the electron is often negligible, so that it is sufficient to include only the spin contribution. The molar magnetic susceptibility $\varkappa_m$ can be determined from the magnetic moment:

$$\varkappa_m = \frac{\mu^2}{3kT}\mu_0 N_A\qquad(14\text{-}18)$$

REFERENCES

1. Selwood P. W.: *Magnetochemistry.* Interscience, New York 1956.
2. Hutchison C. A., Jr.: in *Determination of Organic Structures by Physical Methods* (ed. Braude E. A. and Nachod F. C.). Academic Press, New York 1955.
3. Salem L.: *The Molecular Orbital Theory of Conjugated Systems.* Benjamin, New York 1966.
4. Dunn T. M., McClure D. S., Pearson R. G.: *Some Aspects of Crystal Field Theory.* Harper & Row, New York 1965.

15. THERMOCHEMICAL PROPERTIES AND MOLECULAR STABILITY

15.1 Heats of formation and atomization

The enthalpy of formation of one mole of compound $A_m B_n C_o$ from its elements under standard conditions is called *the heat of formation* (ΔH_f):

$$mA + nB + oC \rightarrow A_m B_n C_o \dots \Delta H_f \tag{15-1}$$

The enthalpy of elements in their standard state is taken to be zero by convention. The standard state is the stable state of elements at a pressure of 101 kPa at the reaction temperature. For the formation of carbon dioxide from carbon and oxygen [Eq. (15-2)], molecular oxygen is considered to be the standard state of oxygen and graphite is the standard state of carbon. If the temperature is not stated, the data are considered to refer to 25 °C.

$$C(s) + O_2(g) \rightarrow CO_2(g), \qquad \Delta H_f = -389.79 \text{ kJ/mol}, \tag{15-2}$$

where (s) denotes the solid and (g) the gaseous state.

Care must be taken when comparing these experimental quantities with the theoretical values, as the quantities obtained directly from the theoretical data are *the heats of atomization* (bonding energies). These heats are obtained from the total energy of the system by subtracting the energies of the isolated atoms. It is obvious, therefore, that the calculated heat of atomization must be reduced by the energy required for atomization of elements in the standard state; in other words the heats corresponding to the formation of the elements from the atoms must be subtracted.

Theoretical calculation of the heats of formation is one of the most important tasks of quantum chemistry. The heats of formation of substances of different structural types can not yet be calculated with "chemical accuracy", i.e. accuracy which permits calculation of reaction enthalpies which are suitable for calculation of sufficiently accurate equilibrium constants (a plausible assumption is adopted that

384

the accuracy of the entropy values is not critical). The situation is, however, not as unfavourable as it might seem. All valence electron methods which take into account electronic and nuclear repulsion must be considered primarily. The EHT method is not suitable for these purposes; adapted versions of the CNDO method, on the other hand (for example, the Dewar MINDO/2 and MINDO/3 methods), afford promising results. For illustration, Table 15-1 compares a number of calculated and experimentally determined heats of formation.

Satisfactory agreement between theory and experiment has been achieved with a number of *conjugated* compounds. For the heat of atomization of the conjugated hydrocarbon at temperature T [K] the relationship[1]

$$-\Delta H_a^T = N_C D_{CC}^T + N_H D_{CH}^T + W_b \tag{15-3}$$

is valid, where W_b is the π-electron contribution to the total binding energy of the molecule, N_C and N_H are the number of $C-C$ and $C-H$ bonds in the hydrocarbon and D_{CC}^T and D_{CH}^T are the empirical energies of the $C-C$ and $C-H$ σ-bonds (i.e., σ-bonds in conjugated compounds). It is thus assumed that the $C-C$ and $C-H$ σ-bonds can be described by universally applicable values, in agreement with the older finding that the heats of formation of aliphatic hydrocarbons can be calculated very accurately using the group contribution method. The empirically determined values of these energies amount to 3.812 eV for D_{CC}^{298} and 4.432 eV for D_{CH}^{298}. In more accurate calculations, however, the hybridization of the carbon atoms must not be ignored, because, for example, the difference in the energies of the $C(sp^2)-H$ and $C(sp^3)-H$ bonds amounts to more

Table 15-1

Heats of Formation (in kJ/mol) Obtained in Experiments and by Calculation Using the Dewar Method MINDO/2 (25 °C)

Substance	Experiment	Calculation	Relative error
			%
H_2O	−242	−247	−2.0
CO_2	−394	−389	+1.3
NH_3	−46.1	−46.9	−1.7
CH_4	−74.9	−67.8	+10.5
Ethane	−84.6	−90.9	−6.9
Ethylene	52.3	61.1	+14.4
Acetone	−216	−237	−8.9
Nitromethane	−51.1	−62.0	−17.6
Aniline	87.1	99.7	+12.6

than 8 kJ/mol. Some authors have calculated the energies of $C-C$ bonds using the empirical relationship between this energy and the length of the $C-C$ bond.

For the π-electronic binding energy, W_b, the relationship

$$W_b = -(W - nU_\mu - E_{cr}) \qquad (15\text{-}4)$$

is valid, where W is the total energy of the system of n π-electrons, $-U_\mu$ is the ionization potential of carbon atom in the valence state and E_{cr} is the repulsion energy of the positively charged cores. The enumeration of Eq. (15-4) is extremely simple on the HMO level, because the ionization potential has the value α and E_{cr} is neglected; for the total energy $(q_\mu \equiv P_{\mu\mu})$ we have

$$W = \sum_{\mu=1}^{n} q_\mu \alpha + 2\sum\sum_{\mu<\nu} P_{\mu\nu}\beta = n\alpha + 2\sum\sum_{\mu<\nu} P_{\mu\nu}\beta \qquad (15\text{-}5)$$

and for the binding energy we have

$$W_b = -(n\alpha + 2\sum\sum_{\mu<\nu} P_{\mu\nu}\beta - n\alpha) = -2\sum\sum_{\mu<\nu} P_{\mu\nu}\beta \qquad (15\text{-}6)$$

The corresponding SCF expression (Pople) is, of course, more complicated, but can be derived in a similar manner:

$$W_b = -\left[2\sum\sum_{\mu<\nu} P_{\mu\nu}\beta + \frac{1}{4}\sum q_\mu^2\gamma_{\mu\mu} + \sum\sum_{\mu<\nu}(q_\mu - 1)(q_\nu - 1)\gamma_{\mu\nu} - \right.$$

$$\left. - \frac{1}{2}\sum\sum_{\mu<\nu} P_{\mu\nu}^2\gamma_{\mu\nu}\right] \qquad (15\text{-}7)$$

The meaning of the symbols is the same as in Chapter 10.

When the HMO estimation of the π-electron energy is sufficiently accurate (planar, conjugated, alternant systems, mainly benzenoid hydrocarbons) the $-\Delta H_a$ values calculated by the HMO and SCF methods differ only slightly. In systems with a single Kekulé structure (for instance, in polyenes or fulvenes) calculation of W_b by both the HMO and the Pople SCF method is dubious. In such cases the molecular geometry must be duly considered and only methods in which the dependence of $\beta_{\mu\nu}^c$ on the bond length is properly taken into account lead to good results.

15.2 Delocalization energies of conjugated compounds

The delocalization energy is a quantitative measure of the extent of delocalization of the π-electrons and thus also an indicator of "aromaticity". The question of aromaticity will not be discussed in detail here. However,

386

it should be mentioned that for years incorrect and confused ideas have been held in this field; one of them will be mentioned later.

For calculation of delocalization energies from experimental data one of the two methods described below is usually used. The first (more accurate but seldom applied) compares the heat of hydrogenation of the conjugated molecule with the heat of hydrogenation of a fictitious molecule which has the same number of double bonds which, however, are not conjugated. The second method is based on comparison of the heat of formation calculated from the heat of combustion with that calculated by summing up the bond energies. The difference between the two heats of formation gives the experimental delocalization energy.

Before the MO delocalization energies (see Chapter 11) are compared with the values obtained from experimental data, it must be noted that the experimental delocalization energy (for example, for benzene) characterizes the difference between the energy of benzene (I) and localized (fictitious) structure II.

I   II   III

This follows from the fact that the empirical contributions used in the calculation were obtained from the data for paraffinic and oleofinic hydrocarbons. *The MO delocalization energies*, in turn, characterize the energy difference between form I and fictitious form III, which is cyclohexatriene with a uniform (benzene) bond length. The delocalization energy of the form III is called *the vertical energy*. In order to calculate these energies from experimental delocalization energies, it would be necessary to know the energy which would have to be supplied to system II for its conversion to system III. This is the energy required for shortening the single and elongating the double bonds of structure II to a uniform (benzene) bond length. This is called *the distortion energy*. Its values are not commonly available but there appears to be a close correlation between the MO and the experimental delocalization energies, as the distortion energies increase roughly parallel to the delocalization energies. Correlations between calculated and observed values afford the possibility of using the HMO delocalization energies for systems in which the approximations of the HMO method are justified. Generally, however, methods in which the molecular geometry is considered in more detail than in the HMO method and in the simple SCF method, i.e. methods containing the β^c variation (Fig. 15-1), must be used.

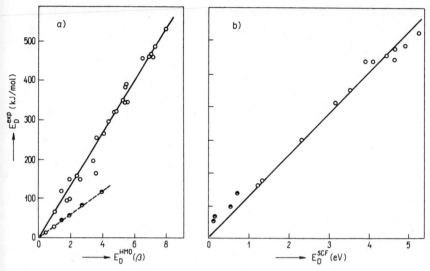

Fig. 15-1. Dependence of experimental delocalization energies on (*a*) HMO quantities, (*b*) SCF quantities (including β^c variation). Symbols: ○ alternant hydrocarbons and their derivatives; ◒ non-alternant hydrocarbons with only one Kekulé structure.

For years, the opinion that the delocalization energy is a measure of the aromaticity of a molecule has prevailed. Chemists generally understand the concept of aromaticity to refer to the stability of conjugated compounds in the sense of being a sort of "unwillingness" to react (in the sense which is characteristic for benzene). This "unwillingness" depends, of course, on the kinetic stability and not on the thermodynamic stability. The essence of this controversy lies in the fact that the delocalization energy is a thermodynamic characteristic which, in principle, gives no information on the height of the energy barriers (activation energies) for different reactions. It has been found empirically that systems with large delocalization energies often have large localization energies (cf. Section 16.5.1) in the individual positions and are thus also kinetically stable. Consequently, delocalization energies have frequently yielded correct estimates of the kinetic stability of molecules.

15.3 Stabilization of coordination compounds

The difference between the energy of the free ion and the ion in the complex is called the stabilization energy (CFSE, *crystal field stabilization energy*). Accurate calculation using the method of configuration inter-

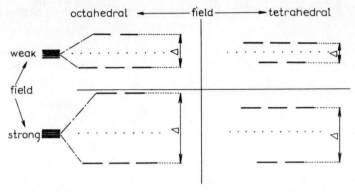

Fig. 15-2. Splitting of degenerate d levels in fields of different symmetry and different strength.

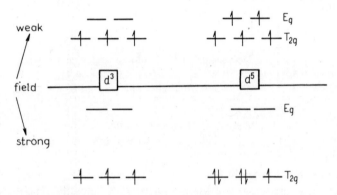

Fig. 15-3. Occupation of the T_{2g} and E_g levels by electrons in weak and strong octahedral fields.

actions is computer-time-consuming; however, the one-electron approach in terms of the simple crystal field method also leads to useful results.

The most important types of complexes, octahedral and tetrahedral complexes, will be discussed (Fig. 15-2), along with complexes with strong and weak ligand fields (complexes with low and high spins). In complexes with a strong field the splitting of the d orbital levels ($\Delta = 10Dq$) is relatively large and thus the levels with lower energy are occupied first. Conversely, if the splitting of the levels is small, they are occupied, as far as possible, by only one electron per orbital, corresponding to a high spin complex (cf. Section 10.6.2). For illustration, octahedral d^3 and d^5 complexes will be discussed (Fig. 15-3). In both d^3 complexes the stabilization energy amounts to $-12Dq$. In the second example, in the d^5 complex the conditions are quite different. If the $10Dq$ value is large, configuration T^5 is preferable, corresponding to a doublet; the stabilization energy amounts to $-20Dq$ (5×4). In turn, if the value of $10Dq$ is small then the complex exists in the T^3E^2 (sextet) configuration and the stabilization energy equals zero: the energy gain corresponding to three electrons located in the T_{2g} orbitals is compensated by the loss of energy induced by the occupation of the E_g orbitals by two electrons.

This conclusion can easily be generalized and the stabilization energy in octahedral complexes with n d electrons, of which x are in the T_{2g} orbitals, can be expressed by the relationship

$$\Delta E(\text{CFSE}) = -[4x - 6(n - x)]\,Dq \tag{15-8}$$

Analogously, for tetrahedral complexes it holds that

$$\Delta E(\text{CFSE}) = [6(n - x) - 4x]\,Dq \tag{15-9}$$

In Tables 15-2 and 15-3 the stabilization energies of octahedral and tetrahedral complexes are given. The more correct procedure requires that, in complexes with low spin, an additional term be considered in the expressions for the stabilization energy, corresponding to the spin pairing energy.

The stabilization energies obtained can be employed in the solution of thermodynamic and kinetic problems. The former can be illustrated by the heats of hydration of the ions of the transition metal elements; the latter is mentioned in the next chapter.

Table 15-2

Stabilization Energies (CFSE) of Octahedral Complexes Expressed in Multiples of Dq

Number of d electrons	Complexes				
	with high spin		with low spin		
	configuration	CFSE	configuration	CFSE	
1	T^1	-4	T^1	-4	
2	$T^{2\ a}$	-8	T^2	-8	
3	T^3	-12	T^3	-12	
4	T^3E^1	-6	T^4	-16	
5	T^3E^2	0	T^5	-20	
6	T^4E^2	-4	T^6	-24	
7	$T^5E^{2\ a}$	-8	T^6E^1	-18	
8	T^6E^2	-12	T^6E^2	-12	
9	T^6E^3	-6	T^6E^3	-6	

[a] The more accurate CI calculation leads to a value of -6 to $-8Dq$; this more correct description shows that further configurations play a role in the ground state.

Table 15-3

Stabilization Energies (CFSE) of Tetrahedral Complexes Expressed in Multiples of Dq

Number of d electrons	Complexes			
	with high spin		with low spin	
	configuration	CFSE	configuration	CFSE
1	E^1	-6	E^1	-6
2	E^2	-12	E^2	-12
3	$E^2 T^{1\,a}$	-8	E^3	-18
4	$E^2 T^2$	-4	E^4	-24
5	$E^2 T^3$	0	$E^4 T^1$	-20
6	$E^3 T^3$	-6	$E^4 T^2$	-16
7	$E^4 T^3$	-12	$E^4 T^3$	-12
8	$E^4 T^{4\,a}$	-8	$E^4 T^4$	-8
9	$E^4 T^5$	-4	$E^4 T^5$	-4

[a] See the note to Table 15-2.

By plotting the heats of hydration of divalent or trivalent ions, for example, in the series $Ca^{2+} \ldots Zn^{2+}$ against the number of d electrons, in place of the linear dependence a curve with two extremes is obtained. If, however, the stabilization energies of the individual complexes are subtracted from the heats of hydration, the expected simple dependence is obtained.

REFERENCE

1. Chung A. L. H., Dewar M. J. S.: *J. Chem. Phys.* **42**, 756 (1965).

16. CHEMICAL REACTIVITY

16.1 Introductory comments

Although theoretical studies of various physical properties of inorganic and organic compounds have been of particular interest in recent years, it is nevertheless evident that the theory of chemical reactivity is (in a narrower sense) the most important theory in chemistry[1-11].

Chemical reactivity includes both equilibria and the rates of processes. It should be noted that this part of the theory of the chemical bond has not yet been treated in sufficient detail. Although a great many experimental studies are published every year on equilibria and the rates of chemical processes, the amount of suitable and well-defined experimental data available for theoretical purposes is still small.

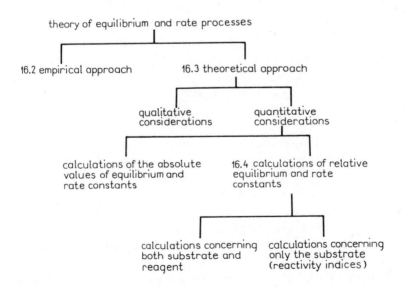

Fig. 16-1. Survey of approaches discussed (the numbers denote the sections of this chapter).

392

Analysis of the present state of the theory of chemical reactivity is comparatively complex and therefore the individual aspects will be discussed according to the arrangement in Fig. 16-1.

Most topical from the theoretical point of view are attempts to calculate the absolute values of equilibrium and rate constants. In order to give a relatively wide survey of present possibilities and trends, also the calculation of relative values will be discussed in greater detail and, for the sake of completeness, empirical procedures and more recent theoretical methods, which might find wider use in the future, will also be mentioned.

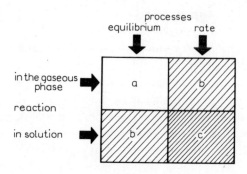

Fig. 16-2. State of theoretical studies in the field of chemical reactivity. a — Absolute values of equilibrium constants can be calculated. b — Regions which look promising for simple reactions. c — Calculations of absolute rate constants will be feasible only in the future; progress depends on advances in regions sub b.

Before discussing the individual theories, it should be noted that the difficulties encountered when dealing with the rates of processes are qualitatively greater than with equilibria. The situation is similar when passing from the gaseous phase into solution; it has so far been very difficult to calculate solvation energies, but now the outlook in this field is relatively favourable. The most topical field in chemistry, the kinetics of reactions in solutions, is, unfortunately, the most complicated from the theoretical point of view. The situation is surveyed in Fig. 16-2. The growing difficulty of theoretical processing is expressed by a darker shade for the corresponding field.

Whereas the procedure to be employed in the calculation of equilibrium constants in the gaseous phase (see below) is now clear, the calculation of the rate of processes is rather complicated. There are several theoretical possibilities and it is not always easy to decide which to choose. The procedures used so far can be divided into two large groups; those which use *energy hypersurfaces* and those which do not use them. At present only methods of the first group can be employed for numerical calculations. They can be divided according to whether an equilibrium between the reactants and the activated complex is assumed (relatively long-lived complexes; *the theory of absolute reaction rates*) or not (short-lived intermediate complexes; *direct processes*). Although there have been certain objections to the theory of absolute reaction rates, it is at present the only theory suitable for quantitative studies of the reactions which are of the great interest to chemists.

In the following sections the above-described classification will be employed.

16.2 Empirical approach

A relatively large number of empirical relations have been described, which, for some reactions in compounds of certain structural types, enable correlation of experimental equilibrium and rate constants, or of another quantity connected with them[12-18]. Relations of this type are only marginally interesting here; nevertheless, they are worth mentioning since they have played a very interesting role in the last three decades, especially in organic chemistry, and they cannot be considered to have been surpassed in application even from the point of view of recent, theoretically better founded, procedures. They have not much in common with the actual, physically based theory, however.

One of the oldest and probably also most important of these empirical equations is the Hammett equation; Hammett observed that a straight line is obtained when the logarithms of equilibrium (or rate) constants for one series of *m*-substituted and *p*-substituted benzene derivatives are plotted against data for another series. *Functional groups* and *substituents* which influence the property of the functional group will be distinguished. These are, of course, relative concepts; when studying, for example, the acidity of *m*-substituted and *p*-substituted benzoic acids, the carboxyl group is the functional group and the second group bonded to the benzene nucleus is a substituent. In Fig. 16-3 the original Hammett dependence[14] between logarithms of the rate constants for the hydrolysis of the esters of the benzoic acids [alkaline hydrolysis of ethyl-benzoates, 87.8% ethanol, 30° C; Eq. (16-1)] and for the dissociation of the benzoic acids [water, 25° C; Eq. (16-2)] is depicted.

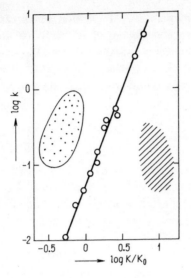

Fig. 16-3. Dependence of the logarithms of the rate constants for the hydrolysis of the esters of substituted benzoic acid on the logarithm of the relative dissociation constants of the corresponding benzoic acids: *m*- and *p*-substituted derivatives (O), *o*-derivatives (hatched area), aliphatic compounds (dotted area).

$$\text{X}\!-\!\!\left\langle\!\!\!\bigcirc\!\!\!\right\rangle\!\!-\!COOC_2H_5 \quad \xrightarrow{\;OH^{\ominus}\;} \quad \text{X}\!-\!\!\left\langle\!\!\!\bigcirc\!\!\!\right\rangle\!\!-\!COO^{\ominus} \;+\; C_2H_5OH \tag{16-1}$$

$$\text{X}\!-\!\!\left\langle\!\!\!\bigcirc\!\!\!\right\rangle\!\!-\!COOH \quad \rightleftharpoons \quad \text{X}\!-\!\!\left\langle\!\!\!\bigcirc\!\!\!\right\rangle\!\!-\!COO^{\ominus} \;+\; H^{\oplus} \tag{16-2}$$

The reactions studied occur on the side chain. For the dependence given in Fig. 16-3, the relationship

$$\log k_i = \varrho \log K_i + A \tag{16-3}$$

is valid, where k_i (K_i) denotes the rate (equilibrium) constant of reaction (16-1) or (16-2) for the i-th substituent and k_0 (K_0) are similar quantities for reference substances which are unsubstituted derivatives, i.e. benzoic acid and its ethyl ester; ϱ and A are constants. The expression $\varrho \log K_0$ is added to and subtracted from the right-hand side of Eq. (16-3):

$$\log k_i = \varrho(\log K_i - \log K_0) + (A + \varrho \log K_0) \tag{16-4}$$

The expression $(A + \varrho \log K_0)$ is equal to $\log k_0$ and it is therefore valid that

$$\log k_i - \log k_0 = \varrho(\log K_i - \log K_0), \tag{16-5}$$

where k_i refers to the rate of ester hydrolysis. Hammett demonstrated that Eq. (16-5) is satisfied (with different values of constants ϱ) for many series of equilibrium and rate data; apparently a reaction series can be chosen and then employed as reference series; in all further correlations the data for this series will serve as "universal constants". The dissociation constants of the benzoic acids (water, 25° C) were chosen for this purpose; the expression $\log K_i - \log K_0$ is then denoted σ. Thus, Eq. (16-5) can be written in the form

$$\log \frac{K_i}{K_0} = \varrho\sigma \qquad (16\text{-}6)$$

Quantity σ is called the substituent constant and its value depends upon the nature of the substituent and is independent of the reaction. ϱ is referred to as the reaction series constant and applies for all substituents and depends on the given reaction series. Examples of values of these constants are given in Tables 16-1 and 16-2.

The Taft equation[17] is applied for similar correlations in aliphatic series and has the same form as the Hammett equation:

$$\log \frac{K_i}{K_0} = \varrho^*\sigma^* \qquad (16\text{-}7)$$

Table 16-1

The Hammett Substituent Constants [Ref. 15]

Substituent	σ_m	σ_p	Substituent	σ_m	σ_p
$-N(CH_3)_2$	-0.211	-0.83	$-Cl$	0.373	0.227
$-NH_2$	-0.16	-0.66	$-Br$	0.391	0.232
$-OCH_3$	0.115	-0.268	$-CHO$	0.381	1.13
$-CH_3$	-0.069	-0.170	$-CN$	0.678	0.660
$-H$	0	0	$-COOCH_3$	0.37	0.45
$-F$	0.337	0.062	$-NO_2$	0.710	0.778

Table 16-2

The Hammett Constants of Reaction Series (Equilibrium and Rate Data) [Ref. 14]

Reaction	ϱ	$\log K_0$
Alkaline hydrolysis of ethylbenzoates, 87.8% ethanol, 30 °C	2.498	-3.072
Ionization of phenols, H_2O, 25 °C	2.008	-9.941
Acid hydrolysis of benzamides, H_2O, 100 °C	0.118	-3.513
Solvolysis of benzoylchlorides, ethanol, 0 °C	1.529	-4.071
Reaction of anilines with dinitrochloronaphthalene, ethanol, 25 °C	-3.690	-1.641
Reaction of phenyldiethylphosphines with ethyliodide, acetone, 35 °C	-1.088	-3.286

The meaning of the quantities is similar to those in Eq. (16-6); σ^* characterizes the polar influence of the substituent. An additional term is introduced into Eq. (16-7) in order to describe the steric effect.

A linear relation[18] has also been used for the correlation of biological activities; the "$\alpha\beta$ equation" is used for correlation in aliphatic series when a physical process determines the activity (for example, the partitioning of biologically active substances among various components of the biophase, passage through the cell wall):

$$\log \frac{\tau_i}{\tau_0} = \alpha\beta, \tag{16-8}$$

where τ_i is the biological activity of the i-th member of the series, β is the constant of the i-th substituent and α is a constant characterizing the respective biological system.

16.3 Theoretical approach

Here procedures which lead, at least in principle, to absolute values of equilibrium and rate constants will be treated. Initially, however, a procedure which determines qualitatively (or semiquantitatively) whether or not a particular reaction can be realized under the given conditions will be discussed. Application of such a procedure also indicates the changes which have to be made in the reactant in order to make the reaction (which under the originally chosen conditions seemed impossible) more probable.

16.3.1 Qualitative considerations

If a given chemical transformation which does not occur spontaneously sufficiently rapidly under normal laboratory conditions (laboratory temperature, pressure of about 100 kPa and dilute solutions) is to be carried out, an attempt can be made to influence it by increasing the temperature (thermal initiation) or by changing the solvent. In some cases this purpose is achieved, generally when the orbital and the spin parts of the wave functions of the reactants and products (or the activated complex) are correlated.

With very many processes, however, this condition is not fulfilled. for example

$$N_2O\,(^1\Sigma^+) \quad \rightarrow \quad N_2(^1\Sigma_g^+) + O\,(^3P) \tag{16-9}$$

$$N_2O\,(^1\Sigma^+) \quad \rightarrow \quad N\,(^4S) + \cdot NO\,(^2\Pi) \tag{16-10}$$

$$\text{(16-11)} \quad \underset{H_2C}{\overset{H_2C}{\diagup}}\hspace{-6pt}\diagdown C=O \longrightarrow \begin{array}{c} CH_2 \\ \| \\ CH_2 \end{array} + CO$$

$$2\ C_2H_4 \longrightarrow \begin{array}{ccc} H_2C & \!\!\!-\!\!\! & CH_2 \\ | & & | \\ H_2C & \!\!\!-\!\!\! & CH_2 \end{array} \quad \text{(16-12)}$$

Reactions (16-9) and (16-10) are spin-forbidden; reactions (16-11) and (16-12) are spin-allowed; however, they are orbitally forbidden (cf. Section 11.2.6).

Nevertheless, if the transformation is to be realized, it can frequently be made possible by introducing certain agents which do not change the catenation of the atoms in the studied molecule and whose effect is reversible and is often manifested by a change in the symmetry of the frontier orbital and also by a change of the multiplicity of the state. Thus the principal obstacle in realizing the reaction by thermal initiation is supressed and the reaction often occurs. This effect can most frequently be produced by particles such as photons, electrons, protons or electronically excited atoms.

The most important products of these changes are radical ions (cations and anions) and electronically excited states which usually have

Survey of Studied Types of Reactions: Table 16-3
Parallelism between Non-catalytic and Catalytic Reactions

Formation of	Reactions	
	non-catalytic	catalytic
radical cation	1. Action of oxidants (e.g. Lewis acids) 2. Electrooxidation 3. Electron impact	1. p-type semiconductor: depletive chemisorption 2. n-type semiconductors: cumulative chemisorption
radical anion	1. Action of reducing agents (e.g. alkali metals) 2. Electroreduction 3. Absorption of thermal electrons	1. n-type semiconductors: depletive chemisorption 2. p-type semiconductors: cumulative chemisorption
electronically excited state	Interaction with energy-rich particles: photons, electrons, electronically excited atoms (photosensitization)	Interaction of the reacting substance with a transition element involving back donation

398

the character of biradicals. Influencing the reactivity of substrates or reagents in this respect has a long tradition in chemistry (radical and photochemical reactions). Quantum chemists have recently paid a great deal of attention to both of these types of reactions, as well as to the electronic structure of radical ions and electronically excited states.

A further extensive and important field of chemical dynamics is *catalysis*, in which the transfer of electrons (with formation of radical ions or doubly charged ions, as well as states similar to electronically excited states) has a key position; it is therefore opportune to utilize experience with radicals and excited molecules for studies of catalytic processes. In Table 16-3 a survey of the types of processes studied is given.

Before discussing these processes in greater detail, the requirements which must be fulfilled if the studied process is to proceed relatively easily should be noted: *a*) conservation of the total angular momentum, *b*) conservation of the total spin and *c*) conservation of orbital symmetry in all the concerned elementary steps of the reaction[19].

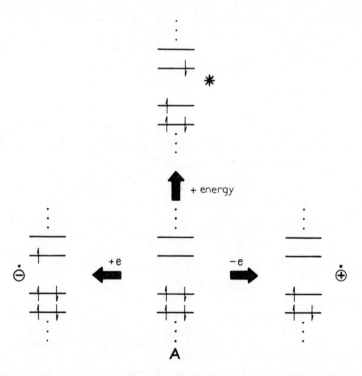

Fig. 16-4. Electron configuration of initial system A, of the system after ionization (radical cation), of the system after acceptance of an electron (radical anion), and after electronic excitation (*).

Changes accompanying ionization, acceptance of an electron and electronic excitation

It is usually not important whether simple (HMO, EHT) or more complicated (PPP, CNDO/2) methods are employed to obtain qualitative and semiquantitative information.

In Fig. 16-4 the formation of radical cations (ionization), radical anions (acceptance of an electron) and excited states (the lowest possible, corresponding to the $N \rightarrow V_1$ excitation) from system A are schematically represented:

$$A \rightleftarrows A^{\overset{\cdot}{\oplus}} + e \qquad (16\text{-}13)$$

$$A + e \rightleftarrows A^{\overset{\cdot}{\ominus}} \qquad (16\text{-}14)$$

$$A + h\nu \rightleftarrows A^* \qquad (16\text{-}15)$$

Table 16-4

Common Properties of Radical Cation, Radical Anion and Electronically Excited State in Relation to the Singlet Ground State
(EA is electron affinity, IP is ionization potential)

Radical cation	Electronically excited state $(N \rightarrow V_1)$	Radical anion
High EA (easy reducibility)	High EA (easy reducibility)	—
—	Low IP (easy oxidizability)	Low IP (easy oxidizability)
$\longleftarrow$ The longest wavelength transition is frequently located in the near IR region[a] $\longrightarrow$		
$\longleftarrow$ Numerous transitions in the visible (or near IR) region $\longrightarrow$		
Weakening of bonding in molecule	Weakening of bonding in molecule	Weakening of bonding in molecule
Elongation of some bonds[b]	Elongation of the majority of bonds	Elongation of some bonds[b]
Great increase in reactivity towards nucleophilic reagents	Great increase in reactivity towards electrophilic and nucleophilic reagents	Great increase in reactivity towards electrophilic reagents
Paramagnetism	In T state – paramagnetism	Paramagnetism
Change of symmetry of the state	Change of symmetry of the state	Change of symmetry of the state
Change of multiplicity	In T state – change of multiplicity	Change of multiplicity

[a] In the frontier orbital theory, condition for an easy monomolecular decomposition.
[b] Different bonds are elongated as a rule in anions and cations.

The orbital schemes given in Fig. 16-4 demonstrate the reasons for the similarity between radical ions and excited states. For instance, a radical anion has an electron located in the lowest antibonding MO, similarly for the lowest electronically excited state. The similarities are surveyed in Table 16-4.

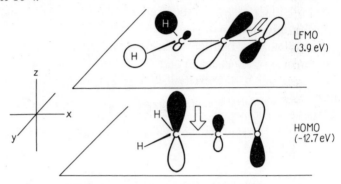

Fig. 16-5. Graphical representation of the CNDO/2 frontier orbitals in ketene, CH_2CO: the highest occupied (HOMO) and the lowest free (LFMO) molecular orbital. The light and dark parts of the orbital denote positive and negative signs of the orbital, respectively. A bond which is weakened by ionization (HOMO) or by acceptance of an electron (LFMO) is indicated by an arrow.

For illustration, the reactivity of ketene and of fulvene and of their radical ions, and excited states (including certain catalytic reactions[19]) will be discussed. The frontier orbitals of ketene are depicted in Fig. 16-5. It is known from experimental studies that the different "forms" of ketene decompose to give carbene and carbon monoxide (or their ions) or atomic oxygen and C_2H_2 (an isomer of acetylene or its ion):

$$CH_2CO^{\oplus}(D) \rightarrow \ \overset{\displaystyle H}{\underset{\displaystyle H}{\diagdown \diagup}}C^{\oplus}(D) + CO(^1\Sigma) \qquad (16\text{-}16)$$

$$CH_2CO^{\ominus}(D) \rightarrow \ \overset{\displaystyle H}{\underset{\displaystyle H}{\diagdown \diagup}}C=C^{\ominus}(D) + O(^3P) \qquad (16\text{-}17)$$

$$CH_2CO^*(S) \rightarrow \ \overset{\displaystyle H}{\underset{\displaystyle H}{\diagdown \diagup}}C(^1B_1) + CO(^1\Sigma) \qquad (16\text{-}18)$$

In Fig. 16-5 it can be seen that the removal of an electron from the HOMO of ketene is connected with weakening of the C—C bond [cf. reaction (16-16)] and the acceptance of an electron into the LFMO, on the other hand, weakens the C=O bond [cf. reaction (16-17)]. The photochemical decomposition of ketene into carbene and CO suggests that the N→V$_1$ excitation is connected with a greater weakening of the C=C bond than of the C=O bond.

The conditions for the decomposition of fulvene can be similarly examined, although the experimental data are not complete here. Whereas fulvene in the ground state can be considered to contain three weakly interacting double bonds (this feature might be important in thermally iniciated decomposition which could lead to two molecules of acetylene

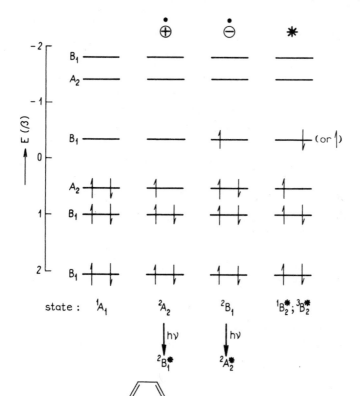

Fig. 16-6. HMO energy of fulvene (C_{2v}), of its radical ions, and of the N→V$_1$ excited state. In the figure the symmetry of the molecular orbitals and symmetry of the corresponding states are given.

and vinylidene), a different structure and another decomposition path are to be expected with radical ions and in the $N \rightarrow V_1$ excited state (Fig. 16-6) of fulvene. This is due to the fact that the changes discussed appear to be connected with a significant decrease in the bond orders of the bonds that, in the parent substance, have the character of double bonds (Fig. 16-7). The same decomposition can be expected on interaction of fulvene with a catalyst containing atoms of a transition element; the expected decomposition would lead to acetylene and methylenecyclopropene (Fig. 16-8; cf. Fig. 16-7):

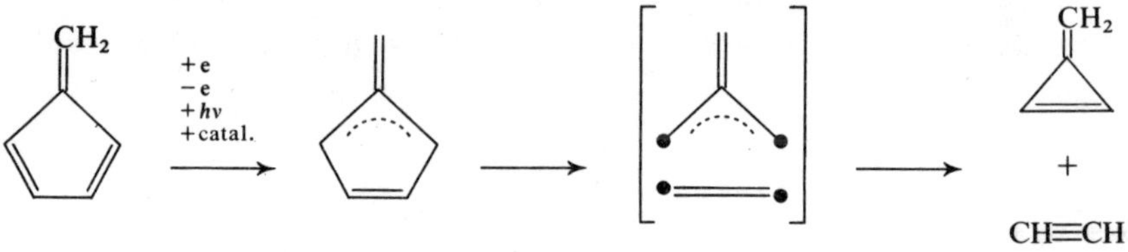

(16-19)

Next, the influence of various changes will be discussed semi-quantitatively. Under normal experimental conditions the *cis-trans* isomerisation of butene-2 [Eq. (16-20)]

(16-20)

is generally accompanied by migration of the double bond leading to butene-1 [Eq. (16-21)]:

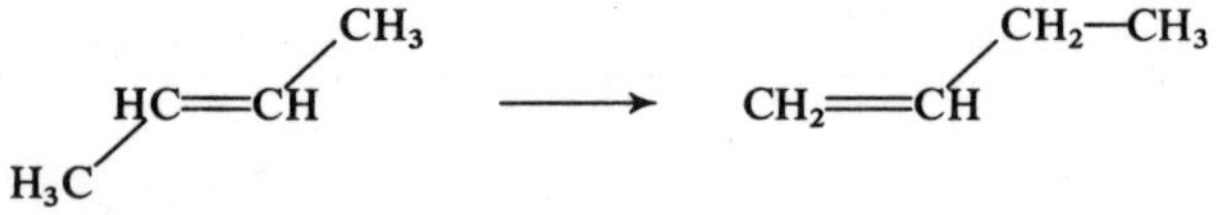

(16-21)

The activated complex in the model reaction (16-20) corresponds to "perpendicular" butene-2

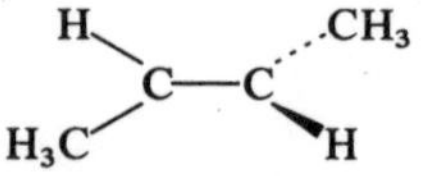

In Fig. 16-9 the results of EHT energy calculation for the *cis* and *trans* isomers as well as for the activated complex are given. The model considered for the interaction of the two forms of butene-2 with a metal atom can be seen in Fig. 16-10a. Course (*a*) corresponds to the

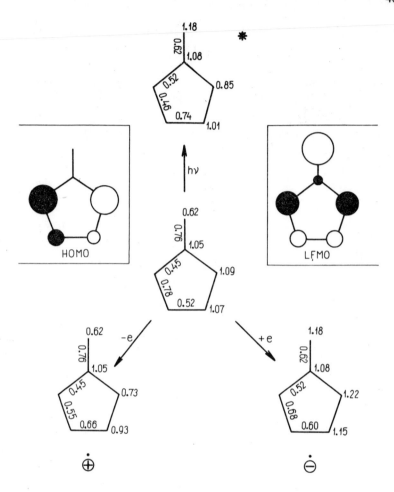

Fig. 16-7. Molecular diagrams of the systems in Fig. 16-6 and electron distribution in frontier orbitals (HOMO, LFMO).

thermally initiated reaction, in courses (b) and (c) the radical ion is the reactant, in course (d) we have the electronically excited state ($N \rightarrow V_1$, transition $12 \rightarrow 13$) and finally (e) and (f) correspond to the reaction catalyzed by iron (cf. Fig. 16-10b) and to the reaction occurring in the presence of Ag^+. The results are also characteristic for numerous other processes: the transition from the parent substance to the radical ion causes a substantial decrease in the activation energy for isomeration and the transition to the excited state is connected with an energy minimum on the potential energy curve, which corresponds to "perpendicular" butene-2. This is in agreement with the experimentally found

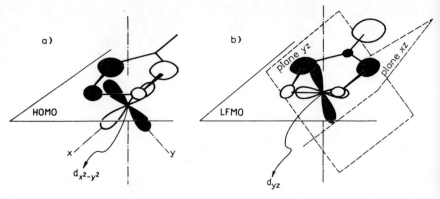

Fig. 16-8. Interaction of the frontier orbitals of fulvene with the d orbitals of a transition element located below the fulvene plane. (*a*) The $d_{x^2-y^2}$ atomic orbital is free and accepts electrons from the HOMO of fulvene. (*b*) The d_{xz} and d_{yz} orbitals are occupied and electrons pass from them into the LFMO of fulvene. For convenience, the d_{xz} orbital is not depicted in the figure but the *xz* plane is marked.

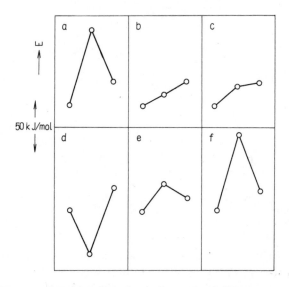

Fig. 16-9. EHT energy of *cis*-butene-2 and *trans*-butene-2 and of the corresponding activated complex: *cis* ⇌ [ ]‡ ⇌ *trans*. (*a*) Thermal initiation, (*b*) radical cation, (*c*) radical anion, (*d*) N→V$_1$ excited state, (*e*) effect of Fe, (*f*) effect of Ag.

perpendicular arrangement in the first excited state of ethylene. Transition elements do not have such a strong influence; they have merely a catalytic effect or, more accurately, lead to a decrease in the energy barrier from 104 to 42 kJ/mol. It should be added that the migration of the double

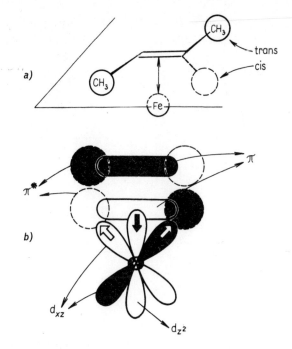

Fig. 16-10. Model describing catalysis of the *cis-trans* isomerization of butene-2 by iron. (*a*) Mutual orientation of the components, (*b*) flow of electrons from the π-MO of the hydrocarbon into the d_{z^2} orbital of iron and back donation from the d_{xz} orbital of iron into the π^*-MO of the hydrocarbon. The direction of electron flow is indicated by thick arrows. The positive and negative parts of the orbitals are light and dark, respectively.

bond, which usually accompanies *cis-trans* isomerisation, can be catalyzed in a similar manner, but this reaction has not yet been investigated semiquantitatively.

The next example is also interesting from the catalytic point of view. This is the cycloaddition of ethylene leading to cyclobutane. As demonstrated by Hoffmann, this process[20] is forbidden by symmetry in the ground state. It becomes allowed either through excitation or through catalysis[21] (Fig. 16-11). There is no doubt that the barrier for the thermally initiated process is decreased if one of the reactants is transformed into a radical ion or into the $N \rightarrow V_1$ excited state. The application of semiempirical methods encounters certain difficulties and thus only a qualitative description (Fig. 16-11) based on the theory of the frontier orbitals will be given. In the upper part of Fig. 16-11 a diagram of orbital energies, for both ethylene molecules (1, 2) consisting of the π and π^* levels, is given. It is evident from the figure that, when both molecules are in the ground state, significant donor-acceptor interaction cannot occur. If, however, one of the molecules is in the form of a radical ion or in an electronically excited state, the situation changes considerably, due to the fact that a strong donor, a strong acceptor or even a system which is simul-

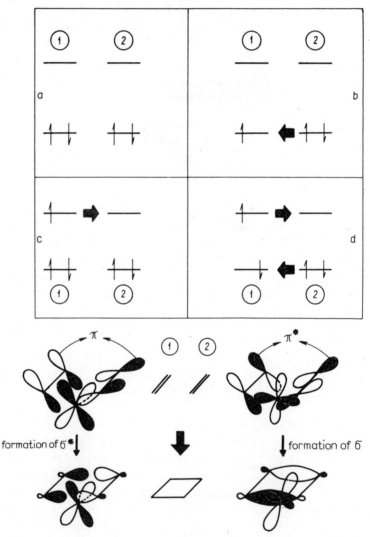

Fig. 16-11. Interaction between two molecules of ethylene (denoted by ① and ②) in different forms leading to different forms of cyclobutane: (*a*) thermal initiation (orbitally forbidden reaction), (*b*) radical cation, (*c*) radical anion, (*d*) excited state, (*e*) catalysis by an atom of a transition element. Formation of σ* orbitals in cyclobutane depends on participation of the d orbital lying in the plane perpendicular to the C=C bond of ethylene, formation of the σ orbital is mediated by the d orbital located in the plane parallel to the C=C bond.

taneously a strong donor and a strong acceptor is involved in the interaction. The direction of the electron flow is represented by an arrow. In the bottom part of Fig. 16-11 the catalytic effect of an atom of a transition element with an occupied d_{yz} orbital and an unoccupied $d_{x^2-y^2}$ orbital is depicted graphically. It can be seen in the left (lower) part of the figure that the electron flow from the occupied π orbitals of the two ethylene molecules into the empty $d_{x^2-y^2}$ orbital is connected with the formation of a σ^* orbital in cyclobutane; in the right-hand (lower) part of the figure the formation of bonding σ orbitals in cyclobutane by electron flow from the d_{yz} orbital into the antibonding π^* orbitals of the two ethylene molecules is depicted.

The last example is the decomposition of nitrogen monoxide[19]. N_2O is a relatively stable oxide; its decomposition into N_2 and O begins only at temperatures of 800 to 900° C. The dissociation of the $\dot{N}-O$ and $N-N$ bonds in the ground state (thermal initiation) is spin-forbidden:

$$N=N=O\,(^1\Sigma^+) \quad \rightarrow \quad N_2\,(^1\Sigma_g^+) + O\,(^3P) \qquad (16\text{-}22)$$

$$N=N=O\,(^1\Sigma^+) \quad \rightarrow \quad N\,(^4S) + \cdot NO\,(^2\Pi) \qquad (16\text{-}23)$$

Decomposition in the sense of reaction (16-23) is also orbitally forbidden.

The spin and orbital forbiddenness can be overcome by electronic excitation or by changing the molecule to the radical ion form.

Before discussing the decomposition of electronically excited forms and of radical ions, the structure of the molecular orbitals in N_2O must be described (Fig. 16-12). The following Wiberg indices* correspond to nitrogen monoxide:

$$N \underline{\quad 2.42 \quad} N \underline{\quad 1.49 \quad} O$$

The semiempirical LCI-SCF theory of the CNDO/2 type aptly describes both excited singlet states at 150 and 120 nm; the CI data together with information on the nature of the individual MO's (Fig. 16-12) can be used to interpret the structure of the isolated decomposition products. It must be borne in mind that the removal of an electron from the bonding MO and the introduction of an electron into the antibonding MO leads to a weakening of the respective bond.

The decomposition of N_2O photosensitized by mercury will now be discussed:

$$Hg(^3P_1) + N_2O(^1\Sigma^+) \quad \rightarrow \quad N_2(^1\Sigma_g^+) + O(^3P) + Hg(^1S_0) \quad (16\text{-}24)$$

---

* In the framework of the CNDO/2 method, this index is defined by the expression [cf. Eq. (10-26)]

$$P_{MN} = \sum_{\mu\in(M)} \sum_{\nu\in(N)} P_{\mu\nu}^2$$

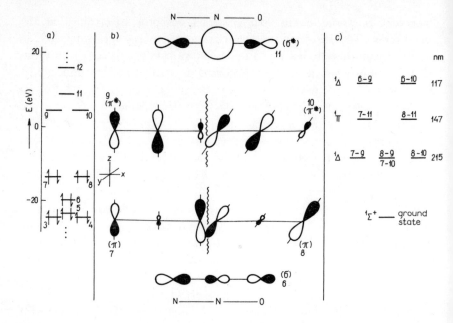

Fig. 16-12. CNDO/2 characteristics of N_2O. (*a*) Orbital energies, (*b*) the 6th to 11th molecular orbitals (π states are degenerate), (*c*) LCI-CNDO/2 energies of the ground state and of excited singlet states. Symmetry data are given for the individual states; the positions of the excited states are characterized by the wavelength of theoretical transitions from the ground state into the excited states.

This is a spin-and-orbital allowed process; the decomposition proceeds smoothly.

Three experimentally studied N_2O decomposition paths will be examined in their excited states.

(*i*) Excitation in the 180 nm region. A transition into the first triplet state occurs. It can be expected from experience that the electron distribution will be similar as in the first excited singlet state and that similar products can therefore also be expected. It has been experimentally determined that the decomposition takes the following course:

$$N_2O(^3\Pi) \Big\langle \begin{array}{l} \nearrow N_2(^1\Sigma_g^+) + O(^3P) \qquad\qquad (16\text{-}25) \\ \searrow N(^4S) + \cdot NO(^2\Pi) \qquad\qquad (16\text{-}26) \end{array}$$

The $N \rightarrow V_1$ excitation is connected with a somewhat greater weakening of the $N-N$ bond (cf. Fig. 16-12), which explains the disappearance of both the $N-N$ and $N-O$ bonds.

(*ii*) Excitation in the 147 nm region. This band corresponds to transitions $7-11$ and $8-11$ ($^1\Pi \rightarrow {}^1\Sigma^+$). It can be seen from the nature of the respective MO's (Fig. 16-12) that excitation affects the $N-N$ bond a little more than the $N-O$ bond; thus both decompositions can also be expected in this case, in agreement with experimental results:

$$N_2O(^1\Pi) \begin{cases} \nearrow N_2(^3\Sigma_u^+) + O(^3P) & (16\text{-}27) \\ \searrow N(^2D) + \cdot NO(^2\Pi) & (16\text{-}28) \end{cases}$$

(*iii*) Excitation in the 124 nm region leads to the decomposition:

$$N_2O \text{ (the excited S-state, probably } {}^1\Delta) \rightarrow \cdot NO(^2\Pi) + N(^2P) \quad (16\text{-}29)$$

Theory attributes the band in this region to excitations $6-9$ and $6-10$ (transition $^1\Delta \leftarrow {}^1\Sigma^+$). This excitation corresponds in both MO's to more significant weakening of the $N-N$ bond, in agreement with experimental results.

With reaction on a catalyst, which acts simultaneously as an electron acceptor and donor (this type of donation ability is called "back donation"), breaking of the $N-N$ and $N-O$ bonds can be expected.

The $N_2O^{\dot\oplus}$ cations that are formed in the source of a mass spectrometer decompose according to the reaction

$$N_2O^{\dot\oplus}(^2\Pi) \rightarrow NO^\oplus(^1\Sigma^+) + N(^4S) \quad (16\text{-}30)$$

This is a spin-forbidden process, which is apparently made possible by predissociation. It has also been found that decomposition according to reaction (16-31) does not occur:

$$N_2O^{\dot\oplus}(^2\Pi) \nrightarrow N_2(^1\Sigma_g^+) + O^{\dot\oplus}(^4S) \quad (16\text{-}31)$$

The observed dissociation course can be interpreted qualitatively from the form of the highest occupied MO (Fig. 16-12), which is also in agreement with the experimental and theoretical course of the ion-molecular reaction

$$O^\oplus + N_2 \rightarrow [N_2O^{\dot\oplus}(^4\Sigma^-)] \rightarrow NO^\oplus + N \quad (16\text{-}32)$$

Ab initio CI calculations have shown that the $N_2O^{\dot\oplus}(^2\Pi)$ decomposition which occurs via the $^4\Sigma^-$ excited state (crossing of surfaces) to give $NO^\oplus$ and N requires 46 kJ/mol less than decomposition to N_2 and $O^\oplus$.

No such detailed calculations are available for the decomposition of radical ions which have been studied in protic and aprotic solvents by flash photolysis as well as catalytically on oxides acting as *p*-type semiconductors. Molecular nitrogen is formed in all cases:

$$N_2O^{\dot\ominus}(^2\Pi) \rightarrow N_2(^1\Sigma_g^+) + O^\ominus(^2P) \quad (16\text{-}33)$$

This is a spin-allowed decomposition, obviously energetically more favourable than further spin-allowed processes:

$$N_2O^{\ominus}(^2\Pi) \begin{cases} N^{\ominus}(^3P) + \cdot NO(^2\Pi) & \text{(16-34)} \\ N(^4S) + NO^{\ominus}(^3\Pi) & \text{(16-35)} \end{cases}$$

It is worth noting that the Wiberg index for the $N-O$ bond in the radical anion approaches values which are typical for single bonds:

$$N\underline{\quad^{2.42}\quad}N\underline{\quad^{1.49}\quad}O \quad \pm \quad N\underline{\quad^{1.87}\quad}N\underline{\quad^{1.20}\quad}O^{\ominus}$$

16.3.2 Quantitative considerations. Calculations of absolute values of equilibrium and rate constants

Reactivity theory attempts to calculate absolute values of equilibrium and rate constants when the molecular geometry of the reactants, products and activated complexes are known, using universal constants and the methods of statistical and quantum mechanics. The expressions for the calculation of these constants have already been available for more than three decades, so that the lack of completed calculations is surprising. Analysis of the respective expressions can explain this apparent paradox.

The results of chemical and statistical thermodynamics and reaction kinetics are used in the study of this entire problem. Consequently, for example, the expressions for equilibrium and rate constants are not derived here in terms of partition functions and enthalpy changes.

It is interesting to study the problem simultaneously from the thermodynamic and kinetic aspects. It appears that, if the concept of the formation of an activated complex which is in equilibrium with the reactants (and is therefore comparatively long-lived*) is used when studying kinetic problems, then the expressions for calculation of the equilibrium and rate constants are very similar. However, the similarity of these expressions gives no information on a possible parallel between the equilibrium and the rate constants in a series of structurally similar substances. Yet these constants are very often correlated. That the parallel is not trivial follows from the fact that the rate of transformation of the reactants to the products depends on the Gibbs energy of activation ($\Delta G^{\neq}$),

---

* The lifetime of a complex is related as a rule to the time required for rotating the molecule through 2π (of the order of 10^{-12} s). An activated complex is formed when the lifetime allows several rotations.

whereas the establishment of equilibrium is given by the Gibbs energy (ΔG) (cf. Fig. 16-13).

Still more critical is the fact that the theories which are to be compared have different validities. Whereas the expressions for calculation of the equilibrium constant are very generally valid and have a very sound physical basis (they depend solely on the validity of the second law of thermodynamics and statistical mechanics), the validity of the theory of absolute reaction rates is more limited, as it depends on satisfying the rather uncertain assumption of equilibrium between the reactants and an activated complex. It has been demonstrated experimentally with crossed molecular beams (in a high vacuum with particles of defined velocity and quantum state) that there are also direct processes, i.e. reactions in which the lifetime of the complex formed by the collision of the reactants is very short (approximately 10^{-12} s or less). There is, of course, no justification for interpreting these processes in terms of the theory of absolute reaction rates. The discovery of direct reactions has lead to a revival of the classical and quantum collision theories. Expressions for calculation of equilibrium and rate constants will now be given.

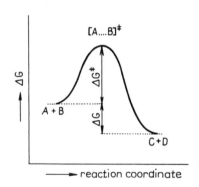

Fig. 16-13. Dependence of ΔG on the reaction coordinate for reactions A + B → C + D.

Equilibrium processes. Reaction (16-36) can serve as an example:

$$a\text{A} + b\text{B} + \ldots \overset{K}{\rightleftharpoons} m\text{M} + n\text{N} + \ldots, \tag{16-36}$$

For equilibrium constant K the usual expression holds where the activities can be replaced by concentrations, assuming ideal behaviour of the reactants and products:

$$K = \frac{[\text{M}]^m [\text{N}]^n \ldots}{[\text{A}]^a [\text{B}]^b \ldots} \tag{16-37}$$

$$\Delta G = \Delta H - T\Delta S = -RT \log K \tag{16-38}$$

412

The equilibrium constant can be expressed in terms of statistical thermodynamics:

$$K = \frac{Q_M^m Q_N^n \cdots}{Q_A^a Q_B^b \cdots} \, e^{-E_o/RT} \qquad (16\text{-}39)$$

(for the significance of the symbols see below).

Similar expressions are valid in the theory of absolute reaction rates (ART) for the rate constant.

Rate processes.

$$aA + bB + \ldots \; \overset{K^{\neq}}{\rightleftarrows} \; [C]^{\neq} \; \overset{\tilde{k}}{\rightarrow} \; \text{products} \qquad (16\text{-}40)$$

For the rate constant $\tilde{k}$ (this denotation is used locally to avoid confusion with the Boltzmann constant) of process (16-40) the following expression is valid:

$$\tilde{k} = \varkappa \frac{kT}{h} \, e^{-\Delta G^{\neq}/RT}, \qquad (16\text{-}41)$$

where $\varkappa$ denotes the transmission coefficient. Furthermore,

$$\Delta G^{\neq} = \Delta H^{\neq} - T\Delta S^{\neq} = -RT \log K^{\neq} \qquad (16\text{-}42)$$

Using statistical thermodynamics for calculation of the activation entropy change, it follows that

$$\tilde{k} = \varkappa \frac{kT}{h} \frac{Q^{\neq}}{Q_A^a Q_B^b \cdots} \, e^{-E_o^{\neq}/RT} \qquad (16\text{-}43)$$

Most of the symbols in Eqs. (16-36)−(16-43) have the usual significance: Q_X in Eq. (16-39) and (16-43) denotes the total partition function of substance X. The energy data in Eqs. (16-39) and (16-43) refer to absolute zero, the symbols without a double cross refer to calculation of the actual equilibrium, symbols with a double cross refer to description of rate processes.

The total partition functions can be expressed as the product of the partition functions for the individual energy forms:

$$Q = Q_t Q_r Q_{ir} Q_v Q_e Q_n, \qquad (16\text{-}44)$$

where symbols, t, r, ir and v denote the translation, rotation, internal rotation and vibration energies. Contributions with indices e and n are related to the energies of electrons and nuclei, respectively.

Before repeating the expressions for calculation of these partial partition functions, known from statistical thermodynamics, it must be noted that their "average" numerical values differ considerably (Table 16-5).

Table 16-5

Order of Magnitude of the Partition Function Values

Energy	Degrees of freedom	Numerical values
translation	3	10^{25}
rotation[a]	3	$10^2 - 10^3$
vibration (for each normal vibrational mode)	1	$1 - 10^b$
hindered rotation	1	$1 - 10$
electronic	–	1
nuclear	–	1

[a] Non-linear molecule.
[b] At lower temperatures (up to 500 K) the values approach unity.

For the individual partition functions it holds that

$$Q_t = \frac{(2\pi mkT)^{3/2}}{h^3} \quad \text{(per unit volume)} \tag{16-45}$$

$$Q_r = (\pi I_A I_B I_C)^{1/2} \left(\frac{8\pi^2 kT}{h^2} \right)^{3/2} * \tag{16-46}$$

$$Q_{ir} = \frac{(8\pi^3 I' kT)^{1/2}}{h} \int_0^{2\pi} \exp\left[-f(\alpha) \right] d\alpha, \tag{16-47}$$

where $f(\alpha)$ is a function parametrically dependent upon the height of the potential barrier for internal rotation and α denotes the angle of rotation of the rotating groups. For free rotation, the integral in expression (16-47) equals unity.

$$Q_v = \prod_i \frac{1}{1 - e^{-hv_i/kT}} \tag{16-48}$$

$$Q_e = \sum_i g_{e,i} e^{-\varepsilon_i/kT} = 1 \tag{16-49}$$

$$Q_n = 1 \tag{16-50}$$

For enumeration of these partial functions it is necessary to know the relative atomic (molecular) masses, the Boltzmann (k) and Planck (h) constants, the product of the moments of inertia with respect to three mutually perpendicular axes, $I_A I_B I_C$ (or the moment of inertia of internal rotation), and the normal modes of vibration, v_i. In the general case, Q_e and Q_n are not equal to unity but have this value in the majority of reactions interesting for chemists; two exceptions, however, are important:

---

* The symmetry number is assumed to equal unity.

first, instances where the first electronically excited state lies relatively close to the ground state (i.e. $\Delta E < 0.5$ eV), so that the excited state can be thermally populated even at "laboratory" temperatures. This is sometimes true for radicals and biradicals. In addition, there are cases where the value of ΔE is so large that the thermal population is not important and the excited state is populated through radiation supplied by an external source.

It can be shown that the product of the moments of inertia, $I_A I_B I_C$, is given by the determinant[22]

$$I_A I_B I_C = \begin{vmatrix} I_{xx} & -I_{xy} & -I_{xz} \\ -I_{xy} & I_{yy} & -I_{yz} \\ -I_{xz} & -I_{yz} & I_{zz} \end{vmatrix} \qquad (16\text{-}51)$$

The diagonal and off-diagonal elements have the form

$$I_{xx} = \sum_i m_i(y_i^2 + z_i^2) \qquad (16\text{-}52)$$

$$I_{xy} = \sum_i m_i x_i y_i \qquad (16\text{-}53)$$

In Eqs. (16-52) and (16-53) m_i denotes the mass of the i-th atom and x_i, y_i, z_i are its coordinates in any system of Cartesian coordinates which has its origin at the centre of gravity of the molecule. The summation is carried out over all the atoms.

The only remaining quantities to be mentioned are the normal modes of vibration. For numerous small molecules (containing 4 to 5 atoms), these quantities can be determined by analysis of infrared spectra. For the majority of molecules, however, whose reactivity is important here, these data are unknown. These values can, of course, be obtained by quantum mechanical calculation, which has been discussed in the section on vibrational spectra. The values of the normal modes of vibration λ_i were obtained there by solving the determinant equation

$$\left| \Sigma\, G_{ij} F_{ji} - \lambda \right| = 0 \qquad (16\text{-}54)$$

[cf. Eq. (13-60)].

The scheme in Fig. 16-14 indicates the procedure for the calculation to be performed for each reactant and product or activated complex. A method is sought which allows calculation in the whole indicated range with reasonable demands on computer time. In principle, semi-empirical and nonempirical methods come into consideration. There is a considerable number of problems still unsolved. The first step indicated in Fig. 16-14 concerns the calculation of minima on the energy hypersurface of each component. Minimization must be carried out with respect to all

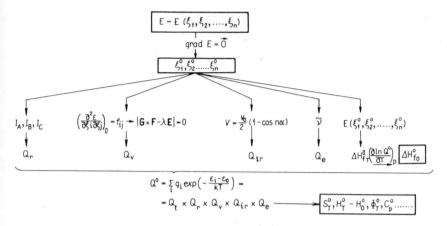

Fig. 16-14. Survey of quantum chemical and statistical thermodynamical calculations leading to absolute values of equilibrium and rate constants. The scheme suggests that the total energy (calculated by a chosen quantum chemical method) of each component is minimized with respect to all coordinates (ξ_i). The obtained optimized coordinates (ξ_i°) serve for calculation of the enthalpies of formation (ΔH_{fT}°), and of partial and total partition functions (Q_i, Q°) and finally also of entropy values (S_T°).

coordinates ξ_i ($i = 1, 2, \ldots, n$). For nonsymmetrical five-atom molecules this is already an extremely difficult task which could not be solved classically (i.e., by stepwise minimization with respect to the individual coordinates in larger molecules). In nonempirical methods data on the experimentally determined geometry can be used for reactants and products. In semiempirical methods, on the other hand, calculations should be performed for optimized geometry found using one of the recently developed procedures, amongst which that introduced by Pancíř[28] appears to be particularly effective. In principle, it consists of a suitable combination of the SCF iterative method with calculation of the gradient of the total energy; at an energy minimum it holds that grad $E = 0$. The ascertained optimal coordinates ξ_i° ($i = 1, 2, \ldots, n$) permit calculation of the moments of inertia (and thus also Q_r), the force constants, the normal modes of vibrations (Q_v) and solution of the Schrödinger equation for the ground state (ΔH_f^0) and for the electronically excited states (Q_e). The total partition function can easily be formed from the partial partition functions, thus giving the thermodynamic functions S_T° and $H_T^\circ - H_0^\circ$. All these considerations are related to equilibrium and rate processes in the gaseous phase. The absolute calculations for reactions in solutions cannot avoid calculations of the solvation energies, which is not easy. For methods which can be used to solve this task see Chapter 17.

416

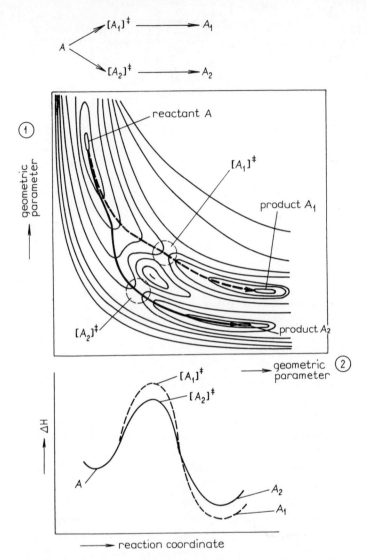

Fig. 16-15. Energy hypersurface corresponding to decomposition of reactants A into products A_1 and A_2 through activated complexes $[A_1]^\ddagger$ and $[A_2]^\ddagger$: (a) two-dimensional representation of the hypersurface, (b) sections through the hypersurface in the direction of the reaction coordinates.

With rate processes it was assumed that saddle points could be found on the potential energy hypersurface (cf. Fig. 16-15). This is numerically a very difficult task and can hardly be solved without chemical experience and a certain amount of imagination. Recently the Pancíř procedure, which has proven satisfactory in the search for minima on energy hypersurfaces, was adapted for studying the reaction coordinates and therefore was also made suitable for studies of activated complexes. Considering the short lifetime of these complexes, only theoretical investigations can be carried out; it is possible, however, that the development of extremely rapid spectroscopies (the most rapid technique so far works in the picosecond region) will make it possible, in some cases at least, to investigate them experimentally.

16.4 Calculations of relative equilibrium and rate constants

In theoretical studies in this field, the same relations for the equilibrium and rate constant are employed as a starting point as in the previous section i.e. Eqs. (16-38) and (16-42).

Considerable simplification will be achieved if the reasonable assumption of a linear dependence of the enthalpy change on the entropy change[10,11] is introduced:

$$\Delta H^{(\neq)} = a\,\Delta S^{(\neq)} + b, \tag{16-55}$$

where a and b are constants and the exponent in parentheses indicates that the relation is assumed to hold for both thermodynamic and activation quantities. The assumption expressed by Eq. (16-55) is more reasonable than the occasionally employed assumption that the (activation) entropy change is constant for all members of the series under study.

Furthermore, for reactions of conjugated systems, the Dewar assumption that the total change in the (activation) enthalpy is composed of contributions originating from the π electrons, the σ electrons and the other electrons is employed:

$$\Delta H^{(\neq)} = \Delta H_\pi^{(\neq)} + \Delta H_\sigma^{(\neq)} + \Delta H_c^{(\neq)} \tag{16-56}$$

A similar assumption is also introduced in studies of the reactivity of transition element complexes; $\Delta H^{(\neq)}$ is considered to consist of contributions originating from the d electrons and all the remaining electrons:

$$\Delta H^{(\neq)} = \Delta H_d^{(\neq)} + \Delta H_o^{(\neq)} \tag{16-57}$$

The justification of assumptions (16-56) and (16-57) is supported by experience from quantum chemistry on the separability of contributions originating from electrons of various types in planar conjugated systems and in some complexes of transitions elements; these systems will be discussed below. Finally, if in conjugated systems quantities $\Delta H_\sigma^{(\neq)}$ and

418

$\Delta H_c^{(\neq)}$ are assumed to be constant with all members of the studied reaction series,* then Eqs. (16-38) and (16-42) can be substantially simplified. First, however, the term "reaction series" should be explained. If the studied process is characterized by the equation

$$R_iX + A \; \rightleftarrows \; R_iA + X, \qquad i = 1, 2, \ldots, n \qquad (16\text{-}58)$$

then a reaction series (series of structurally related substances) is understood to refer to a group of substances R_iX $(i = 1, 2, \ldots, n)$ whose individual members are investigated in relation to reaction (16-58). For conjugated compounds we consider only those R_i which differ (a) in the size of the conjugated skeleton (which must, however, be of a distinct type, i.e., for example, alternant or non-alternant), (b) in the number of heteroatoms in the skeleton, (c) in the number and type of substituents. With complexes, only cases where the members of the series differ in the number of electrons on the central atom are considered.

Assumptions (16-55) and (16-56) allow Eqs. (16-38) and (16-42) to be written in the form

$$\log K = c_1 \, \Delta H_\pi + c_2 \qquad (16\text{-}59)$$

$$\log k = c_3 \, \Delta H_\pi^{\neq} + c_4 \qquad (16\text{-}60)$$

for the reactions of conjugated systems and in the form

$$\log K = b_1 \, \Delta H_d + b_2 \qquad (16\text{-}61)$$

$$\log k = b_3 \, \Delta H_d^{\neq} + b_4 \qquad (16\text{-}62)$$

for the reactions of transiton metal complexes, where c_i and b_i $(i = 1, 2, 3, 4)$ are constants.

The problem is formally simple: it is sufficient to calculate, using a suitable quantum chemical method, the respective changes in the π-electron energy or in the d-electron energy. It appears, however, that interpretation of the results by comparing the theoretical data with experimental data is not quite so simple. In addition, from the broader viewpoint of the theory of chemical reactivity, it must be borne in mind that the problem formulated by the compromise approach concerns a relatively narrow group of reactions, which are, however, important for chemistry.

---

* It is not certain whether this assumption is valid especially with the $\Delta H_\sigma^{(\neq)}$ term; most probably, however, the more restricted assumption of a linear dependence of $\Delta H_\sigma^{(\neq)}$ on $\Delta H_\pi^{(\neq)}$ is fulfilled, which is quite sufficient.

The majority of reactions mentioned in this chapter occur in solution. This is explicitly taken into account only exceptionally. Mostly no attempt is made to calculate the changes in the enthalpy of solvation, and this change is assumed to be either constant in the overall reaction series or to be proportional to the enthalpies calculated for the reactions in the gas phase. In some cases, of course, this assumption is qualitatively incorrect and then the expected correlation is not found.

For equilibrium processes Eqs. (16-59) and (16-61) permit direct treatment of the problem. For rate processes the situation is complicated by the fact that, in the formally preferable approximation, the reagent is included explicitly in the activated complex, whereas in the simpler approximation calculation is confined to the substrate. In view of the nature of the models and approximations employed, the real gain from application of the first method is small; yet this first approach often involves a numerically far more difficult calculation than the second. Thus calculations including a reagent in addition to the substrate are described only superficially.

16.5 Compromise approach:
the quantum chemical treatment

16.5.1 Reactions of conjugated compounds

Some reactions for which the dependence of the equilibrium constants on quantum chemical quantities was studied will be discussed using the given simplifying assumptions. For reasons given below, the three acid-base reactions

$$\text{(16-63)}$$

$$\text{(16-64)}$$

$$\text{(16-65)}$$

will be discussed first.

The free bases are neutral substances so that the conjugated acids are cations. There is, however, a delocalized charge solely for reaction product (16-63); the products of reactions (16-64) and (16-65) are systems with a charge which is essentially localized. It must, of course, be expected that use of a non-bonding AO for bonding purposes [Eq. (16-64)] will be manifested in a change in the electronegativity (and a corresponding change in the Coulomb integral); nevertheless, the protonation of the non-bonding AO in pyridine is not directly connected with the π electrons of the system. [Realize the mutual orientation of the $2p_z$ and sp^2 AO's and, furthermore, the difference between the nitrogens of the pyridine and pyrrole (aniline) types.]

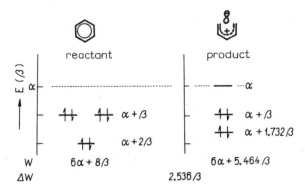

Fig. 16-16. Protonation of benzene: HMO π-electronic energies of the reactant and product.

From the viewpoint of the theory given in the previous section, it is necessary to determine whether the expected linear dependence between the logarithms of the equilibrium constants (or pK) and the π-electronic contributions to the total enthalpy change [cf. Eq. (16-59)], accompanying protonation, in fact, exists. The calculation can, of course, be performed on the HMO or SCF level[23]. For benzene the π-electronic energy for the two systems depicted in Fig. 16-16 must be calculated (HMO data). (These quantities, termed "localization energies", will be encountered when analyzing the rates of chemical reactions.)

The formation of a conjugated acid is manifested in pyridine by an increase in the Coulomb integral for nitrogen (Fig. 16-17).

If N-protonation is to occur in aniline, a change in the hybridization $(sp^2 \rightarrow sp^3)$ must occur, leading to removal of the nitrogen orbital from the conjugation. Unless allowance is to be made for the inductive or hyperconjugative effect of the ammonium group (which appears to be un-necessary), then evidently ΔH_π is given by the difference in the π-electron energies of the amine and of the corresponding parent hydrocarbon; for example, the conditions in aniline are indicated in Fig. 16-18.

The π-electronic contribution to the enthalpy change can be calculated easily for the entire series of substances and the theoretical and experimental data can be compared. The results of measurement of protonation equilibria in hydrogen fluoride and the theoretical characteristics (HMO and SCF) are given in Table 16-6.

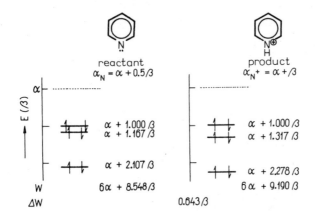

Fig. 16-17. Illustration of calculation of the change in the HMO π-electronic energy of pyridine during protonation.

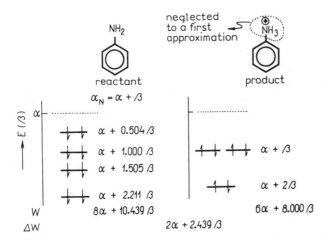

Fig. 16-18. Illustration of calculation of the change in the HMO π-electronic energy of aniline during protonation.

Table 16-6

Basicity of Benzenoid Hydrocarbons (0 °C, HF)

Hydrocarbon	Position	log K^a	$\dfrac{A^{HMO}}{\beta}$	$\dfrac{A^{SCF}}{eV}$
Benzene	–	−5.6	2.536	25.715
Naphthalene	1	0	2.299	24.643
Phenanthrene	1	0.5	2.320	24.510
Phenanthrene	9	0.5	2.299	24.497
Anthracene	9	8.1	2.013	23.502
Triphenylene	1	−0.8	2.374	24.86[b]
Pyrene	1	6.1	2.189	23.855
Chrysene	6	2.6	2.248	24.20[b]
Benz[a]anthracene	7	6.6	2.050	23.49
Tetracene	5	9.8	1.932	23.07
Perylene	3	8.4	2.139	23.64[b]
Benz[a]pyrene	6	11.1	1.962	23.22[b]

[a] E. L. Mackor, A. Hofstra, J. H. van der Waals: *Trans. Faraday Soc.* **54**, 66 (1958).
[b] Interpolated from the graph of the dependence of A_e^{SCF} on A^{HMO}.

Fig. 16-19 depicts these data graphically. The result is interesting and dependences of the type shown in Fig. 16-19 are characteristic for numerous series of equilibrium and (as we shall see below) kinetic data[23]. It is rather interesting that, while the SCF data lead to the expected result (Fig. 16-20), the points obtained using the HMO data are divided into three groups. This division corresponds to the three types of protonated positions in cyclic conjugated hydrocarbons: these are benzene-like, α-naphthalene-like and mezoanthracene-like positions. By plotting the pK values of polynuclear pyridine derivatives, in both cases (using HMO and SCF data) data divided into three groups are obtained. With polynuclear amino compounds, on the other hand, data treatment leads in both cases to a single straight line. For reactions (16-63) and (16-65) the expected result was obtained (i.e. a simple linear dependence) using SCF data. It must first be determined why the HMO method gives reasonable results for reaction (16-65) and fails with reaction (16-63). The following explanation can be given: quantity ΔH_π in reaction (16-65) is given by the difference of neutral particle energies, whereas in reaction (16-63) it is the difference between the energies of the neutral and charged particles. The shortcomings of a method which does not allow for repulsion (HMO) are comparable with reactants and products for reaction (16-65); they are, however, different for reaction (16-63).

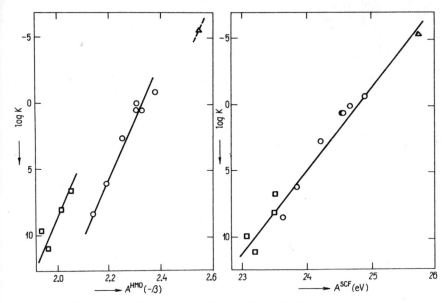

Fig. 16-19. Dependence of the logarithms of protonation equilibrium constants for benzenoid hydrocarbons on the HMO localization energies. Designation (class of the position): △ (0), ○ (1), □ (2).

Fig. 16-20. The same dependence as in Fig. 16-19, but for SCF electrophilic localization energies. Designation (class of the position): △ (0), ○ (1), □ (2).

A factor which plays an important part in many reactions in solution, *the solvation energy*, must now be mentioned. For the reactions of neutral particles in non-polar solvents the contribution of this quantity is, of course, small, but for the ions (or strongly polar molecules) it is considerable in polar solvents. Because of the limited knowledge on the theory of the liquid phase, no accurate calculation of the total solvation energy can be performed, but one of the contributions to the solvation energy can be estimated using the Born formula. For this purpose the electron charges on the atoms (q_ν) of the dissolved substance and the relative permittivity of the solvent (ε_s) must be known:

$$E_s = \frac{1}{4\pi\varepsilon_0} \left[\frac{1}{2} \sum_\mu \sum_\nu \frac{q_\mu q_\nu}{r_{\mu\nu}} \left(1 - \frac{1}{\varepsilon_s} \right) \right] \qquad (16\text{-}66)$$

In Eq. (16-66) $r_{\mu\nu}$ denotes the distance between the μ-th and ν-th atoms; when $\mu = \nu$, the effective atomic radius is used.

It appears that neglecting the solvation energy when calculating the enthalpy of reaction (16-64) leads to the split of the dependence of pK on ΔH^{HMO} and ΔH^{SCF}. By including the solvation energies, the expected simple linear dependence is obtained for the protonation of pyridine.

Still another type of equilibrium reaction, *the dismutation* (dispro-portionation) reaction of radicals warrants closer attention. These, are, for example, the following processes:

$$2 \quad \text{(·CH}_2\text{)} \quad \rightleftarrows \quad \text{(}^\ominus\text{CH}_2\text{)} \quad + \quad \text{(}^\oplus\text{CH}_2\text{)} \tag{16-67}$$

$$2 \quad \text{(}^\ominus\text{)} \quad \rightleftarrows \quad \text{(}^{2\ominus}\text{)} \quad + \quad \text{(} \text{)} \tag{16-68}$$

More specifically, a dismutation reaction is disproportionation of a sub-stance with a medium degree of oxidation (denoted by the symbol Sem) to two related substances with higher (Ox) and lower (Red) degrees of oxidation, i.e. reactions (16-67) and (16-68) proceed from left to right. Reversible reactions are called coproportionation reactions; generally it can be written that

$$\text{Ox} + \text{Red} \quad \rightleftarrows \quad 2\,\text{Sem} \tag{16-69}$$

For the dismutation constant the relationship

$$K_{\text{d}} = \frac{[\text{Sem}]^2}{[\text{Ox}][\text{Red}]} \tag{16-70}$$

is valid. If the assumptions given in Section 16.4 are accepted then the π-electronic contribution to the enthalpy can be written as

$$\Delta H = 2W_{\text{Sem}} - W_{\text{Red}} - W_{\text{Ox}} \tag{16-71}$$

If the energies of the individual forms are calculated using the HMO method, then, after substituting into Eq. (16-71), it follows for all hydro-carbon systems that

$$\Delta H^{\text{HMO}} = 0 \tag{16-72}$$

Obviously this result is incorrect, since, in reality, the dismutation constants for various systems usually have very different values.

This result represents one of the qualitative failures of the HMO method. If W in Eq. (16-71) is replaced by values E_{tot} calculated considering the electron repulsion (SCF method), an interesting result is obtained:

$$\Delta H^{\text{SCF}} = \iint \varphi_m^*(1)\, \varphi_m^*(2)\, g(1,2)\, \varphi_m(1)\, \varphi_m(2)\, d\tau_1\, d\tau_2 \tag{16-73}$$

The integral (16-73) can be easily interpreted as a measure of the repulsion of the electron pair occupying, in the reduced form, the MO which is occupied in the radical by an unpaired electron (m-th MO). It is usually denoted by the symbol J_{mm}. In view of the assumptions already made it obviously holds that

$$\log K_d = aJ_{mm} + b, \qquad (16\text{-}74)$$

where a and b are empirical constants. According to Eq. (16-74) a straight line is obtained if the logarithms of the dismutation constants of a series of structurally related substances are plotted against integrals J_{mm}. In Fig. 16-21 this dependence is shown for a number of radicals containing nitrogen and sulphur, for example I – III.

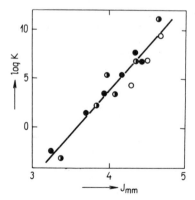

Fig. 16-21. Dependence of the logarithms of dismutation constants of organic redox systems on the value of the electron repulsion integrals, J_{mm}. m denotes the singly occupied molecular orbital in the radical.

The electronic repulsion integrals J_{mm} [Eq. (16-73)] can be easily calculated. If the MO is expressed in the LCAO form, then J_{mm} can be written as

$$J_{mm} = \sum_{\mu} \sum_{v} c_{\mu m}^2 c_{vm}^2 \gamma_{\mu v},\qquad(16\text{-}75)$$

where $c_{\varrho m}$ is the expansion coefficient for the m-th MO occurring at the ϱ-th AO and $\gamma_{\mu v}$ denotes the electronic repulsion integral within the AO basis set. Its empirical expression can be successfully carried out using the formula introduced by Mataga and Nishimoto (see Chapter 10). For estimation of the relative dismutation tendency in a series of structurally related substances it is sufficient to restrict the summation in Eq. (16-75) to one-centre contributions ($\mu = v$) and to contributions corresponding to the interaction between nearest neighbours (μ and v are centres connected by chemical bonds). For one-centre and two-centre γ's the values 10.53 and 5.20 eV, respectively, are obtained, so that, for the simplified form of Eq. (16-75), it holds that

$$J_{mm} = 10.53 \sum_{\mu} c_{\mu m}^4 + 5.20 \sum_{\substack{\mu \quad v \\ (\mu,v:\,\text{neighbours})}} c_{\mu m}^2 c_{vm}^2 \qquad(16\text{-}76)$$

The dependence of the π-electronic energy in a conjugated system on the reaction coordinate will now be discussed in greater detail.[25]

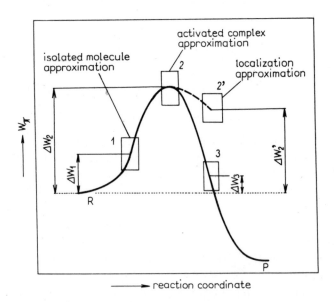

Fig. 16-22. Dependence of the π-electronic energy of a system (W_π) on the reaction coordinate Explanation is given in the text.

Five regions can be distinguished on the curve of the dependence of W on the reaction coordinate (Fig. 16-22). With reactants (R) and products (P) the energy calculation is relatively simple since both systems are completely conjugated. The conjugation is weakened in regions 1 and 3, and markedly weakened (if not completely absent) in region 2, this region corresponding to the activated complex. In this theory only the relative rates are calculated and therefore the difference in the energies of states R and 1, R and 2, R and 2′ or R and 3 can be used as an estimate, unless, of course, there is no crossing of the courses for the individual members of the reaction series. For clarification, it is sufficient to choose two cases and to take into consideration only two members of the reaction series (Fig. 16-23).

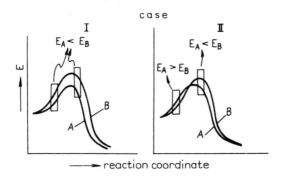

Fig. 16-23. Illustration of the "non-crossing" rule of chemical reactivity: detailed explanation is given in the text.

Case I is favourable because prediction of the reactivity, based on energy difference ΔW_1 and ΔW_2 leads to the same reactivity order for substances A and B. This is not so in case II, where the chemical "non-crossing" rule introduced by R. D. Brown[4] is not fulfilled.

The states of the system in regions 1 and 3 can be considered to be perturbed states of the reactant and product; state 2 corresponds to the actual activated complex and state 2′ corresponds to a certain exaggerated model of the activated complex (see below). Although it is hardly possible to determine the structure of the activated complex experimentally, it is usually not too difficult to obtain a rather realistic idea of its geometry. It therefore seems simplest to calculate energy ΔW_2 in order to estimate the relative reactivity sequence. Due to difficulties with computers in the first half of the fifties and because of some complications which will be discussed below, this method has not been used and predictions have been based on the calculated values ΔW_1

and $\Delta W_{2'}$. Let us start with ΔW_1. It seems to be very easy to estimate energy differences in a system in regions R and 1 using the perturbation calculation. Fig. 16-24 depicts the beginning of the interaction between the substituting chemical agent and the substrate. If the reagent is an electroneutral radical, then the hybridization of the substituted centre changes from sp^2 to sp^3 when the substrate – reagent bond starts to be formed, leading to weakening of the original bonds on the substituted centre. The first-order perturbation theory leads to the following expression for the change in energy connected with the change in the resonance integral:

$$\delta W = 2 \sum_v P_{\mu v} \delta \beta_{\mu v}, \tag{16-77}$$

the summation being carried out over all the bonds on the substituted centre μ. The free valence of centre μ can be defined as

$$F_\mu = \text{const} - \sum_v P_{\mu v} \tag{16-78}$$

So that it follows for δW that

$$\delta W = -2(F_\mu - \text{const}) \delta \beta_{\mu v} = -2F_\mu \delta \beta_{\mu v} + \text{const} \tag{16-79}$$

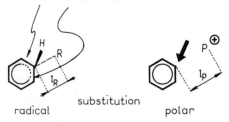

these bonds are weakened

substitution

radical

polar

Fig. 16-24. Models of activated complexes during substitution of conjugated systems considered in application of the perturbation treatment.

Due to the validity of relation (16-79), a linear dependence between $\log k_{\text{exp},\mu}$ and F_μ can be expected; it is also evident that large values of F correspond to relatively large negative changes δW and an increase in the reactivity with increasing free valence can thus be expected.

In polar reactions the effect of the substituent (cation or anion) is manifested at greater distance than in the previous instance because the effect of the reagent charge is effective at a distance where the new σ bond is not yet formed. The reagent induces a change in the Coulomb integral on the atom that it is approaching (Fig. 16-24); the first-order perturbation theory then leads to the following expression for the change in the π-electronic energy:

$$\delta W = q_\mu \delta \alpha_\mu \tag{16-80a}$$

Apparently the increase in the π-electronic energy is smallest for electrophilic (nucleophilic) substitution in the position with the maximum (minimum) π-electron density. The changes in energies induced by bonding of the reagent in different positions or approach to different positions need not be calculated, because of the assumption that the degree of perturbation ($\delta\beta_{\mu\nu}$, $\delta\alpha_\mu$) is constant at all positions; therefore only values of the free valences and π-electron densities, i.e. data found from the molecular diagram, need be known. For azulene, the expected sequence of electrophilic, nucleophilic and radical reactivities is as follows (HMO data):

e: $1 > 2 > 5 > 6 > 4$,
   since $q_1 > q_2 > q_5 > q_6 > q_4$

n: $4 > 6 > 5 > 2 > 1$,
   since $q_4 < q_6 < q_5 < q_2 < q_1$

r: $4 > 1 > 6 > 5 > 2$,
   since $F_4 > F_1 > F_6 > F_5 > F_2$

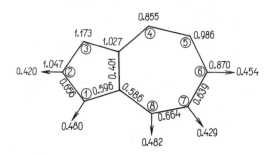

Predictions based on SCF data lead to a somewhat different sequence, but the positions of maximum reactivity are the same in both theories:

n: $4 > 6 > 2 > 5 > 1$
r: $4 > 1 > 6 > 5 > 2$

This entire discussion is particularly important as it has been shown that the use of the electron density and free valence for reactivity estimation is justified because these quantities are proportional to the π-electronic contribution to ΔH ($\Delta H^{\neq}$).

It appears that prediction of the course of polar substitution in alternant hydrocarbons cannot be based on the first-order perturbation theory because the electron densities are uniform and it is thus necessary

430

to employ the second-order perturbation theory:

$$\delta W = \underbrace{q_\mu \delta\alpha_\mu}_{\substack{\text{constant term} \\ \text{for alternant} \\ \text{hydrocarbons}}} + \tfrac{1}{2}\Pi_{\mu\mu}(\delta\alpha_\mu)^2 \tag{16-80b}$$

In practice it is useful to be aware that the sequence of values of self-polarizabilities ($\Pi_{\mu\mu}$) and free valences (F_μ) is the same and that the numerical values of $\Pi_{\mu\mu}$ and F_μ are almost identical (Table 16-7).

Consequently it must be expected that the sequence of polar and non-polar reactivities in non-equivalent positions, for example in benzenoid hydrocarbons, will be the same. This is in agreement with experimental data.

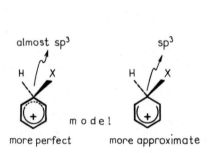

almost sp³   sp³

m o d e l

more perfect   more approximate

Fig. 16-25. Models of an activated complex during substitution of aromatic systems considered in the localization approximation.

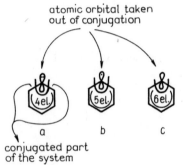

atomic orbital taken out of conjugation

a   b   c

conjugated part of the system

Fig. 16-26. Wheland complexes for electrophilic (a) radical (b) and nucleophilic (c) substitutions in benzene.

Energy change $\Delta W_{2'}$ will now be calculated (Fig. 16-22). First it must be determined in which way the structure of the activated complex is exaggerated. In the activated complex (for example, an activated complex of benzene during electrophilic substitution, Fig. 16-25) the hybridization of the substituted carbon can be expected to approach the sp³ state, so that the conjugation is weakened but not entirely absent. "Exaggeration" of this structure consists of predicting complete transition from sp² hybridization to sp³ hybridization. Experience suggests that such a model is permissible (see below). The numerical calculation then becomes considerably simplified; for $\Delta W_{2'}$ it holds that

$$\Delta W_{2'} = W_{2'} - W_R \tag{16-81}$$

Wheland termed this difference *the atomic localization energy* (A), as quantity $\Delta W_{2'}$ gives the energy that must be supplied to the conjugated system for removal of one of the atomic orbitals from the conjugation, so that it is localized. This AO can be occupied by two, one, or no electrons; the situation with benzene is shown in Fig. 16-26. A complex of type a easily reacts with a positively charged reagent or with a reagent possessing an electron hole; this reagent is termed electrophilic.

Table 16-7

Chemical Reactivity Indices of Positions in Conjugated Hydrocarbons

Hydrocarbon	Structure	Position	q	F	$\Pi_{\mu\mu}$	A_e	A_r	A_n
Benzene	(1)[a]	1	1	0.398	0.398		2.536	
Naphthalene	(2)[a]	1	1	0.453	0.443		2.299	
		2	1	0.404	0.405		2.479	
Anthracene	(3)[a]	1	1	0.459	0.454		2.230	
		2	1	0.408	0.411		2.423	
		9	1	0.520	0.526		2.013	
Phenanthrene	(4)[a]	1	1	0.450	0.439		2.320	
		2	1	0.403	0.403		2.499	
		3	1	0.408	0.407		2.454	
		4	1	0.441	0.429		2.367	
		9	1	0.452	0.442		2.299	
Azulene	(5)	1	1.173	0.480	0.425	1.924	2.262	2.600
		2	1.047	0.420	0.419	2.362	2.362	2.362
		4	0.855	0.482	0.438	2.551	2.240	1.929
		5	0.986	0.429	0.429	2.341	2.341	2.341
		6	0.870	0.454	0.424	2.730	2.359	1.988
Acenaphthylene	(6)	1	1.066	0.477	0.473	2.124	2.124	2.124
		3	0.909	0.469	0.447	2.525	2.262	1.999
		4	1.008	0.397	0.395	2.513	2.513	2.513
		5	0.927	0.478	0.466	2.606	2.311	2.016

[a] In alternant hydrocarbons $A_e = A_r = A_n$ is valid.

(1)   (2)   (3)   (4)

(5)   (6)

432

Similarly, a radical or nucleophilic reagent readily reacts with a complex of type *b* or *c*. Before beginning the numerical calculation of the localization energies, it must be noted that this model of the activated complex is non-specific inasmuch as it does not include more detailed data on the structure of the substituting reagent. The values of the localization energy are quite independent of the reagent performing the electrophilic substitution, e.g. $^\oplus NO_2$, $D^\oplus$, $Br^\oplus$, or $^\oplus Si(CH_3)_3$. As, however, only relative rates are calculated, this fact is not generally important. This approximation is termed *the localization approximation*. This shortcoming is also characteristic for the isolated molecule approximation.

Fig. 16-27. HMO orbital energies of the pentadienyl system. In the (d) cation, two bonding levels are occupied by four electrons.

The electrophilic localization energy of benzene can now be calculated. By definition

$$A_e = W_i - W_R \tag{16-82}$$

W_R denotes the π-electron energy in benzene, $6\alpha + 8\beta$. The energy of localized structure (*d*) consists of two contributions. The first is equal to the energy of a pentadienyl cation; the orbital energies of the pentadienyl system are indicated in Fig. 16-27. Both bonding orbitals are occupied in the pentadienyl cation, so that its π-electron energy is given by

$$W = 4\alpha + 5.464\beta \tag{16-83}$$

The energy of an electron located in the carbon $2p_z$ orbital equals α; the energy of two electrons is 2α. Hence, energy $W_{2'}$ will be given by

$$W_{2'} = 4\alpha + 5.464\beta + 2\alpha = 6\alpha + 5.464\beta \tag{16-84}$$

Substitution of Eqs. (16-83) and (16-84) into the expression for A_e yields

$$A_e = \underbrace{6\alpha + 5.464\beta}_{W_{2'}} - \underbrace{(6\alpha + 8.000\beta)}_{W_R} = -2.536\beta \tag{16-85}$$

Because the π-electronic energy of the localized structure is always higher than the energy of the initial system and because β is a negative quantity, localization energies always have a negative sign. To simplify the treatment of the localization energies they are given in units of $-\beta$ and the calculation is carried out only for binding energies. In agreement with this convention, the electrophilic localization energy of benzene can be written as

$$A_e = 5.464\beta - 8.000\beta = 2.536(-\beta) \tag{16-86}$$

The π-electronic energies of the pentadienyl radical and anion are equal to $5\alpha + 5.464\beta$ and $6\alpha + 5.464\beta$, respectively; the binding HMO energies of

Table 16-8

Calculation of Localization Energies

Parent hydrocarbon	Localized structure			$\dfrac{A}{(-\beta)}$
C⸺C⸺C⸺C 4.472β	C⸺C⸺C 2.828β	C 0β		1.644
  10.424β	 8.000β	C 0β	C 0β	2.424
  4.000β	C⸺C⸺C 2.828β	C 0β		1.172
  13.683β	 11.384β	C 0β		2.299
  13.683β	 11.203β	C 0β		2.480

all three systems derived from pentadienyl are the same, however, and to 5.464β. Consequently, the values of the electrophilic, nucleophilic, and radical localization energies of the "carbon atoms" in benzene are also equal:

$$A_e = A_r = A_n \qquad (16\text{-}87)$$

This equality is, of course, satisfied for all the even alternant hydrocarbons.

The positions of the different alternant hydrocarbons for which the localization energies are to be calculated and the numerical data necessary for this calculation are given in Table 16-8.

16.5.2 Substitution reactions of complexes of the transition elements[26,27]

Substitution reactions have been studied in a great variety of complexes. Nucleophilic reactions occur far more frequently than electrophilic reactions and follow similarly as organic compounds S_N1 or S_N2 mechanisms. In the former case, the monomolecular reaction is the rate-determining step, in the latter case, the bimolecular reaction. In both cases the structure of the transition state (of the activated complex) can be quite accurately predicted. If an octahedral complex undergoes substitution, then this complex is a square pyramid (S_N1) or a pentagonal bipyramid (S_N2) (cf. Fig. 16-28).

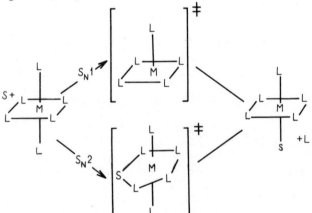

Fig. 16-28. S_N1 and S_N2 substitutions in an octahedral complex.

It should be mentioned that the transition state of the S_N1 mechanism can also be a trigonal bipyramid. The stabilization energy values, however, suggest that a square pyramid is more probable. If only the energy difference between the transition state and the reactant is to be calculated, then the crystal field theory is adequate for semi-quantitative purposes. It is assumed in the calculation that only the differences in the energies of the d electrons are important and that the energy changes of the other electrons are constant in the entire series of studied complexes. In order to calculate the crystal field stabilization energy (CFSE), it is necessary to know (in addition to the structures of the reactant and the transition state) the number of d-electrons and the nature of the crystal field (whether it is strong or weak).

Table 16-9

S_N2 Reactions [Ref. 24]

(1) Reactants: octahedral complex + substituting ligand. (2) Activated complex: pentagonal bipyramid (values are expressed in multiples of Dq)

System	Strong field			Weak field		
	(1)	(2)	ΔE	(1)	(2)	ΔE
d^0	0	0	0	0	0	0
d^1	−4	−5.28	−1.28	−4	−5.28	−1.28
d^2	−8	−10.56	−2.56	−8	−10.56	−2.56
d^3	−12	−7.74	4.26	−12	−7.74	4.26
d^4	−16	−13.02	2.98	−6	−4.93	1.07
d^5	−20	−18.30	1.70	0	0	0
d^6	−24	−15.48	8.52	−4	−5.28	−1.28
d^7	−18	−12.66	5.34	−8	−10.56	−2.56
d^8	−12	−7.74	4.26	−12	−7.74	4.26
d^9	−6	−4.93	1.07	−6	−4.93	1.07
d^{10}	0	0	0	0	0	0

Tables 16-9 and 16-10 summarize the results of calculations of the reactant energies and of the transition states for S_N2 and S_N1 reactions, which permits explanation of numerous experimental data. These data often reflect great differences in the kinetic stability of individual, often formally similar, complexes. Thus, for example, accepting a seventh ligand does not require much energy in a d^5 ion with high spin $[(T_{2g})^3 (E_g)^2]$ (the charge distribution is spherically symmetrical in these ions). Essentially, it suffices for this ligand to approach to a distance comparable with the distance characteristic of the original ligands. With the d^6 ion with low spin $[(T_{2g})^6]$ the situation is different. The charge distribution differs considerably from spherical symmetry. If the substituting reagent is to be equivalent to the bonded ligands, transfer of one electron from orbital T_{2g} to orbital E_g must occur. This process requires a relatively large amount

Table 16-10

S_N1 Reactions [Ref. 24]

(1) Reactant: octahedral complex. (2) Activated complex: square pyramid (values are expressed in multiples of Dq)

System	Strong field			Weak field		
	(1)	(2)	ΔE	(1)	(2)	ΔE
d^0	0	0	0	0	0	0
d^1	-4	-4.57	-0.57	-4	-4.57	-0.57
d^2	-8	-9.14	-1.14	-8	-9.14	-1.14
d^3	-12	-10.00	2.00	-12	-10.00	2.00
d^4	-16	-14.57	1.43	-6	-9.14	-3.14
d^5	-20	-19.14	0.86	0	0	0
d^6	-24	-20.00	4.00	-4	-4.57	-0.57
d^7	-18	-19.14	-1.14	-8	-9.14	-1.14
d^8	-12	-10.00	2.00	-12	-10.00	2.00
d^9	-6	-9.14	-3.14	-6	-9.14	-3.14
d^{10}	0	0	0	0	0	0

of energy. d^3 Complexes are similar $[(T_{2g})^3]$; formation of an activated complex requires either transfer of an electron $(T_{2g} \rightarrow E_g)$ or the coupling of two electrons in the T_{2g} orbital.

It is evident that the calculations described have some features in common with the HMO treatment of the Wheland complex in organic reactions: a) only electrons of a certain type (here d electrons) are explicitly considered, b) the model of the activated complex is not specific for the substituting reagent, c) electronic repulsion is neglected.

REFERENCES

1. Zahradník R.: *Aspects de la chimie quantique contemporaine,* p. 87. Editions du CNRS, Paris 1971.
2. Vlček A. A.: *Struktura a vlastnosti koordinačních sloučenin.* Academia, Praha 1966.
3. Streitwieser A., Jr.: *Molecular Orbital Theory for Organic Chemists.* Wiley, New York 1961.
4. Brown R. D.: *Quart. Rev.* 6, 63 (1952).
5. Glasstone S., Laidler K. J., Eyring H.: *The Theory of Rate Processes.* McGraw-Hill, New York 1941.
6. Veselov M. G.: *Elementarnaya kvantovaya teoria atomov i molekul.* Gos. izd. fiz.-mat. lit., Moscow 1962.
7. Bazilevskij M. V.: *Metod molekularnych orbit i reakcionnaya sposobnost organicheskich molekul.* Izd. Chimiya, Moscow 1969.
8. Jungers J. C. et al.: *Cinétique Chimique Appliquée.* Société des Éditions Techniq. Paris 1958.

9. Johnson F. H., Eyring H., Polissar M. J.: *The Kinetic Basis of Molecular Biology.* Wiley, New York, Chapman, London 1954.

10. Daudel R.: *Théorie quantique de la réactivité chimique.* Gauthier-Villars, Paris 1967.

11. Dewar M. J. S.: *The Molecular Orbital Theory of Organic Chemistry.* McGraw-Hill, New York 1969.

12. Leffler J. E.: *J. Org. Chem.* **20**, 1202 (1955).

13. Exner O.: *Collect. Czech. Chem. Commun.* **29**, 1094 (1964).

14. Hammett L. P.: *Physical Organic Chemistry.* McGraw-Hill, New York 1940.

15. Hine J.: *Physical Organic Chemistry.* McGraw-Hill, New York 1962.

16. Exner O.: *Chem. Listy* **53**, 1302 (1959).

17. Taft R. W., Jr.: *J. Am. Chem. Soc.* **75**, 4231 (1953).

18. Zahradník R.: *Experientia* **18**, 534 (1962).

19. Zahradník R., Beran S.: Unpublished results 1972.

20. Woodward R. B., Hoffmann R.: *Die Erhaltung der Orbitalsymmetrie.* Verlag Chemie. Weinheim 1970.

21. Mango F. D.: *Advan. Catal.* **20**, 291 (1969).

22. Hirschfelder J. O.: *J. Chem. Phys.* **8**, 431 (1940).

23. Zahradník R., Chalvet O.: *Collect. Czech. Chem. Commun.* **34**, 3402 (1969).

24. Čársky P., Hünig S., Scheutzow D., Zahradník R.: *Tetrahedron* **25**, 4781 (1969).

25. Zahradník R.: *Chem. Listy* **66**, 50 (1972).

26. Cf. refs. 73 and 76 in Chapter 10.

27. Basolo F., Pearson R. G.: *Mechanism of Inorganic Reactions.* Wiley, New York 1958.

28. Pancíř J.: *Collect. Czech. Chem. Commun.* **40**, 1112 (1975).

17. WEAK INTERACTIONS

17.1 Introduction

Interactions between atoms, ions and molecules leading to the formation (and breaking) of chemical bonds are of particular importance in chemistry. However, weak and very weak interactions between systems with closed electronic shells have already been studied for more than a century. No reaction in the chemical sense occurs between these systems under "normal" laboratory conditions. The existence of the liquid state and of molecular crystals is a consequence of the existence of attractive intermolecular forces. The equilibrium distance between molecules forming associates in the liquid and solid phases is dependent on the compensation of the attractive and repulsive forces. Repulsive forces have been shown to decrease very sharply with increasing intermolecular distances (approximately with the twelfth power of this distance); the increase in attractive forces, on the other hand, is not as rapid, as the distance decreases (it is roughly proportional to its sixth power). This has important consequences: whereas repulsive forces practically cease to be effective at distances greater than the length of chemical bonds, attractive forces are not negligible even at distances of about 0.4 nm; we therefore speak of *long-range forces*. A very important place is occupied by dispersion forces; in Section 17.2 the quantum mechanical derivation will therefore be discussed using a simple model. Expressions, resulting from the perturbation calculation, suitable for description of the intermolecular effects, will be introduced. Initially a few words will be said on the origin of *Coulomb, induction* and *dispersion forces*. Coulomb interaction is based essentially on the fact that the interacting systems are formally composed of a number of multipoles. Induction forces result from interaction between the permanent and induced multipoles of these systems. Interaction between systems without permanent multipoles characterizes dispersion forces. Multipole moments are also, however, formed in these systems as a result of electron fluctuations and their existence is, of course, limited in time. The interaction between the time-limited multipole and the multipole induced in the second system is important in calculating the dispersion interaction. The energies of these three types of interactions are, in the simplest cases (interaction monopole – monopole, monopole – induced dipole, time-variable dipole – induced dipole) indirectly proportional to the square, fourth and sixth power of the distance between the interacting systems which, ideally, act as systems of point particles. In interaction between systems without permanent dipoles, the potential energy $E(r)$ can be described by the empirical relationship introduced by Lennard-Jones, also called 6-12 potential:

$$E(r) = 4\varepsilon \left[\left(\frac{\sigma}{r} \right)^{12} - \left(\frac{\sigma}{r} \right)^{6} \right], \tag{17-1}$$

where ε and σ are empirical constants with the dimensions of energy and length, respectively. Concretely, ε denotes the depth of the potential minimum and σ is the distance between the interacting systems at which energy $E(r)$ is zero. It follows from the introduction to this section that a term with a positive sign in Eq. (17-1) describes repulsion and a term with a negative sign describes attraction between the systems.

17.2 van der Waals interaction between a pair of linear oscillators

This case corresponds to the interaction of two diatomic molecules[1,2]. Each is characterized by a dipole and, in addition, both will be considered to be harmonic oscillators. In Fig. 17-1 several symmetrical arrangements of interacting oscillators are shown. The linear model (Fig. 17-1a) will be discussed first. The total potential energy of this system which is to be substituted in the Schrödinger equation is composed of the potential energy of the dipole interaction and the energy of the oscillators.

The first contribution, provided the absolute values of the charges in the dipole equal q, has the value

$$V' = \frac{q^2}{4\pi\varepsilon_0} \left(\frac{1}{r} + \frac{1}{r - x_1 + x_2} - \frac{1}{r - x_1} - \frac{1}{r + x_2} \right) \tag{17-2}$$

If we expand the individual terms in parentheses into a series and if r is much greater than x_1 and x_2, we obtain the following simplified expression:

$$V' = - \frac{2q^2 x_1 x_2}{4\pi\varepsilon_0 r^3} \tag{17-3}$$

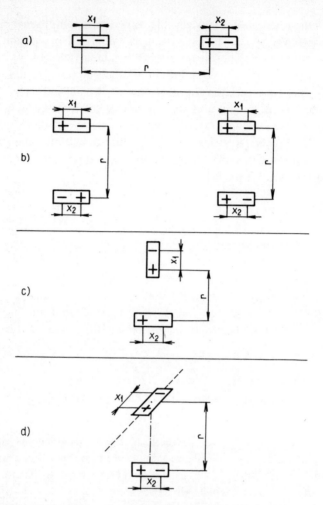

Fig. 17-1. Symmetrical models of two interacting systems. Models *a*) to *c*) are planar, model *d*) is three-dimensional. x_1 and x_2 designate the length of dipoles, r the distance between them.

If the force constant of both oscillators equals f, then it is valid for the second contribution that

$$V'' = \frac{1}{2}f(x_1^2 + x_2^2) \tag{17-4}$$

and the total potential energy has the form

$$V = V' + V'' = \frac{1}{2}f(x_1^2 + x_2^2) - \frac{q^2 x_1 x_2}{2\pi\varepsilon_0 r^3} \tag{17-5}$$

By substituting this expression into the Schrödinger equation and by introducing new variables z_1 and z_2 $[z_1 = (x_1 + x_2)/\sqrt{2}; z_2 = (x_1 - x_2)/\sqrt{2}]$, we obtain

$$\frac{\partial^2 \Psi}{\partial z_1^2} + \frac{\partial^2 \Psi}{\partial z_2^2} + \frac{8\pi^2 \mu}{h^2} \left(E - \frac{1}{2} f_1 z_1^2 - \frac{1}{2} f_2 z_2^2 \right) \Psi = 0, \quad (17\text{-}6)$$

where $f_1 = f - \dfrac{2q^2}{4\pi\varepsilon_0 r^3}$ and $f_2 = f + \dfrac{2q^2}{4\pi\varepsilon_0 r^3}$ and μ denotes the reduced mass.

In Eq. (17-6) it is possible to separate the variables:

$$\frac{\partial^2 \Psi_1}{\partial z_1^2} + \frac{8\pi^2 \mu}{h^2} \left(E_1 - \frac{1}{2} f_1 z_1^2 \right) \Psi_1 = 0 \quad (17\text{-}7)$$

$$\frac{\partial^2 \Psi_2}{\partial z_2^2} + \frac{8\pi^2 \mu}{h^2} \left(E_2 - \frac{1}{2} f_2 z_2^2 \right) \Psi_2 = 0, \quad (17\text{-}8)$$

where $\Psi = \Psi_1(z_1) \Psi_2(z_2)$ and $E = E_1 + E_2$.

The analogy of Eqs. (17-7) and (17-8) with Eq. (3-14) is obvious; therefore, for the total energy of two linear dipole oscillators, the expression

$$E = \left(n_1 + \frac{1}{2} \right) h\nu_1 + \left(n_2 + \frac{1}{2} \right) h\nu_2 \quad (17\text{-}9)$$

can be written, where for ν, using Eq. (3-74) and the definitions for f_1 and f_2, it is valid that

$$\nu_1 = \nu_0 \left(1 - \frac{2q^2}{4\pi\varepsilon_0 fr^3} \right)^{1/2} \quad (17\text{-}10)$$

$$\nu_2 = \nu_0 \left(1 + \frac{2q^2}{4\pi\varepsilon_0 fr^3} \right)^{1/2} \quad (17\text{-}11)$$

Expression (17-9) can now be easily evaluated; for the ground state $(n_1 = n_2 = 0)$ the expression

$$E_0 = \frac{1}{2} h\nu_0 \left[\left(1 - \frac{2q^2}{4\pi\varepsilon_0 fr^3} \right)^{1/2} + \left(1 + \frac{2q^2}{4\pi\varepsilon_0 fr^3} \right)^{1/2} \right] \quad (17\text{-}12)$$

is obtained. If the square roots in Eq. (17-12) are expanded using the binomial theorem and higher terms of the expansion are neglected, then the relationship

$$E_0 = h\nu_0 \left(1 - \frac{q^4}{2f^2 r^6 (4\pi\varepsilon_0)^2} \right) \quad (17\text{-}13)$$

is valid. Since the total energy of the zero point of the two-oscillator system equals hv_0, the difference between E_0 [Eq. (17-13)] and hv_0 indicates the interaction energy of the studied system

$$U(r) = -\frac{hv_0 q^4}{2f^2}\frac{1}{r^6}\frac{1}{(4\pi\varepsilon_0)^2}, \qquad (17\text{-}14)$$

and represents the van der Waals energy of the pair of oscillators. The important feature of this result, i.e. the proportionality of the interaction energy to the sixth power of the reciprocal distance, is also preserved for the generalized three-dimensional case. This term appears in all expressions for the dispersion energy.

17.3 Various means of calculating intermolecular interaction energies

This is, in principle, a very common task which does not differ at first sight from those usual in the theory of chemical reactivity. Two sub-systems, A and B, affect each other and establish a composed system. It is thus a matter of calculating the difference between the energies of system $A-B$ and of the two sub-systems A and B. Two kinds of difficulties arise: quantitative and qualitative.

The first difficulty is connected with the fact that the difference in energy which is to be calculated is usually smaller by roughly one order of magnitude than the analogous difference in chemical reactions. This difference usually amounts to 4 to 35 kJ/mol and is given as the difference of two large numbers. Thus, the demands on accuracy are extraordinarily high. However, the calculation can be based on the perturbation method instead of the variation method.

The qualitative difficulty is related to the fact that the dispersion energy can only be calculated using a method describing the correlation energy. It is therefore impossible, in principle, to calculate the dispersion energy using the Hartree-Fock SCF method. On application of the variation treatment, a satisfactory result can be obtained using the non-empirical SCF method combined with complete configuration interaction; the calculation can also be carried out using a Hamiltonian which includes the interelectronic coordinate. For rougher calculations, inclusion of doubly excited configuration suffices, but even here the calculation is very extensive, particularly for a molecule containing several atoms. Even proper study of the interaction of two hydrogen molecules is a rather difficult task at the present. The results of recent calculations are surveyed in Table 17-1.

Semiempirical SCF methods, in which the parameter values are chosen to cover the greatest possible part of the correlation energy, can also be used. It appears, however, that these methods are too rough for correct description of such fine effects.

Table 17-1

Interaction Energy of Two H_2 Molecules

Method	$\dfrac{-\Delta E}{\text{J mol}^{-1}}$
"ab initio" + CCI[a]	173.3
"ab initio" + limited CI[b]	304.5
perturbation treatment[c]	260.1
experimental (molecular beams)[d]	274.4

[a] C. F. Bender, H. F. Schaefer III: *J. Chem. Phys.* **57**, 217 (1972).
[b] O. Tapia, G. Bessis: *Theoret Chim. Acta,* **25**, 130 (1972).
[c] A. A. Evett, H. Margenau: *Phys. Rev.* **90**, 1021 (1953).
[d] J. M. Farrar, Y. T. Lee: *J. Chem. Phys.* **57**, 5492 (1972).

With the perturbation calculation, the treatment can be carried out as follows: a total Hamiltonian in the form

$$\mathcal{H} = \mathcal{H}^A + \mathcal{H}^B + \mathcal{U} \qquad (17\text{-}15)$$

is introduced, where $\mathcal{H}^A$ ($\mathcal{H}^B$) is the Hamiltonian of subsystem A (B) and $\mathcal{U}$ is *a perturbation term* which contains only *intermolecular interactions*:

$$\mathcal{U} = \sum_{i \in A} \sum_{j \in B} g(i,j) + \frac{1}{4\pi\varepsilon_0} \times \qquad (17\text{-}16)$$

$$\times \left[-\sum_{i \in A} \sum_{I \in B} \frac{Z_I e^2}{|r_i - R_I|} - \sum_{j \in B} \sum_{J \in A} \frac{Z_J e^2}{|r_j - R_J|} + \sum_{I \in A} \sum_{J \in B} \frac{Z_I Z_J e^2}{|R_I - R_J|} \right]$$

(using the symbols introduced in Section 10.2.1).

At large distances, it is assumed that the total wave function can be approximated by the product of the wave functions of the subsystems. At medium and small distances the overlap between the subsystems is not negligible, so that a properly antisymmetrized product of the wave functions must be considered:

$$\Phi = \mathcal{A} \Psi_0^A \Psi_0^B, \qquad (17\text{-}17)$$

where $\mathcal{A}$ is the antisymmetrizer and Ψ_0^A (Ψ_0^B) is the wave function of subsystem A (B).

Application of the antisymmetrizer guarantees that wave function Φ is an antisymmetrical function with respect to the transposition of an arbitrary pair of electrons irrespective of the system they come from.

From the point of view of computation it is expedient to discuss three cases separately, according to the distance between the interacting subsystems. If the distance between subsystems A and B is sufficiently large, it is possible to work with the product function and the first-order perturbation treatment leads to the classical Coulomb interaction E^C, which is either attractive or repulsive. The second order perturbation treatment leads to the induction (E^I) and the dispersion (E^D) interactions; both these energies reduce the energy of the total system. The perturbation calculations of the first and second order (cf. Section 4.6) lead to the following expressions for the energy:

$$E^{(1)} = \langle \Psi_0^A \Psi_0^B \mid \mathcal{U} \mid \Psi_0^A \Psi_0^B \rangle \tag{17-18}$$

$$E^{(2)} = \sum_{K \in A} \frac{\langle \Psi_0^A \Psi_0^B \mid \mathcal{U} \mid \Psi_K^A \Psi_0^B \rangle^2}{E_0^A - E_K^A} + \sum_{L \in B} \frac{\langle \Psi_0^A \Psi_0^B \mid \mathcal{U} \mid \Psi_0^A \Psi_L^B \rangle^2}{E_0^B - E_L^B}$$
$$+ \sum_{K \in A} \sum_{L \in B} \frac{\langle \Psi_0^A \Psi_0^B \mid \mathcal{U} \mid \Psi_K^A \Psi_L^B \rangle^2}{(E_0^A - E_K^A) + (E_0^B - E_L^B)}, \tag{17-19}$$

where summations K and L include all the excited configurations of subsystems A and B. One of the possible means of evaluation Eqs. (17-18) and (17-19) consists of the direct utilization of the perturbation Hamiltonian [Eq. (17-16)] and of replacing the respective Slater determinants for the wave functions; these wave functions are the result of MO calculations for subsystems A and B. Within the framework of the CNDO/2, CI method (including only singly excited configurations) the following expressions are valid:

$$E^C = \sum_{I \in A} \sum_{J \in B} \left[\gamma_{IJ}(Q_I Q_J - Q_I Z_J - Q_J Z_I) + \frac{1}{4\pi\varepsilon_0} \frac{Z_I Z_J e^2}{|\mathbf{R}_I - \mathbf{R}_J|} \right] \tag{17-20}$$

$$E^I = -2 \sum_{r \in A}^{(occ.)} \sum_{s \in A}^{(unocc.)} \frac{\left(\sum_{\mu \in A} \sum_{J \in B} c_{\mu r} c_{\mu s} q_J \gamma_{\mu J} \right)^2}{\Delta E_{r \to s}^A}$$
$$-2 \sum_{t \in B}^{(occ.)} \sum_{u \in B}^{(unocc.)} \frac{\left(\sum_{v \in B} \sum_{I \in A} c_{vt} c_{vu} q_I \gamma_{vI} \right)^2}{\Delta E_{t \to u}^B} \tag{17-21}$$

$$E^D = -4 \sum_{r \in A}^{(occ.)} \sum_{s \in A}^{(unocc.)} \sum_{t \in B}^{(occ.)} \sum_{u \in B}^{(unocc.)} \frac{\left(\sum_{\mu \in A} \sum_{v \in B} c_{\mu r} c_{\mu s} c_{vt} c_{vu} \gamma_{\mu v} \right)^2}{(\Delta E_{r \to s}^A + \Delta E_{t \to u}^B)} \tag{17-22}$$

Here Q_I denotes the electron density on atom I (cf. Section 11.2.4),

q_I ($\equiv Z_I - Q_I$) is the charge on atom I, γ is the repulsion integral in the AO ·basis set, Z is the charge of the core, c is the expansion coefficient, $\Delta E_{r \to s}$ is the excitation energy corresponding to the transfer of an electron from molecular orbital r into molecular orbital s; the summations $\sum\limits_{\mu \in A}$ and $\sum\limits_{v \in B}$ are carried out over all the AO's of subsystems A and B; the terms $\sum\limits_{r \in A}^{(occ.)} \sum\limits_{s \in A}^{(unocc.)} \sum\limits_{t \in B}^{(occ.)} \sum\limits_{u \in B}^{(unocc.)}$ denote summations over the occupied and unoccupied virtual MO's of subsystems A and B and finally $\sum\limits_{I \in A} \sum\limits_{J \in B}$ are summations over all the atoms of subsystems A and B.

The first order perturbation calculation using Eq. (17-17) leads to the expression for repulsion energy (E^R); this contribution is predominant at small distances. At medium distances an interesting effect can become important, namely, the transfer of an electron from one subsystem (the donor) to the second subsystem (the acceptor); this is referred to as charge-transfer (E^{CT}). The two last contributions are of great importance; however, since the respective expressions are even more complicated than the previous expression [Eqs. (17-20) to (17-22)] they will not be discussed here; they can be found in the literature[4,5].

It should be noted that expressions (17-20) to (17-22) permit calculations with a high degree of accuracy. The only difficulty is of a technical nature: these calculations, owing to the multiple summations involved in all the contributions, impose tremendous demands upon computation time for polyatomic molecules. Thus the majority of the calculations performed so far concern small molecules.

It is often necessary for practical reasons to estimate the extent of the interaction energy in relatively large systems. Various empirical formulae have long been used for these purposes; one of them has already been mentioned [cf. Eq. (17-1)]. Buckingham introduced another important formula:

$$E(r) = b \exp(-ar) - cr^{-6} - dr^{-8}, \qquad (17\text{-}23)$$

where $E(r)$ is the interaction energy corresponding to the distance r between the point particle subsystems and a, b, c, d are constants. Both the given empirical formulae are applicable, however, only for non-polar spherically symmetrical molecules. Calculation of interaction energies in more complicated systems requires suitable modification of the empirical potential. Most frequently this is performed using the simplified Buckingham potential [Eq. (17-23)], in which the term containing r^{-8} is neglected and the total interaction is, in addition, considered as the sum of the interactions among all the atoms of subsystems A and B:

Table 17-2

Constants in Eq. (17-24)

Atom I	$\dfrac{C_1}{\text{kJ mol}^{-1}}$	$\dfrac{C_2}{\text{kJ mol}^{-1}}$	C_3	$\dfrac{\varrho_I}{\text{nm}}$
H	0.3957	145.61	13.587	0.120
C	1.0078	370.95	13.587	0.170
N	0.8952	329.35	13.587	0.150
O	1.0928	402.15	13.587	0.140

$$E^{\text{INT}} = -C_1 \sum_{I \in A} \sum_{J \in B} \left(\frac{|R_I - R_J|}{\varrho_I + \varrho_J} \right)^{-6} +$$

$$+ C_2 \sum_{I \in A} \sum_{J \in B} \exp\left(-C_3 \frac{|R_I - R_J|}{\varrho_I + \varrho_J} \right), \tag{17-24}$$

where C_1, C_2, C_3 are constants following from thermochemical or conformational data (Table 17-2), ϱ_I and ϱ_J are the van der Waals atomic radii. This potential is applicable to interactions between uncharged non-polar systems. The description of charged and polar systems must contain terms allowing for the Coulomb and induction interactions.[4]

17.4 Application of weak interactions from the point of view of physical chemistry

In Fig. 17-2 the regions are schematically represented in which weak intermolecular interactions are effective to a significant degree. It would not be reasonable to try to decide the sequence of importance of these interactions. It is, rather, important to be aware that Fig. 17-2 includes both cases in which, in essence, pair interaction occurs (for example, charge-transfer complexes in the gas phase), and cases in which interaction of a large group of molecules takes place (for example, the cohesion of molecules in a molecular crystal). It appears, however, that also in this second case it is possible (for example, in the process of molecular crystal melting) to estimate the change in the energy from knowledge of the pair interaction energy and from the coordination number of the individual molecules (see below). The complete theoretical treatment of the systems depicted in the scheme in Fig. 17-2 comprises not only calculation of the enthalpy term (the interaction energy following from theoretical or empirical expressions is the internal energy) but also calculation of the entropy term using the methods of statistical thermodynamics. It remains to be added that a gradual transition exists between interactions leading to chemical bonds and weak intermolecular interactions: this can be demonstrated by the existence of the hydrogen bond. It can be assumed that in this case there is a significant contribution of the dispersion forces to the bonding effect.

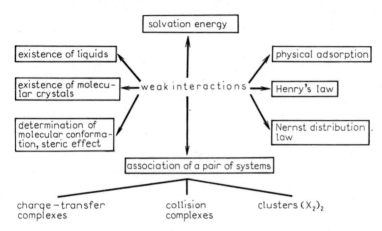

Fig. 17-2. Survey of the regions in which weak interactions play an important role.

For illustration, a few specific cases can be mentioned, beginning with pair interactions between homonuclear diatomic molecules. The results for hydrogen have already been given (Table 17-1). In heavier molecules (for example N_2, O_2, F_2, I_2) the interaction energies can be estimated most easily using the empirical potential (Table 17-3). This is the process:

$$2\,X_2 \;\rightleftarrows\; (X_2)_2 \tag{17-25}$$

Table 17-3

Enthalpy of Dimerization of Diatomic Molecules [Eq. (17-25)]:
the Buckingham potential, Eliel parametrization [Eq. (17-24)]

Product	$\dfrac{T}{K}$	$\dfrac{-\Delta H}{\text{kJ mol}^{-1}}$	
		experiment	calculation
$(O_2)_2$	70.8	1.13	1.817
$(Br_2)_2$	408.2	10.9	5.950
$(I_2)_2$	605	12.1	8.077

Optimization of the geometry in dependence on the mutual orientation leads to an "elongated" tetrahedron (the main axes of the two diatomic molecules lie in mutually perpendicular planes). Calculations by the more

Table 17-4

Experimental and Calculated[6] Values of the Energy of Vaporization at the Boiling Point

Substance	$\dfrac{-\Delta U}{\text{kJ mol}^{-1}}$	
	experiment	calculation
Methane	8.00	7.24
Ethane	13.98	13.19
Propane	16.92	16.87
Benzene	25.00	27.84

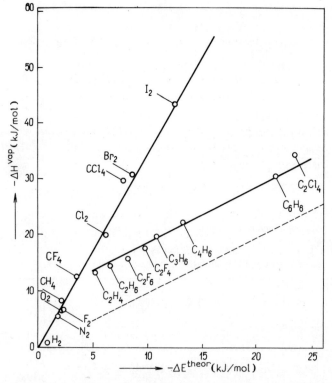

Fig. 17-3. Graph of the dependence of experimental enthalpies of vaporization on pair interaction energies.

accurate perturbation method lead to a linear configuration. The difference in the results is caused by the fact that only the atomic distances are included in the potential term and thus the dispersion energy is calculated using an expression in which it is proportional to the sixth power of the reciprocal distance between the atoms. In reality, "the centres for dispersion energy" are located in the region of the bond centres, this being well characterized by the perturbation method but not by the empirical potential.

The second case concerns heats of vaporization. In Table 17-4 the experimental and calculated vaporization energies for several hydrocarbons are given. This calculation requires a certain knowledge of the structure of liquids. The problem[6] has been treated as the interaction of one molecule with a continuous medium and the numerical calculations were performed using spherical integrals. However, good agreement between the calculated and experimental data could not be achieved without the introduction of a scaling factor. The fact that there is a linear relationship between the pair interaction energies and the vaporization energies is useful not only from a practical point of view. The different values of the slopes of these dependencies are probably connected with the coordination numbers of the molecules (Fig. 17-3).

In conclusion, the great possibilities of applying the formalism introduced in this chapter for studying *solvation effects* and energies should be mentioned. The calculation of these quantities represents a difficult point in theoretical studies of reactivity in solutions. Studies in this field have, however, already been initiated. (For additional reading see Ref. 7−9.)

REFERENCES

1. Lennard-Jones J. E.: *Proc. Phys. Soc.* (London) **43**, 461 (1931).
2. Glasstone S.: *Theoretical Chemistry*. D. van Nostrand Comp., New York 1944.
3. Margenau H., Kestner N. R.: *Theory of Intermolecular Forces*. Pergamon Press. London 1971.
4. Hobza P., Zahradník R.: *Chem. Listy* **68**, 673 (1974).
5. Fueno T., Nagase S., Tatsumi K., Yamaguchi K.: *Theoret. Chim. Acta* **26**, 43 (1972).
6. Huron M. J., Claverie P.: *J. Phys. Chem.* **76**, 2123 (1972).
7. Pullman B. (Ed.): *Environmental Effects on Molecular Structure and Properties*. Reidel, Dordrecht, Holland 1976.
8. Hirschfelder J. O., Curtiss C. F., Bird R. B.: *Molecular Theory of Gases and Liquids*. Wiley, New York 1954.
9. Hobza P., Zahradník R.: *Weak Intermolecular Interactions in Chemistry and Biology*. Elsevier, Amsterdam 1980.

INDEX